THE
MEATHEAD
METHOD

THE
MEATHEAD
METHOD

A BBQ Hall of Famer's Secrets and Science on BBQ, Grilling, and Outdoor Cooking with 114 Recipes

MEATHEAD

Foreword by Alton Brown

Photographs by Meathead

With Recipe Inspirations from Brigit Binns

HARVEST

An Imprint of WILLIAM MORROW

THE MEATHEAD METHOD. Copyright © 2025 by AmazingRibs.com. All rights reserved. Printed in Malaysia. No part of this book may be used or reproduced in any manner whatsoever without written permission except in the case of brief quotations embodied in critical articles and reviews. For information, address HarperCollins Publishers, 195 Broadway, New York, NY 10007.

HarperCollins books may be purchased for educational, business, or sales promotional use. For information, please email the Special Markets Department at SPsales@harpercollins.com.

FIRST EDITION

Designed by Tai Blanche
Photography by Meathead unless otherwise noted
Illustrations by Lisa Kolek unless otherwise noted
Culinary Assistants: Brigit Binns and Richard Longhi
Food Styling by Richard Longhi

Library of Congress Cataloging-in-Publication Data has been applied for.

ISBN 978-0-06-327284-2

25 26 27 28 29 IMG 10 9 8 7 6 5 4 3 2 1

TO MARY LOU TORTORELLO

For half a century,

my wife,

my best friend,

my taster and editor,

my food safety expert,

and the best cook in the house.

You can have your kitchen back now.

And apologies for all the failed experiments.

CONTENTS

3. COOKING SCIENCE 33

4. SOME MORE IMPORTANT TOOLS 116

13. PORK 243

14. CHICKEN AND TURKEY 257

FOREWORD

I am seriously not happy with the author of this book. When he sent me the most recent version, I opened it and, after reading about ten pages, slunk to my office, retrieved the first three chapters of my own manuscript on outdoor cooking, and shredded it.

Meathead got there first.

To suggest that this sad state of affairs explains the presence of a drone, hovering high over my backyard, during my own backyard recipe testing would be paranoid at the least, and at the most, a serious disrespect to a man who has written more great barbecue content than, near as I can tell, anyone on earth. *The Meathead Method* is his grand opus, and honestly probably the only book on outdoor cookery that you'll ever need.

The problem with barbecue literature is that lore, myth, heaps of "historical" hyperbole, and tall tales of various dimensions have passed into canon. Plainly stated: BS abounds. Heck, just drop by any American barbecue festival or competition and you can literally marinate in a litany of lies, all told with the kind of tent-revival conviction that you know you just ought not to waste time arguing with.

Somehow, I sense that Meathead has heard all the lies and, having carefully considered the canards, lined them up and tested each and every one. Now, I'm not going to pretend that I agree with him on every single point of science, and I could swear I invented a couple of the processes described herein, but I contend that most everything here is righteous and true, and reasonable, and put to paper in such a clear and cogent manner that anyone with a piece of meat and a match could profit from its full and careful consideration.

My only suggestion is that they publish this book in a waterproof/heatproof edition, because I feel certain that I'm going to keep it, like my last half-dozen or so digital thermometers, out at the smoker.

Alton Brown

FOODIST, TV CHEF, MEATHEAD'S HERO

INTRODUCTION TO THE MEATHEAD METHOD

"Almost anything you can cook indoors can be cooked outdoors. Only better."

Without question the most important discovery of all time was fire. This awe-inspiring magical phenomenon kept our ancestors warm, illuminated the dark, and scared away predators. Huddling around the hypnotic campfire transformed herds of humans into communities and bred cooperation. The most destructive power on earth became the most constructive power on earth.

Fire also melted metals for tools and weapons and it produced enegerific charcoal. But most important, fire enabled cooking. Humans are the only animals that cook and fire placed us atop the food chain. Cooking had massive benefits to primitive humans. It killed bacteria, made food tastier, made food juicier, and made food more nutritious by unbinding beneficial compounds. Harvard primatologist Richard Wrangham wrote in his landmark 2009 book *Catching Fire: How Cooking Made Us Human* that cooking made meat easier to chew, releasing its calories faster so eating took less time, and allowed humans more time for activities such as inventing things, building things, and mating. The surfeit of calories fed their brains so they grew larger, and, because digestion was easier, their bellies got smaller so they could walk upright and run faster on the hunt. In short, fire made us human.

The Meathead Method is how I meld science and art to create a toolbox with which you can gain inspiration and elevate your cooking. Although this book is a manual for outdoor cooks, there is also plenty for indoor cooks, because the core principles and science of the culinary arts are the same indoors and out.

Along the way I will be telling you that much of what you have been taught is out of date, even if you learned it in culinary school, from a famous chef on TV, from a revered cookbook, or even the USDA. We have new tools and we know a lot more about the chemistry and the physics of cooking. The reason for the accolades for my last book is simple: For years I have been busting myths by debunking "Old Husbands' Tales." I continue herein and along the way share many of my tricks. I have also tried to up your game by getting outside the box of standard Southern Barbecue and by using ingredients you don't likely use often.

I have tried not to repeat what was in the last book and on my website, but there are some foundational concepts I need to make sure you understand. The recipes are creative, aspirational, nontraditional dishes that are meant to inspire you to experiment with methods, ingredients, and recipes not covered in other barbecue books. This book takes you where barbecue goes next. Science melded with art is the Meathead Method.

MEET MEATHEAD

I am honored to be, at press time, one of only forty-one living Barbecue Hall of Famers. I am also the author of the *New York Times* bestseller *Meathead: The Science of Great Barbecue and Grilling* written with science guidance from Professor Greg Blonder of Boston University.

It was named one of the "100 Best Cookbooks of All Time" by *Southern Living Magazine*, one of the "25 Favorite Cookbooks of All Time" by Christopher Kimball's Milk Street, one of the "22 Essential Cookbooks for Every Kitchen" by SeriousEats.com, and an "indispensable guide" by Helen Rosner of *The New Yorker*. Jacques Pépin, the world's most famous French chef and the author or the landmark book *La Technique*, called it "*La Technique* of barbecue." It is available on Amazon and at better bookstores and is an essential companion to this tome.

I studied journalism at the University of Florida and have a Masters in Fine Arts from the School of the Art Institute of Chicago. In a previous life I was a syndicated wine critic for the *Washington Post* and *Chicago Tribune*,

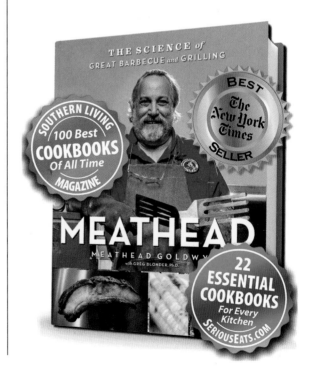

Meathead's
AMAZINGRIBS.COM
The Science & Art Of BBQ & Grilling Since 2005

founded the Beverage Testing Institute, published the magazine *International Wine Review*, lectured on wine at Cornell University's School of Hotel Administration for ten years, and taught for three years at Le Cordon Bleu in Chicago.

I live in the Chicago area with my wife, a PhD microbiologist and a food safety expert with many scientific publications under her belt, recently retired and now a certified Master Gardener. She also edits me, so if you dine at our house, you will eat and drink well, and safely.

In 2005 I founded Meathead's AmazingRibs.com, called "by far the leading resource for BBQ and grilling information" by *Forbes* and "the gold standard of foundational smoking recipes" by *The New Yorker*. Now under the direction of champion BBQ cook Clint Cantwell, it is a growing compendium about the science and art of all forms of outdoor cooking. With more than two thousand free pages, the site offers tested recipes, tips on technique, original science research, myth-busting, and unbiased equipment reviews.

It has numerous extraordinary features, among them a searchable database of detailed reviews of hundreds of grills and smokers by the world's only full-time grill and smoker tester, Max Good. There is also the world's largest collection of thermometer reviews by electrical engineer Bill McGrath. You can trust our reviews because when we recommend a product it is because we really like it, not because someone has paid us to say so. Suppliers are never charged to have products reviewed. Advertising is sold by third-party ad networks so we never know who is buying and we don't do sponsored articles or accept junkets.

Our Pitmaster Club is the world's largest barbecue association, with more than 15,000 paid members who enjoy the politics-free give and take of a lively community forum, exclusive content including more than

120 broadcast-quality videos, and many other benefits. If you bought this book, join the fun and take a free 60-day trial membership. No credit card necessary. Go to AmazingRibs.com/pitmaster and enter "Meathead Method" in the coupon box. New members only, limit one per person.

STAY IN TOUCH

Alas, there just isn't enough room for everything in this tome, so occasionally I will refer you to my last book and to my website for more. I have set up a web page to be an auxiliary to this book with links to related info, reviews, recipes, any corrections, and a place to ask questions at Amazingribs.com/mm or just scan this QR code.

If I make any new discoveries or find any errors in this book, I will share them in our free email newsletter, "Smoke Signals." You can subscribe on any page of Meathead's AmazingRibs.com or Amazingribs.com/newsletter. No spam!

In addition, we post a cooking tip almost every day at x.com/meathead and facebook.com/AmazingRibs. Stay in touch!

SENSORY SCIENCE

"The culinary arts employ all of the senses, more than any other art form."

I magine your TV gets only five channels: NBC, FOX, CNN, ESPN, and MTV. Everything you know about the world is what comes into your brain through those five channels. But you spend most of your time watching ESPN and MTV. So everything you know about the world comes through those two channels. Well, that's most of us. We have five senses—sight, hearing, touch, smell, and taste—yet most of us go through life using only two channels, seeing and hearing. But we know so little about smell, taste, and touch. Food is the antenna that can tune us into our senses.

FLAVOR, SMELL, AND AROMA

"Flavor" and "taste" are among the most frequently used words in this book. We often use them interchangeably because they are our reactions to chemicals in foods and they work so closely together, but scientifically, they are different.

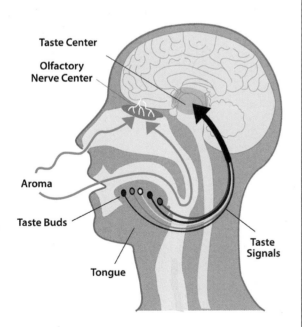

Taste Center

Olfactory
Nerve Center

Aroma

Taste Buds

Tongue

Taste
Signals

Technically, flavor is largely the result of what we smell, sometimes called aroma. It is in green in the illustration here. Aroma is due to volatile, aerosolized molecules that come into contact with the olfactory nerve center at the back of the nasal cavity, the only part of the brain exposed to air.

It is estimated that 80% of what we call taste is really flavor/smell/aroma. While in the mouth, warmth volatilizes aromatic compounds and they enter the nasal cavity through a passage at the back of the mouth where they make contact with the wet surface of the olfactory nerve center.

The jelly bean test illustrates this. Pinch your nose and eat a fruit-flavored Jelly Belly. All you can taste is the sweetness and maybe some tartness. Then let go of your nostrils and you can sense the fruitiness. That's because chewing releases volatile compounds that mix with air and they flow to the back of the mouth and upward into the nasal cavity.

Smell can often trigger memory. The smell of salt air as you approach the beach might take you back to a fishing trip with Dad, or the scent of a horse transports you to the first time you went riding with Mom, or a bowl of pasta might put you back in Grandma's kitchen on a Sunday afternoon.

Furthermore, smell and taste sensations may be perceived differently by each of us. That's why some of us love cilantro while others think it smells like soap. That's why taste is a matter of taste.

Alas, we often lack the vocabulary to discuss the sensations of smell and taste. If I say "red," we visualize a color that is in the red wavelength range, but many of us are visualizing slightly different shades. If I say something smells "earthy," chances are we each are imagining a wide range of sensations. Even "sweet" is relative. Something that is sweet to you might not be sweet to me. And just what are the proper terms for the sensations of eating a truffle? If you've never tasted one, there's no way I could describe it to you.

The best way to describe a scent is to say that it reminds you of something. That's why wine lovers often describe wines as *redolent* or *reminiscent* of raspberries, bell peppers, cheeses, barnyard, cat pee, all fragrances caused by chemicals that also appear in food. We call these *descriptors* and they are often more poetic than descriptive.

Here's a fun experiment: Blindfold yourself and have a friend go to the spice rack and pull the bottles of six common herbs and spices. Have your friend take the lid off the bottles and hand them to you one at a time. Try to name them. You will likely recognize all of them but have great difficulty naming them. Then have your friend tell you their names in random order. Once you know who the players are you will have little difficulty in naming most of them. This is why blind tasting wine and guessing its name is such a difficult task.

TASTE

Scientists tell us that we have 8,000 to 10,000 taste buds when young, mostly on our tongues, but there are some on the roof of our mouth and in our throat. As we age the number decreases to 5,000 or so. Our preferences also change. We love sweet and hate bitter when young but as we age most of us learn to appreciate a bit of bitterness and prefer things less sweet.

Most taste buds operate on what is called the "lock and key" model, meaning that the sensors in the buds are like a padlock with a specific keyhole shape, and the compounds that make up much of taste fit only into the locks designed for them where they send a signal to the brain.

The role of saliva is crucial. Saliva is an enzyme-laden solvent that dissolves food, releasing and creating new compounds and transporting it to all parts of the mouth in order to trip the greatest number of buds.

Each bud contains receptor cells that are sensitive to sweet, acid, salt, bitter, and umami. Scientists think there may be others sensitive to fat and metallic sensations. The tongue is also sensitive to texture, temperature, astringency, and spicy hot, but they are not considered tastes because taste buds don't seem to react to them.

SWEET. Sweet detectors sense sugars and sometimes their cousins, carbohydrates. We like sweet the moment we are born because breast milk is sweet and it is laden with calories.

ACID, SOUR, TART. Most fruits gain sugar and lose acidity as they ripen. When something is sour it usually means it needs more time in the sun.

SALT. Sodium and chloride ions carry charges (Na+ Cl–) needed to keep our electrical/nerve/brain systems running. Salt amplifies flavors without altering them. The taste of almost everything benefits from just a little salt. A pinch can even amp up a cocktail. Since our bodies do not make salt, we need to ingest it.

BITTER. Plants can't run from predators so they produce chemical and physical deterrents such as bitterness. Many poisons are bitter, so humans have developed a detection system in our taste buds. Capsaicin is the active ingredient in hot peppers, an irritant that makes your mouth feel like it is on fire.

UMAMI. Umami is the Japanese word that means roughly "deliciousness." It is a puzzlement to many people because it is not as easy to identify as the other tastes. It is created by an amino acid called glutamic acid, common in breast milk, most meats, mushrooms, cheese, and other protein-laden foods, as well as tomatoes and fermented foods such as soy sauce. On the tongue it appears as a rich, resonant, low sensation, like the bass fiddle in a jazz band. Some people describe it as earthy or meaty. Helen Rosner of *The New Yorker* calls it "rocket fuel" for chefs.

METALLIC. Some minerals create a metallic sensation that some scientists believe are a taste.

FAT. This one doesn't show up in any of the textbooks as a taste yet, but I think it should be. Fat is also a carrier of flavor. We need fat because it is a source of energy and we need fatty acids.

The Truth about MSG

Monosodium glutamate (MSG) is a first-rate source of umami. As a food additive, it is harmless in normal use and is an effective flavor enhancer, although a very, very small number of people may be sensitive to it. I keep a jar next to the stove.

Some people think it causes headaches. That started in 1968 when a doctor wrote a letter to the *New England Journal of Medicine* stating that he experienced a hangover-like effect when he ate Chinese food. He never mentioned MSG and there was no research involved. But "Chinese Restaurant Syndrome" went viral and there has been no undoing the harm.

When people who think they are sensitive to it are brought into a lab and fed foods with MSG, there is usually no reaction. The only credible research showing harm was when it was used is large quantities on an empty stomach. Among the many foods loaded with glutamate are Doritos, Pringles, Campbell's Chicken Noodle Soup, and parmesan cheese, so if you can eat them, then you are not sensitive to MSG. Food critic Jeffrey Steingarten once wrote an essay about MSG titled "Why Doesn't Everyone in China Have a Headache?"

Ac'cent and Aji-no-moto are commercial powder forms of MSG that can amp up umami. You can find Ac'cent in most spice sections of the grocery. I recommend ½ teaspoon on a pound of food just before cooking. Sprinkle it on meats, add it to stews and soups. Try it on mac and cheese and popcorn. I use it in savory dishes such as my Paris Chicken (page 271), but not sweets.

TOUCH

In addition to the taste buds, the tongue has sensors for tactile/touch sensations called mouthfeel. Among them are:

TEXTURE. The tongue can sense textures. The graininess of grits, the thickness of hot cocoa, the crunch of croutons, and the tenderness of foie gras.

JUICINESS. This is a complicated sensation caused by water, collagen, melted fats, and even saliva. I discuss juiciness in depth in Keeping Food Juicy (page 101).

VISCOSITY. Our tongue can sense the difference in thickness between milk, cream, and a milk shake.

TENDERNESS. Our tongue can tell us when a food is soft and needs little chewing like an oyster, or if a food is dense and need serious mastication like a steak.

FAT. Our tongue can feel the slippery oiliness of lipids.

TEMPERATURE. The tongue is like Goldilocks, it can tell us when food is hot or cold or just right.

ASTRINGENCE. This is the dry, dusty, puckery sensation found in many teas, young red wines, and underripe bananas. It comes from compounds called tannins, the same ones used to tan animal hides. They glom onto the protein in our saliva and our tongue, and the sensation can accumulate.

SPICY HOT. This is the stinging sensation and there are special receptors on the tongue for them. It can be cumulative and even painful. There are several sources:

CHILE PEPPERS. The heat in chile peppers comes from capsaicin, a large molecule that binds to protein on your tongue and hangs on for too damn long.

GINGER. The heat in fresh ginger comes from a compound called gingerol.

PEPPERCORNS. Black pepper gets its zing from a compound called piperine. It irritates tissues in the mouth and nose and causes us to sneeze.

HORSERADISH, WASABI, AND MUSTARD. The heat in horseradish, wasabi, and mustard comes from allyl isothiocyanate. This stuff packs a double whammy because it is volatile, so it can float up your nose to the olfactory bulb where it delivers a wallop.

SICHUAN PEPPER. This produces a tingling, numbing effect caused by hydroxy-alpha sanshool.

SOUND

Think of the crunch of an apple, the fizz of champagne or beer, the gurgle of pouring wine, the snap of chocolate, the squeak of a squash, the music of the shattering crust of fried chicken, and the disgusting sound of a teenager chewing with his mouth open.

Saliva starts flowing when the steak hits hot metal and sizzles. The popping of water leaving bacon is almost as bewitching as the scent. Experienced cooks know that when deep-frying, food makes a lot of noise at first as the moisture evaporates, but when it gets quiet, it is a sign that it is approaching done.

SIGHT

While cooking we can see when onions are translucent and therefore less harsh, when broccoli is al dente and when it is about to become mush, when the mahogany bark on ribs is just right, and then there is GBD, "Golden Brown and Delicious,"as in when fried chicken is *à pointe*.

But the role of sight is most evident in presentation. Everyone knows the old saw that you eat first with your eyes, and certainly a beautiful presentation can make a meal feel more special. Great culinary art should have the visual artistry of a painting and sculpture. There is no celebration in just slapping food on a plate. You killed the fish, now bring back its beauty. Beautiful presentation shows respect for the sacrifice.

BALANCE AND HARMONY

Some dishes are a simple flute solo, direct, pure, elegant, uncomplicated, soaring. Think grilled steak, with salt and pepper only. Then there's the string quartet, complex, harmonious, yet simple. Think grilled salmon, lightly salted, with tarragon and a brown butter sauce. Some dishes are the whole orchestra, graceful, luxurious, swelling with sumptuous depth and compatible notes. Think grilled Cornish game hen, dry-brined for hours, with a velvety mole sauce fashioned from chiles, onions, pureed almonds, herbs and spices, unsweetened dark chocolate, and knitted together with a rich broth. Some dishes are a brass band, loud, boisterous, attention getting. Think Southern-style pork ribs, salted, rubbed with half the spice rack and sugar, smoked, unctuous with rivers of fat, crowned with two layers of a succulent sweet tomato-based sauce amped up with chiles, vinegar, and molasses.

Chefs refer to the combo of flavors as a dish's *flavor profile*. Like music, our food has instruments with distinctive characteristics, and some seem to harmonize effortlessly, they complement each other, while others conflict violently, they contrast each other. Both can work. So let's make a dish that has all the sensory components, but let's make it balanced. Chefs call this "layering" or "building."

In *Seinfeld* episode 88, Elaine asks George to get her The Big Salad from Monk's Café. "What's in The Big Salad?" George asks and Jerry replies, "Big lettuce, big carrots, tomatoes like volleyballs." So that's a start. The lettuce brings cold water and some bitterness. The carrots bring crunch and a bit of sweetness.

Ripe tomatoes are sweet, tart, and savory, loaded with umami, and the jelly in the center is thick and viscous. In Southern France the classic salade Niçoise often has potato, hard-boiled egg, lightly boiled green beans, thinly sliced red onion, olives, bell pepper, crumbled tuna, and anchovies for umami. That's some serious complexity.

And we haven't even started to make the dressing yet. We could go with the most basic vinaigrette: olive oil and white vinegar. But that vinegar is really tart and it is really overpowering. We can bring it to heel with something sweet, like honey. The sugar in the honey doesn't reduce the acidity in the vinegar, but its sugar harmonizes with it and reduces the *impression* of acidity making it less oppressive.

In fact, sweet-acid balance is one of the most important core concepts in creating dishes. They are the Tweedledee and Tweedledum of contrasts that work. 1 + 1 = 3. It varies from person to person, but there does seem to be an, ahem, sweet spot where the two coexist in perfect harmony, and finding that consonance is a big step to making you a great cook.

"Food contains more than ingredients. It contains memories."

The most obvious examples of sweet-acid balance are Chinese sweet-and-sour dishes. Take the sugar out of sweet-and-sour pork and it is inedible. Take the vinegar out, and it makes your teeth hurt. This sweet-acid tango is what makes most barbecue sauces a treat, sweetness with vinegar.

Let's turn up the amp on our salad dressing and add a pinch of salt. It plays off the sugar and may cause a chemical reaction that impacts sugar sensors. Think salted caramel candies. Sugar does the delicate dance with salt that makes sweet ketchup love salty French fries, and an apricot glaze the perfect pair for salty hams. This combo, called *sucré-salé* (sugary-salty) by the French, is another culinary classic.

Acid and fat is another classic. Fat coats the tongue, and acid washes it away. Take a bite of fried chicken and your mouth is coated with the rich fat from the chicken and the fry oil. Then take a sip of a tart Sauvignon Blanc and it cuts the fat, washes it away, and the effect is like brushing your teeth. You are ready for another bite of chicken and it tastes as good as the first bite. Then more wine, then more chicken . . .

So it is handy to remember that when the tomato sauce is too tart, when your hand slips as you are pouring the salt into the pot, a sweetener can't remove the excess acidity or salt, but it can bring them to heel and bring harmony to the world.

What we have done here is make sure this dish has it all. It will look great, smell great, and taste great with sweetness, acidity, sourness, bitterness, spiciness, fat, saltiness, umami, and soft as well as crunchy texture. We have the brass band.

BASIC
FOOD SCIENCE

"Every time you step in front of your grill or set foot in your kitchen, you commence a physics and chemistry experiment."

Something that has become clear in recent years is that too many of us do not understand science. This is boggling in a world where we can fly from continent to continent in a metal tube, when men have walked on the moon, we hold more computing power in our pockets than went with them to the lunar surface, life expectancy has gone from sixty to eighty in the past century, and medical science allows organs to be transplanted from dead people to live people. Yet many fear vaccinations.

Alas, the culinary world is rife with misinformation about food, cooking, and how they affect us. It is shocking how many recipes in major publications and books never mention thermometers, demonstrate ignorance of basic food safety and food science, and parrot common ignorance. Not to mention the multifarious ridiculous health claims I see every day on social media. I am often left breathless when reading articles about food in major publications and comments on social media.

People cling to their myths despite overwhelming facts. It is both hilarious and sad when they hurl colorful language and porous logic at me when I explain how a beer can up a chicken's butt can't magically moisturize the bird or that the fat cap on a brisket doesn't melt and penetrate the meat.

In Western culinary arts, Classicism/Haute Cuisine was followed by Nouvelle Cuisine, California Cuisine, Fast Food, Farm to Table, Tweezer Food, Fast Casual, and now, Nerdism. OK, Nerdism isn't officially a moniker yet. I made it up. But when you look at the chefs fiddling with Modernist Cuisine and Molecular Gastronomy, and more important, when you look at the food writers and influencers who are nerds, we are seeing the spread of Nerdism. There's J. Kenji López-Alt at the *New York Times*, Dan Gritzer and the gang at SeriousEats.com, Alton Brown of Food Network, Christopher Kimball and the gang at Milk Street, Dan Souza and the gang at *Cook's Illustrated*, Shirley Corriher, and of course the godfather of us all, Harold McGee, author of the indispensable food science book *On Food and Cooking: The Science and Lore of the Kitchen*.

"Both art and science begin with curiosity."

The one thing we all have in common is that we are endlessly curious and are constantly asking the same question scientists ask: "How do you know that to be true?"

WHAT IS FOOD?

All foods are made of chemicals. Yet many of us have an inordinate fear of chemicals. We need to get over it.

1. Water: 75%
2. Carbohydrates and sugars: 12%
3. Amino acids and proteins: 10%
4. Fatty acids, flavors, minerals, vitamins, natural colors: 3%

1. Water: 75%
2. Proteins and amino acids: 18%
3. Fats and fatty acids: 5%
4. Minerals, vitamins, carbohydrates: 2%

WATER

Mammalian meats average 70 to 75% water. Fish can be more than 80% water. Many vegetables are up to 90% water. Lettuce can be 95% water. Shipping lettuce from California to New York is a pretty expensive way of shipping water.

Water can have subtle flavor of its own from impurities dissolved in it, usually minerals from the ground from which it was pumped. Those impurities can also make normally neutral water, pH 7 on a 14-point scale, slightly acidic (below pH 7) or slightly alkaline (above pH 7). The AmazingRibs.com Science Advisor, Professor Greg Blonder of Boston University, says, "Water is a polar molecule, shaped like a little wedge, able to pry open other chemicals. Practically a universal solvent. That's why we wash our hands and dishes with it."

Water turns flour into dough and wakes up dormant yeasts. It can dissolve herbs and spices. With the help of enzymes it breaks down proteins, making them more digestible, and breaks complex carbs into sugars. Boiling water is a pretty destructive form of cooking. It is good for some foods (potatoes) and disastrous for others (most meats).

Although water is a good conductor of electricity, it is a poor conductor of heat. If someone drops a hair dryer into the bathtub, you'll be electrocuted immediately. But if you put a pot of water on the stove, it takes a while to heat up. Likewise, water stores heat well. Fill up the sink with hot water to do dishes and it will stay warm until you're done.

SOLID WATER

When chilled below 32°F, water expands slightly and forms ice crystals. You can see that expansion when you accidentally freeze a full water bottle, and who hasn't done that? It can push off the top or even explode the bottle. Freezing preserves food because most microbes require liquid water to thrive and ice crystals destroy them.

Likewise, when meats, fruits, and vegetables freeze, the expansion of the water and the formation of sharp ice crystals breaks through cell walls so that when they are thawed the liquids can run out easily. It's called "drip loss" or "purge." That's why frozen foods, although they have plenty of flavor and nutrition, can dry out faster than fresh foods when subjected to heat. However, and this is important, when frozen very rapidly at very low temps, the ice crystals are smaller and less damage is done. Many frozen foods are now "flash frozen" to preserve moisture. I would rather have a turkey or fish that was flash frozen two months ago, essentially putting it into a fixed state, than a "fresh" one that has never been frozen but was killed a month ago.

GASEOUS WATER

Water can also become a vapor, which is also important in cooking. Water can evaporate at room temperature but when heated to 212°F it starts to seriously evaporate in the form of steam.

A lot of energy is required to vaporize water. Although water vapor droplets are 212°F, they have absorbed extra "latent energy of vaporization," some of the energy used to make the water vaporize. Also, the air surrounding the droplets can be hotter than 212°F. So steam can have more energy than the water below.

When the droplets encounter something cooler, let's say some broccoli, they condense, make the phase change back to water, and transfer a lot of that extra latent energy to the broccoli. That's why steam cooks faster than hot air and causes such severe burns. But surprisingly, according to fascinating experiments conducted by Professor Blonder, steam and boiling water cook at exactly the same rate! "The latent heat of condensation may be huge, but the density of water is six hundred times higher than steam."

Steaming is a great way to break down tough cellulose fibers in vegetables, such as green beans, broccoli, and cauliflower, and quickly tenderize tough meats, such as pastrami, without dissolving flavor and nutrition. And I prefer the texture of steamed vegetables to boiled. So there.

THE IMPORTANCE OF 212°F

The temperature of liquid water cannot go higher than 212°F at sea level. Let's say you toss a hot dog into a pot of boiling water. It will heat up to 212°F. But it can't go any higher until all the water evaporates and the meat touches the bottom of the pan where it can heat beyond 212°F because it is now in contact with hot metal.

In a pressure cooker with a tight lid, water is energized and evaporates, but it is trapped. The pressure inside the cooker goes up. As the pressure increases, so does the boiling point because that added pressure prevents the water from escaping the surface. If you increase the normal atmospheric pressure of 14.7 pounds per square inch (psi) at sea level by

> **HOT TIP**
> ## Boiling point at altitude
> Most recipes are written for use at sea level so you need to adapt them when you cook at altitude. Water boils at a lower temperature at higher altitudes because the column of air sitting on the surface of the water is shorter and therefore lighter, making it easier for vapors to escape. The boiling point goes down about 2°F for every 1,000 feet above sea level. So the boiling point of water in Denver, which is about one mile above sea level, is about 203°F. This means that water evaporates faster so you need to add more water to things like rice. Lower air pressure also means that it takes a minute or two longer to make hard-boiled eggs.

a measly 4 psi to 18.7 psi in the pressure cooker, the boiling temperature goes up to 224°F. As a result, the food can cook faster in a pressure cooker.

PROTEINS

Proteins are the most complex and sophisticated molecules in the known universe. They are machines that can build, interact, and transform other molecules. They are the bricks, mortar, and masons of life.

Proteins are large, long chains of molecules made from other complex molecules called amino acids. Humans have twenty amino acids. Our bodies can produce some and others we must eat. Proteins are folded and bunched into lumps with hills, valleys, and even holes. If they don't fold correctly, the result is disease, like Alzheimer's.

Proteins make up about 18% of lean meat. They are also prominent in beans, eggs, dairy, nuts, tofu, and wheat. "Complete proteins" contain all nine of the amino acids that we cannot produce ourselves, while "incomplete proteins" lack one or more of them. Animal proteins are complete while most plant proteins are not.

Heat and salt alter the shape of proteins, a process called denaturing. They shrink and squeeze out water as they are heated, which is why well-done steaks are drier than medium-rare. Proteins can link to each other (that's why burgers don't fall apart and why clear egg whites turn opaque); they form gels like cheese curds, mayonnaise, hollandaise, omelets, and custards, and custards are the basis of two of my favorite things: Ice cream and crème brûlée. Here are some important proteins with which pitmasters and all cooks must deal.

ACTIN AND MYOSIN make up muscle cells that contract and are largely responsible for muscle motion.

COLLAGEN surrounds muscle fibers and holds them together in bundles. It can make meat tough. Collagen is partially soluble in water and it can form luscious tender gelatin when heated gently to temps north of 160°F. The more collagen in meat, the tougher it is until it melts, *if* it melts. Hardworking muscles and weightbearing muscles have lots of collagen. The meat between ribs has lots of collagen while filet mignon has much less. As animals age, the bonds between collagens are harder to break up, so young animals are more tender. Skin and tendons have a lot of collagen. The amount of collagen in meat determines the temperature at which we cook and how long.

Beef Tenderloin — Silverskin — Muscle

ELASTIN is another protein that acts as a connective tissue. It forms tendons, silverskin, and gristle.

MYOCOMMATA. Because fish float in water, their muscles don't need to fight gravity so they don't have much collagen. Their muscle fibers are short and held together by connective proteins called myocommata.

HEMOGLOBIN is a dark-red-colored protein in red blood cells. It contains iron and transports oxygen to muscles and organs and transports some CO_2 away. It amounts to about 20% of the volume of your blood.

MYOGLOBIN is an oxygen-binding protein found inside muscle fibers. It helps oxygen distribution in muscle tissue. Myoglobin contains iron, which gives muscle its red or pink color. When heated it turns tan or brown. The amount of myoglobin in muscle varies greatly and there are other factors that influence the change of color. That's why you can't rely on color to tell doneness of meat.

"Every time you call it blood, a bell rings in Bedford Falls and Zuzu becomes a vegan. Let's just call it juice from now on, OK?"

Oh, and by the way, that pink juice on your plate when you cut into a steak? It's myoglobin in water called myowater. It's not blood. If it was blood, it would be dark, almost black, thick, and coagulate, just like your blood. But the bright pink liquid in your plate is thin, fluid, and flavorful. Mostly water.

Those boogers coming out of your burgers? They are mostly myowater. When burgers are ground, plump muscle fibers are sheared open. As the meat begins to heat, protein and collagen shrink and squeeze out the proteinaceous fluids, which are pink at first, and then they gel and turn tan just like the meat.

ALBUMIN is a group of proteins, among them the main ingredient of egg whites. Salmon "boogers" are goopy albumin pushed to the surface by shrinkage caused by heat. Brining helps minimize it, but not always. They can usually be wiped off with a paper towel or a brush. Another good method is to paint the surface with a simple wash of sweet wine, mirin, or a glaze just before you take it off the heat.

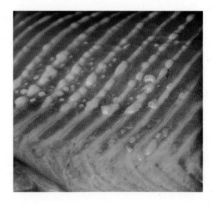

GLUTEN is the much-maligned class of proteins that makes bread and cake possible. Gluten is an elastic network of proteins that forms when flour from wheat, barley, or rye is worked vigorously with water. Gluten is what gives dough its structure.

If you are among the 2% of the population with celiac disease or the 2% with non-celiac gluten sensitivity, you must avoid gluten or suffer serious intestinal inflammation. If you are among the 96% of us who don't have medically confirmed gluten issues, avoiding gluten could have drawbacks. Gluten-free products may lack fiber and nutrients that you need, like iron and folic acid, and gluten-free products often have extra sugar, fat, and salt to improve taste. Plus, gluten-free products are usually more expensive.

ENZYMES

Enzymes put on their hard hats and some combine with other proteins while some split them up. One might think them to be unionized because most do only one task. They are especially important in digestion. Some break complex carbs into sugars, others break proteins into amino acids, others break down cellulose.

CALPAINS AND CATHEPSINS are enzymes that tenderize muscle fibers and connective tissues after slaughter and during aging. They are temperature sensitive. Raise the temperature and they work faster. But not too high. They like it when you start cooking, but at about 105°F calpains shut down and at about 122°F cathepsins take the day off. When you cook low and slow, meat hangs out in the enzyme happy zone longer and your meat will be more tender.

BROMELAIN, PAPAIN, AND ACTINIDIN are enzymes extracted from fruits and they love protein as much as you do. They can take a tough cut of meat and tenderize it in just a few hours. Fresh pineapple has the enzyme bromelain, fresh papaya has papain, and fresh kiwifruit has actinidin. Adolph's Tenderizer, found in the spice section of your grocery, has bromelain. Marinate a steak in pineapple puree for up to 2 hours, rinse it, grill it. It will tenderize the meat without flavoring it more than a hint. Canned fruit or bottled juice won't work.

RENIN is the enzyme in rennet from the stomach of calves and is essential in making many cheeses such as Parmigiano-Reggiano, Gruyère, Manchego, and Emmental.

LIPASES are enzymes that are essential in the production of cheeses such as blue cheeses, Brie, and Camembert.

TRANSGLUTAMINASE. You can buy it as a powder and if you sprinkle it on a piece of meat and lay another piece on top, they will bond into one piece. This "meat glue" is used in binding scraps of chicken into chicken "nuggets," in making sausages, in fish cakes, and in the medical field to help repair wounds. Transglutaminase is found in plants, animals, and bacteria and is perfectly safe, so no, it won't glue your hands together.

BROWN: THE MAILLARD REACTION

Named after a French scientist, the Maillard reaction, the "Golden Brown and Delicious" reaction, aka GBD, is one of the great marvels of cooking. This complex chemical reaction is responsible for much of what we love about the flavor of cooked food. The short explanation: Subjected to heat, water evaporates from the surface, and at up to about 350°F, proteins, amino acids, and sugars react chemically and form a crust made of hundreds of new brown compounds that add depth of flavor and complexity, especially in meats. We love the Maillard reaction. Beyond 350°F, the meat can start to carbonize.

This is the chemistry that makes the crust of meats so desirable, that gives stout its darkly mysterious flavor, that transforms boring beans into coffee and chocolate. It is not the same as caramelization (page 24), which is primarily due to chemical changes in sugars and carbohydrates. Sadly, many TV chefs and authors confuse the two.

To maximize the Maillard reaction, pat food dryish with paper towels before cooking. This is one of the problems with marinating and wet brining. The food is wet. The surface can only go to about 212°F until the water evaporates, too cool to do much browning. Time is wasted boiling off the water that should be spent making Maillard flavors. That's why I prefer dry rubs over marinades. Raising the cooking temperature speeds the evaporation but cannot raise the temperature much above 212°F because foods are mostly water. Immediately below the surface the temp is also about 212°F, the boiling zone.

For optimum browning, you need to give food time and space. Leave plenty of room between food chunks so steam can escape or you won't get browning.

You can make poultry and pork skins browner and crispier with the addition of small amounts of baking soda. It raises the pH, lowering the acidity of the surface, and speeds up the Maillard reaction.

Another way to maximize Maillard is to cook cold foods. They take longer to cook and that allows more time for the surface to brown.

BLACK

Maillard accelerates rapidly at about 300°F and continues up to about 350° to 375°F when it starts to burn, carbonize, pyrolyze, or blacken. Incidentally, that temperature range is also the temperature range for deep-frying.

The Maillard reaction starts at 300°F

BUSTED. The Maillard reaction begins at a low temperature. There is browning in sous vide baths, in braises, and in brown butter, all well below 200°F. Toasting bread, roasting nuts, and sautéing vegetables all take place at about 250°F.

BURNED. In general, black food is bad. It happens when food burns (pyrolyzes). When that happens it becomes dry, has lost some nutritional value, and the flavor can be unpleasant, bitter, and taste like an ashtray. Think burnt toast.

CHARRED. Charring is similar to burning. *Char* is the root word for charcoal and in chemistry it is what is left when something is pyrolyzed. That's why I never eat at The Char House. They're telling me right up front they're going to burn my steak. But sometimes a little char adds a dimension. There's the small black "leopard spots" on the underside of a pizza. Slightly charred carrots, green beans, and asparagus are delicious because of their burnt sugars. Ditto for lemons, peaches, pineapple, and other grilled fruits. The line is narrow. It is easy to cross from the slightly charred sugar flavor to carbon. Do you like your campfire marshmallows all black, black spotted, or brown?

BLACKENED. *Blackened* is a term out of New Orleans probably coined by Chef Paul Prudhomme. Blackened foods are *just this side of burned* with perhaps just a little charring. Blackening produces a smokiness, a hint of pleasant bitterness, complexity, intriguing character. That's what makes the spice mix in Prudhomme's blackened redfish work. It has onion in it and when onions cook they get sweet. I love onions sautéed or grilled and even when a few turn black. So sometimes a little burnt sugar is a burnt offering. The black crust on a beef brisket that makes it look like a meteorite is made of a spice rub that usually has sugar mixed with fats and juices that are *just this side of burned,* but not.

FATS AND OILS

Despite their bad press, fats and oils are essential to support life. We need them. Fats and oils, called lipids, are, from a culinary standpoint, best classified as animal fats or vegetable oils. Although the terms *fats* and *oils* are often used interchangeably by cooks, there are complex differences chemically. Loosely defined, however, fats are solid at room temperature and oils are liquid at room temperature, but there are exceptions.

Animal fats used in cooking include butter, lard, bacon fat, duck fat, chicken schmaltz, and beef tallow. In the vegetable world, nuts such as peanuts, walnuts, and almonds, and seeds such as corn, sesame, canola, or sunflower, and even fruits like olives and avocado have enough oil that it can be extracted. Social media is aflame with misinformation over which are healthier. I have not seen enough solid evidence for me to offer an opinion.

In meat, fats can be as little as 5% of seldom used muscles, such as beef tenderloin, and as much as 30% of hardworking muscles such as those in ribs.

Surely you've heard that "fat is flavor." Steaks with more marbling—fat flecks and threads between the muscle fibers—have much better flavor than lean meats. Some lipids have little flavor, like corn oil and canola oil, while others are rich, like duck fat, extra-virgin olive oil, and toasted sesame oil. Most people can taste the difference between cattle finished on grain and those finished on grass because of the fats.

Lipids also carry heat to food in frying and they prevent foods from sticking to metal. Lipids can get hotter than water so they are much better at browning foods, but at certain temps, their structure changes, called cracking, and they can smoke and burn. Some lipids smoke at low temperatures, like butter (about 300°F) and others can absorb much more heat before they start to smoke, like safflower oil (about 500°F). This temperature is called the smoke point and it matters if you are choosing an oil for frying, which is usually done at 350° to 375°F. If an oil is overheated or reused too often, it can crack and change color and taste. There is a table of important temperatures impacting fats in the appendix of this book.

Lipids combine easily with oxygen to form other aromas. When lipid oxidation occurs rapidly during cooking the result is usually positive. If it occurs slowly over weeks or months in the fridge or freezer, things become rancid, a rank smell that can be quite unpleasant to Westerners. In some cultures, rancid fat can be desirable, as in aged sausages, aged cheeses, and aged meats. You can often get a whiff of this funky smell in Asian restaurants and groceries, and in butcher shops all over Italy.

Like water, liquid lipids are a solvent but they don't dissolve the same things as water and vice versa. Lipids can become tainted with the taste of foods. Oil used to fry fish will taste like fish. Potatoes fried in that same oil will taste fishy.

Cooking with fats changes their chemistry and can thus alter their health impact as well as taste. Depending on how many carbon and hydrogen atoms they have and how they are bonded, some fats are classified as saturated, unsaturated, monounsaturated, polyunsaturated, omega-3 fatty acids, omega-6 fatty acids, trans fats, low-density lipoprotein (LDL), and high-density lipoprotein (HDL).

Current wisdom is that saturated fats and trans fats are bad, while unsaturated fats are more or less good, especially omega-3 fatty acids. But how bad or how good depends on a lot of things such as your health, age, gender, genetics, etc. There is so much conflicting information on the subject of fats and health that I will steer a wide path around the discussion. I wish other food writers would do the same.

> *"All oils are high in calories, about 110 to 120 calories per tablespoon or about nine calories per gram."*

TRIMMING FAT

If a steer eats sagebrush, the fat will absorb the flavor of sage. But it is important to keep in mind there are two kinds of fat, *marbling or intramuscular fat,* the tiny threads of fat weaving through and within the muscle bundles (as seen

on page 207), and *surface fat or intermuscular fat* (in the red circle in the photo opposite), the thick layers of fat that lie on top of the meat and between muscle bundles. When meat cooks, the marbling gets soft, and when we eat it, it mixes with the protein to give us great flavor.

If you must leave on some fat, leave only a thin layer, ⅛ to ¼ inch thick depending on how long and how hot it will cook. Then most of it will melt off and there will be a tasty thin layer of fat with your spices that people will usually eat and love.

Nobody's going to eat that fat and when they trim it off, there go your tasty spices. Trim the fat before cooking and they'll get to enjoy the rub.

Also, get rid of silverskin when you can. It contracts during the cooking and can cause the meat to curl and squeeze out juices. It's also as chewy as Saran Wrap.

EMULSIONS

Fat molecules like to hang out with their own tribe, but they are all hydrophobic, meaning they don't mix with water. That's why oil floats on top of vinegar in a salad dressing. Vinegar is about 94% water and oil is less dense than water. Sounds odd, but it is true.

But oil and water can be forced to mix by making the fat droplets very small and surrounding them with something that prevents them from clumping together called emulsifiers, surfactants, or stabilizers. Emulsifiers can bind with both fat and water. Butter, peanut butter, and chocolate syrup are all emulsions of fat and water. Emulsions are usually thicker than their components and they stick better to foods like salad greens than the components do on their own.

Mustard, raw egg yolks, and some starches are common kitchen emulsifiers. Raw egg yolks can be mixed with oil and vinegar to make emulsified sauces such as mayonnaise, Béarnaise, and hollandaise.

Xanthan gum is a popular emulsifier in commercial food processing and it is often found in salad dressings. It is produced by fermenting the sugars sucrose and glucose with bacteria, nothing to fear.

CARBOHYDRATES, STARCHES, AND SUGARS

They are called carbohydrates because they are made from long chains of carbon, hydrogen, and oxygen. They include all starches and sugars. Carbohydrates are the building blocks of most fruits, veggies, grains, and dairy products. They make up the cell walls, the scaffolding that gives veggies rigidity, they enclose their water, and veggies can be up to 95% water. Cell walls are what makes celery, carrots, and peppers crunch. When heated during cooking and when chewed, they break down, get soft, and form sugars, starch, and fiber.

Nutritionists have divided the category into simple carbohydrates and complex carbohydrates and have long warned us against simple carbs. But the dividing line is fuzzy. The distinctions become less useful because different cooking methods alter their chemistry.

Carbs can conceal flavors and aromas by binding with them and making them unavailable to taste buds. That's why bread can dull the flavor of meat in a sandwich.

FIBER

Fiber, the stuff your doc tells you to eat more of, is mostly cellulose, a complex carbohydrate, just like the paper in this book. It comes from plants not animals. Some of it ferments in the gut and some passes right through. Docs say we need it for digestion and keeping our gastrointestinal biome happy. It helps make you feel sated after eating and helps keep a feller regular. Beans, legumes, nuts, and grains are great sources of fiber as are fruits and veggies, especially their skins and seeds. Whole grains that still have their exterior coat, the bran, are also high fiber.

STARCH

Plants connect sugar molecules in long chains called starches. They can get as large as the thickness of a hair and they are essentially energy storage systems, like batteries.

Roots such as potatoes, carrots, onions, jicama, turnips, and yams are energy storage systems loaded with starches. Peas, corn, beans, rice, barley, and oats are seeds that store sugars to support germination before new plants have a chance to sprout leaves and produce sugar by photosynthesis. Refined grains are seeds that have the exterior layer, the bran, removed and are especially high in starch.

When mixed with water, starch granules such as flours absorb the water and swell up, soften, and change shape, a process called gelation. In a liquid, starches can trap water and thicken the fluid, making them important for thickening sauces. When cooled, wet starches can dry and crystallize. That's how breads get stale.

There are two major types of starch molecules, amylose and amylopectin. When heated with water, amylose forms gels better than amylopectin. As a result, Burbank Russet potatoes, which are high in amylose, are great for mashing and absorbing cream, while Yukon Gold potatoes, high in amylopectin, are better for potato salad chunks.

This is also why thickeners like cornstarch, tapioca, arrowroot, and flour produce different results, and why Arborio rice is better for making risotto than other rice. Cornstarch and tapioca are almost all starch, so they form clear gels, making them ideal for pies, while flour has protein, which makes it opaque, so it is better suited for things like gravy for mashed potatoes.

SUGARS

Sugars are water-soluble carbohydrates packed with potential energy for muscles and brains. Sugars can do much more than just sweeten food, and there are numerous forms of sugars and sweet syrups used in cooking. They can act as a preservative and microbial inhibitor. They can caramelize when heated, a change in chemistry that creates hundreds of deep, rich, tasty flavors. Sugar can also make a shiny glaze. If too much sugar is used, it can turn an entrée into a dessert, a common problem with many barbecue sauces. On the other hand, a very light sprinkling of sugar—as found in Meathead's Memphis Dust (page 165)—on the surface of a pork chop can break down and react with the juices and proteins and help the Maillard reaction turn the surface brown to create a marvelous crust.

Plain old granulated white sugar is flavorless in its raw state, but most other forms, such as brown sugar, honey, or molasses, have their own unique song to sing. Sometimes substituting one for another in cooking can yield delightful results. Other times, disaster. Especially in baking. For example, brown sugar and turbinado have molasses in them and they have more moisture than granulated white sugar. They taste different and react differently to heat. And the moisture can throw a baking recipe off. All sugars are not the same in sweetness. For example, honey can be a lot sweeter than granulated white sugar and it reacts differently to heat.

Sugar is the favorite food of yeasts, which produce carbon dioxide and that makes dough rise, impacting texture. That's why I put a pinch in my pizza dough recipe, although *pizzaiolos* in Naples would throw me in Vesuvius if they knew.

Of course, sugar is crucial in making pastries. When making a cake, muffins, or cookies, sugar crystals trap air when beaten with butter (called creaming), making bubbles that capture gases released by leavenings such as yeast and baking powder. In baking, these bubbles form air pockets called "crumb" that affect tenderness. The more tiny bubbles in your batter, the lighter the texture and finer the crumb. Sugar is also essential for thickening and stabilizing beaten egg whites.

In large quantities, all forms of sugar can be unhealthy, but at this time there is no hard evidence that any form is significantly worse than another in moderate quantities. Yes, raw sugar has a few more nutrients than granulated white sugar, but so few as to be negligible. It is doubtful that high fructose corn syrup in barbecue sauce will hasten your death one minute.

CARAMELIZATION

Different levels of caramelization of granulated white sugar.

Caramelization is the controlled oxidation of sugar in the presence of heat. Large sugar molecules break down, water is released, and the remaining atoms recombine into rich new compounds with buttery caramel flavors. Unlike the Maillard reaction, no proteins, amino acids, enzymes, or alkaline conditions are needed in the process.

Go into your kitchen and pour ½ cup granulated white sugar into a white or shiny saucepan and turn the heat to medium. Gently move a wooden or silicone spoon through the sugar until it all liquefies (wear gloves because liquid sugar can stick to skin like napalm and a tiny bit on your hand will make you scream like a baby). Keep stirring for about five minutes and turn the heat to low as it really starts melting. The energy will take each white odorless crystal and break it into thousands, yes thousands, of new molecules. It will fill the house with seductive, sweet aromas, and turn the white mass into a seething golden syrup. Then it will turn amber, then copper, then brown, then mahogany, then blackish. Along the way the process creates a huge range of rich, complex flavors. Scoop some out at each stage and pour it into a small bowl or onto a silicone mat. When they cool, taste them.

When we brown vegetables like onion or corn, their sugars caramelize and add depth to their flavor. That's why grilling vegetables makes them sweeter.

Different sugars caramelize at different temperatures, and the exact temperature can vary depending on how pure the sugar is. That is why some recipes call for more than one type of sugar. In descending order of how sweet they taste, here are the most common sugars and the temperature at which purified versions caramelize.

FRUCTOSE (the sugar in fruits) caramelizes at about 230°F.

SUCROSE (granulated white sugar) caramelizes at about 320°F.

GLUCOSE (made by plants and animals, the sugar in blood) caramelizes at about 320°F.

MALTOSE (two glucose molecules bound together) caramelizes at about 356°F.

LACTOSE (the sugar in milk) caramelizes at about 397°F.

Professor Blonder did an interesting demonstration of the concepts. He placed a half a bagel, a bowl of granulated white sugar, and a bowl of grape juice in a 225°F oven for several hours. The bagel toasted brown from the Maillard reaction because it contains protein and some reducing sugars, the fructose in the grape juice caramelized, but the sugar remained snowy white.

So what does this mean to a barbecue cook? Says the scientist: "A glaze sweetened with granulated white sugar won't caramelize at the typical low and slow temperature of 225°F. But it will contribute to the Maillard reaction and the bark. On the other hand, a fruit syrup will caramelize and glaze."

Notice he is talking about a sugary glaze not a barbecue sauce. Barbecue sauces are complex with acidic tomato, herbs, spices, and a variety of sugars. Still, because honey is mostly fructose, it makes a good addition to barbecue sauces because it caramelizes at low temperatures, which is why it is in many of my barbecue sauce recipes. However, it can also burn easily. The lesson is that you want to be careful making sugar substitutions.

When they are done roasting, I like to take my ribs out of the low and slow smoker, paint them with a sweet sauce, and hit them with high heat on my gas grill just before serving. This is also why sauces are hard to judge when they are straight out of the bottle and I don't pay much attention to sauce competitions (unless mine win).

If you want a nice dark surface on grilled meats, and all the flavor that comes with it, a few pinches of sugar in a marinade or rub or a light sprinkling of sugar before cooking and allowing it to dissolve in the meat juices can do wonders. But beware: If you apply a sweet sauce too early in the cook, it can mean a black disaster. For long low and slow cooks, a spritz with honey or other fruit syrup like apple juice concentrate will help. If you are at low temps and you use the right sugar, you can amp the taste up a notch.

MINERALS

Foods contain small quantities of many minerals, but small quantities can deliver a lotta bang. Although they are odorless, they are major players in taste and color. Iron is a profound presence in red meats. Calcium is the building block of bones and teeth and helps with blood clotting and muscle contraction. Iodine helps the thyroid function. Phosphorus is required for energy, DNA replication, and tissue growth. We also eat aluminum, copper, magnesium, selenium, zinc, and many other minerals in trace amounts. They conduct various functions including aiding digestion, reproduction, nerve function, blood pressure regulation, wound healing, enzyme production, immune system function, and more. There is also a long list of vitamins in food that aid with all our bodily functions. But by far the most important mineral is salt.

Salt Conversions

Different Sizes and Shapes Determine the Air in Salt and Thus Its Strength

1.0 volume of Morton Salt and Morton Iodized Salt (aka table salt) equals

1.0 volume of Morton Canning & Pickling Salt equals

1.0 volume of Morton All-Purpose Natural Sea Salt equals

1.2 volumes of Morton Coarse Kosher Salt equals

1.6 volumes of Diamond Crystal Kosher Salt equals

1.9 volumes of Maldon Sea Salt flakes

These are conversions only if you are measuring by volume such as teaspoons or cups. Even within one box of salt there will be some slight variation. Larger crystals tend to float to the top like big fluffy popcorns float to the top of the box while the crumbs settle to the bottom. There is also variation from batch to batch from the same manufacturer. Measuring spoons are often inaccurate and measuring cups depend on your eye.

SALT: THE MAGIC ROCK

Salt is a natural mineral, sodium chloride, NaCl, a crystal made of one ion of sodium (Na) and one ion of chloride (Cl). Salt is in all our bodily fluids—blood, saliva, tears, urine, and semen. Without salt our electrical system, our nerves, cannot work. In order to communicate, nerve cells allow sodium to flow between them. Incoming sodium tells nerve cells to fire. Salt can be present in tiny amounts in some foods but most of the salt in our food is added. And we need it added because our bodies can't make salt.

It is also the single most important flavor enhancer of them all. *Nolo contendere.* This tiny water-soluble rock opens your taste buds and really wakes up the flavor of meat and vegetables. It is necessary in almost all recipes. Small amounts in baked goods and desserts, and even a pinch in cocktails, improves them.

Salt reduces bitter sensations and counteracts sweetness, and just a small amount can amplify a dish without changing its basic flavor profile. It changes the physical structure of proteins, which helps it hold on to moisture even under the stress of heat. It tenderizes by dissolving the tough muscle protein myosin and can also act as a preservative and antimicrobial, which is at the core of pickling and is why so many meats and vegetables were brined, pickled, or packed in salt before refrigeration was invented. It can help crisp chicken and turkey skin. Salt also lowers the temperature at which water freezes, prevents pasta pieces from sticking to each other when boiling, and is also a pretty good stain remover.

On the other hand, the fastest way to ruin a dish is to oversalt it. You can always add salt, but you can't take it away. It can sneak up on you. Throw some Italian sausage into a cheesy lasagna, and you've added a lot of salt.

One teaspoon of Morton table salt, which is made of small cubic-shaped grains, contains less air than one teaspoon Morton Coarse Kosher Salt, which is a larger flake and has more air between the grains, or Maldon Sea Salt, which is larger still. So if the recipe calls for a teaspoon (a volumetric measurement) of Morton Coarse Kosher Salt and you use a teaspoon of Morton table salt, the food will be saltier than it is supposed to be. If a recipe calls for kosher salt and doesn't specify which brand, it is a bad recipe because Diamond Crystal Kosher Salt has less salinity by volume than Morton Coarse Kosher Salt. There is a handy interactive conversion calculator at AmazingRibs.com/conversion.

But this is cool: If a recipe calls for salt by weight, let's say eight ounces, it doesn't matter which salt you use, the volume may be different because of the air between the grains, but the amount of sodium and chloride will be the same. For this reason, it is far better to measure salt (as well as sugar and flour) by weight rather than volume *if you have an accurate scale, and you do, don't you?*

"Recipe writers who don't specify which salt to use are guilty of malpractice."

Large grains can provide pops of flavor when used at the table, but beware: Large grains don't dissolve easily and can feel gritty, even rocky, between your teeth. In the same bag of large grains there can be fine powder and that is one of the reasons why I don't use it in cooking. It is impossible to regulate quantities. I use large-grain salt only as garnish at tableside.

There are a number of other ways to bring salt to the party when cooking: Pickles and pickle juice (also good for brining), olives, sauerkraut, soy sauce, fish sauce, oyster sauce, Maggi seasoning, bean pastes, potato chips, salted nuts, pretzels, cured meats (like bacon, ham, prosciutto), salt-packed meats (like sardines and anchovies), many cheeses (especially Parmigiano-Reggiano, Grana Padano), miso, stocks, bases, and canned or instant soups.

WHY BRINE?

We like brining because it amplifies flavor without altering it and it helps proteins to retain moisture. Lean white meats like pork loin, chicken, and turkey especially benefit from brining.

Brining works because salt is only two atoms and they get electrically charged when wet. Sodium gets a positive charge because it has sacrificed a negatively charged electron to the chloride atom, which becomes negatively charged after it gets the electron from the sodium. These electrical charges help it move deep into food. Other spices or herbs can't do this. About 0.5% salt content in food is ideal.

WET BRINING

Wet brining is a good, but not perfect way to get the benefits of salt in food. Typically cooks mix salt in a bucket of water, dunk the meat in, and refrigerate it, and the meat slowly absorbs the salt and all its benefits. Many people wet brine their Thanksgiving turkeys because cooking them to 160°F or higher makes them as dry as cardboard. This requires a huge bucket or beer cooler and you need to keep it cold with ice bags to keep it safe.

A typical wet brine is 6% salt and you cannot leave meat in it too long or it will absorb more than 0.5%. To make a typical 6% wet brine: Mix 1 liter (1 kilogram) water with 60 grams of any salt. A small amount of water goes into the meat but tests have shown that because this water is not tightly bound, almost all of it drips out during cooking.

HOT TIP

Beware of Double Salt Jeopardy. Nowadays a lot of meat processors inject a salt solution into poultry and pork so you don't have to brine. Injecting can also add weight so they get to sell salt water for the price of meat. So check the nutrition label carefully before you brine or you could end up with meat that is way too salty. If your meat is labeled "basted," "self-basted," "enhanced," or "kosher," don't brine.

Many wet brine recipes call for using apple juice instead of water, lots of black pepper, garlic, etc. None of these things can penetrate because their molecules are too large. More on this in my section on marinades on page 169. I much prefer dry brining (below) in which I have better control over the amount of salt penetration.

EQUILIBRIUM BRINING

When you do a typical wet brine, you have no idea how much salt has gotten into the meat. Equilibrium brining is a clever way to wet brine and also be assured the meat gets just the right amount of salt. But it takes more time, 24 hours minimum and as long as 100 hours depending on the thickness of the meat. To do an equilibrium brine, put an empty container on the scale and press the TARE button so the scale ignores the weight of the container. Then put the meat in the container, weigh it, and cover it with double its weight in water. Now calculate 0.5% of the weight of the water and meat combined and add that much in salt. Stir to dissolve. Chilling it for 48 hours should be enough for most meats. You should end up with meat that is about 0.5% salt, ideal salinity.

DRY BRINING

"Dry brining" is an expression I started using in the late 2000s along with "wet brining" and I'm pretty sure I coined both terms. But I did not invent the concepts. Dry brining was championed by beloved San Francisco chef Judy Rogers, who died in 2013. To dry brine you

How Much Salt Do You Need to Dry Brine?

For dry brining, the rule of thumb is ½ teaspoon of Morton Coarse Kosher Salt (or equivalent) per 1 pound of trimmed meat, not including bone. Apply it more heavily on thick sections so that, for example, a pear-shaped turkey breast will get more salt on the thick end than the thin end.

simply sprinkle salt on the food. The water on the food's surface dissolves the salt, the Na and Cl split into electrically charged ions, and they set out on their journey to the center of the meat.

The thicker the meat the longer it needs to dry brine, but you are not likely to oversalt with a dry brine because the food cannot absorb more than you apply. Salt moves very slowly in the fridge. During cooking, salt moves faster. Even if you can only dry brine a short time, it is worthwhile. The salt will penetrate a fraction of an inch. But when you chew, it will be distributed throughout and bring the average to a palatable level. It just won't have time to denature the proteins in the center. Here are some rules of thumb for dry brines, wet brines, and brinerades regardless of how strong the salinity:

½-inch-thick meat should be brined for about ½ hour in the refrigerator

1-inch-thick meat should be brined for about 1 hour in the refrigerator

MYTH

Salt enters meat by osmosis

BUSTED. Gosh I am tired of reading in cookbooks, magazines, and newspapers that salt penetrates food via osmosis. It is not osmosis. It is primarily diffusion, an entirely different process. Salt diffuses into the meat by going through pores, sliced muscle fibers, and capillaries, and mixing with intracellular myowater. Here's what to remember: Diffusion is about the movement of salt. Osmosis is about the movement of water.

DIFFUSION
Salt moves from high concentration to low

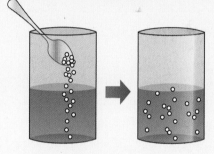

OSMOSIS
Water moves from low concentration to high

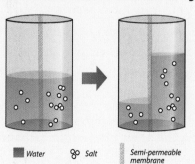

▪ Water ⚇ Salt ▮ Semi-permeable membrane

2-inch-thick meat should be brined for about 4 hours in the refrigerator

3-inch-thick meat should be brined for about 12 hours in the refrigerator

5 inches or more of thickness and the meat should be brined for about 24 hours in the refrigerator.

BRINERADES

A brinerade is a marinade with enough salt to act as a wet brine. You can use salt or a salty solution like soy sauce or miso. When I marinate, I make sure there is enough salt so it acts like a wet brine. Be aware that most brinerades are too salty to be made into sauces.

SALT BLANCHING

Many vegetables benefit from a quick swim in salty water before grilling or frying to tenderize and bring up color. In a big pot, make up a wet brine that is about 6% (see Wet Brining on page 28). Get it simmering, not boiling, and dunk green beans, pea pods, cauliflower, broccoli, asparagus, and okra in there. Thin leafy foods need less than a minute, thicker woodier foods need 2 to 3 minutes. Immediately move them into an ice bath (a pot with half ice and half water), to stop them from cooking. Then grill or fry. Salt blanching is also an essential step before freezing. It kills bacteria and deactivates enzymes that cause them to discolor and turn to mush.

MYTH

Salt will suck moisture out of the meat

BUSTED. At first a small amount of liquid will come to the surface and dissolve the salt. But only a small amount. Then the salt gets wet, dissolves, ionizes, and uses its electrical charges to move toward the center of the meat as you can see in the time-lapse photos below taken over about an hour. If you dry brine with a commercial rub that has both salt, sugar, herbs, and spices, the other ingredients may pull out liquid, but don't blame the salt.

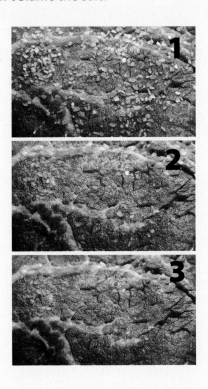

PICKLING

While some recipes may refer to brining as "pickling," technically, this is not accurate. Pickling is the use of high concentrations of salt and/or vinegar for taste and preservation. Brining involves low concentrations of salt.

It is remarkable how quickly some things pickle, especially veggies with their wide vascular system. You can quick-pickle onions and thin slices of cucumber in salt and vinegar in just 30 minutes (see Quick Pickles, page 342). Cucumbers cut into spears can be delightful after only a day in the fridge and hit their stride after 2 to 3 days. Check out my pickling recipes on the website. Meats take longer. Days. Depends on fattiness and thickness.

Pickling fruits and veggies is simple and pretty safe if you follow a recipe and keep it in the fridge. Preserving it so it can be stored at room temp in a pantry is more complicated, but well worth learning. The Ball jar company has several books on pickling. My favorite is *The Preservation Kitchen* by Chef Paul Virant. If you know how to pickle, you can eat well all winter for a lot less money than buying fresh fruits and vegetables from another hemisphere. And your great-granny will be proud of you.

CURING

Some recipes call brining "curing." Technically, curing is preserving with a curing salt with sodium nitrite and/or sodium nitrate added. They are preservatives. Because doing this properly requires working with ingredients that can be harmful if you overdo it, we don't have room to get into it in depth here, but we have a great deal on the subject at AmazingRibs.com/mm. There are recipes and a calculator created exclusively for us by Professor Blonder. It allows you to enter the weight of the meat and its thickness and calculate the correct amount of sodium nitrite, salt, and other variables to make delicious, safe, cured meats, such as bacon, Canadian bacon, ham, corned beef, Disney turkey legs, and more.

Sadly, a long time ago, a single research paper was published that called curing salts carcinogens. Please note that 95% of the nitrites we consume come naturally from vegetables such as celery, lettuce, spinach, carrots, and even some drinking water. Spinach can have more nitrites than an equivalent weight of bacon. Let's give the last word to the American Medical Association: "The risk of developing cancer as a result of consumption of nitrites-containing food is negligible."

But the stigma has not been expunged and many people think hot dogs, bacon, corned beef, pastrami, and ham are potentially dangerous. As a result, many companies have introduced cured products labeled "uncured" because they are treated with extracts from celery and other vegetables that are high in nitrites rather than with a man-made powder.

"Uncured" bacon or "naturally cured" bacon is, in fact, cured. The terms are deceptive and intended to fool you. USDA should ban

The big letters say "UNCURED BACON" and "NO NITRATES OR NITRITES ADDED • NOT PRESERVED," but the fine print says "Minimally processed. No artificial ingredients. Except as naturally occurring in cultured celery powder and sea salt." Well, cultured celery powder is loaded with nitrites.

them. Marketers get away with it by treating the meat with powdered celery powder and other "natural" compounds that are high in sodium nitrite rather than using pure sodium nitrite. Nitrite is nitrite, no matter where it comes from. Amazingly, the law requires these products to be labeled "Uncured Bacon, No Nitrates or Nitrites added." That is pure unadulterated pork excrement.

There is an important difference. When the nitrite comes from sodium or potassium nitrite, precise quantities in food are regulated by law. But there are no regs for nitrite from celery powder and some tests have shown "uncured" foods to be higher in nitrite than the permissible amounts for "cured" foods.

INJECTING

If you want to add salt, flavor, or liquid to meat, brute force works. Injecting with a hypodermic is fast and effective. One can simply inject a stock or broth that contains salt. Competition cooks almost all use injections, often special concoctions that contain ingredients designed to enhance moisture retention and amplify flavors. Butcher BBQ and Kosmos are two brands that are popular. Butcher BBQ Pork Injection contains hydrolyzed soy protein, corn protein, salt, partially hydrogenated cottonseed, soybean oil, monosodium glutamate, sodium phosphate, yeast extract, natural flavors, maltodextrin, and xanthan gum. Sounds scary but they are all food-grade and believed to be safe.

You will also need an injector. The best ones are stainless steel needles with holes on the sides, not the tip, so they won't clog. You can buy them in sporting goods stores or online.

The technique is to insert the needle parallel to the grain, as deep as you can, and slowly press the plunger while slowly pulling the needle out. Then repeat every inch or so. It is a good idea to let the meat rest an hour or even overnight so the injection can disperse a little.

COOKING SCIENCE

3

"A grill line on a piece of meat is a sign of failure."
—LENNOX HASTIE, FIREDOOR RESTAURANT, SYDNEY, AUSTRALIA

Technically, cooking is the transformation of the molecular configurations of food by the application of energy and or chemicals. There are many methods of cooking and I use most of them in this book.

ROASTING AND BAKING. This is what we do in an oven, and grills and smokers are just ovens. Both methods cook food mostly with convection airflow. We tend to think of baking as what you do when you make bread and pastries. We tend to think of roasting as what we do with meats, fruits, and vegetables. But from a physics standpoint, there is no difference.

SMOKING. Roasting or baking with warm convection air infused with smoke, usually from wood.

BROILING. Cooking with direct dry heat either coming from above or below. Indoor ovens usually broil from above; grills broil from below.

FRYING. Cooking food with hot oil usually in the 350° to 375°F range. Cooking in oil is faster than in boiling water because the temperature is much higher. Frying in a very thin layer of oil is called **SAUTÉING.** Frying by using enough oil to cover the food only about halfway is called **SHALLOW-FRYING.** Completely submerging the food in very hot oil is called **DEEP-FRYING.** Food submerged in oil at a low temperature, about 200°F, is called **CONFIT.**

WOKKING OR STIR-FRYING. Woks are shallow bowl-shaped pans and they are used for a type of shallow-frying or sautéing called stir-frying and can be used for deep-frying, braising, and more. Common in China and Chinese restaurants around the world, woks, usually made of carbon steel, are excellent pans. The best wok sessions involve very high heat beneath the wok. There is a section on wok cooking on charcoal chimneys on page 91.

AIR-FRYING is not frying. It is misnamed. Air-frying is roasting. What makes an air fryer neat is that a fan makes the air flow rapidly around the food, moving the boundary layer (more on this on pages 67–68) of cool air surrounding the food away so warm air can make contact.

BRAISING, STEWING, AND POACHING. For braising and stewing, food is put in a container with watery liquids at about 170° to 180°F. Water dissolves some of the food and extracts flavor and moisture. For poaching, the water temperature is lower, usually about 150° to 170°F, perfect for more delicate foods such as eggs and fish.

BOILING. Cooking food by submerging it in a water-based liquid at 212°F.

STEAMING. Cooking food in a vapor of both water droplets and hot air.

GRIDDLING. Food is cooked on a flat hot surface, usually a steel plate. It is similar to cooking in a skillet, only the cooking surface is much larger, giving room for maneuvering and manipulating foods through different temperature zones. Some restaurants call this a plancha or flattop and some even call it a "grill," but it is not.

MICROWAVING. Microwave ovens use electromagnetic waves, which causes molecules to vibrate and generate heat.

MOLECULAR GASTRONOMY AND MODERNIST CUISINE. Molecular Gastronomy and Modernist Cuisine cover a range of newish methods and ingredients that apply food science. Practitioners apply this knowledge to make delicious and creative combinations, including sous vide (see Sous-Vide-Que, page 107), use of liquid nitrogen, hydrocolloids for forming gels, enzymes (such as transglutaminase), infusions, MSG, calcium alginate, calcium chloride, agar-agar, centrifuges, lasers, and syringes. Then again, maybe they are not so modern. Think about soufflés, sauces, and dressings that rely on emulsifiers (such as hollandaise); fermentation, without which we have no wine or beer; molds that make cheese; dehydrators for dried fruits and chiles; baking powder and baking soda; smoking, and so on. They've been around for a long time. Let's just say that chefs have for centuries been applying their scientific knowledge to alter foods.

WHAT ARE BARBECUE AND GRILLING?

"Unless you invented fire you didn't invent barbecue and you don't own it."
—MIKE MILLS, BBQ HALL OF FAMER

Devotees debate the definition of barbecue in the most strident terms and these arguments can end in fisticuffs. Then writers parrot their silliness. We can't even agree on how it is spelled. Is it barbecue, barbeque, barbaque, BBQ, B-B-Que, Bar-B-Q, Bar-B-Que, or Bar-B-Cue? For the record, linguists, historians, and I generally agree that the proper spelling is *barbecue* with a c.

Chauvinists say true barbecue is an American invention. Well, they've been making barbecue in China since long before humans set forth in North America.

A Native American barbacoa at the De Soto National Memorial in Bradenton, Florida, near the spot where Spanish explorer Hernando de Soto came ashore in 1539.

if a word's definition relies heavily on common usage, as many linguists contend, then let's go barbecue some wieners.

The word *barbecue* is used around the world and covers a vast range of cooking methods. When American snobs pontificate about "true barbecue," what they are really discussing is properly called "Southern barbecue," one of many styles of barbecue. The fact is that barbecue is a biiiiig umbrella encompassing many cooking methods. Here is what I have learned: Smoke is what differentiates barbecue from other types of cooking.

The origin of the word was probably *barbacoa*, a word that was first used by Caribbeans to describe an elevated wooden grate that was used to cook lizards, fish, birds, and even dogs. A barbacoa was also used as a sleeping bed and to store grains out of reach of animals. Actually, we don't know for sure what word the Taino tribes used, but Spanish explorers first put the word in print when they got home. Maybe it was the actual Taino word, but barbacoa sure sounds Spanish to me.

Just what the heck is barbecue, anyway? Most people think barbecue is hamburgers and hot dogs on the gas grill. Snobs cringe at the concept. But

BARBECUE

Grilling
Caribbean Barbacoa
Competition Barbecue
Closed Pit Barbecue
Restaurant Barbecue
New Zealand Hangi

Spit Roasting & Rotisserie
Mexican Barbacoa
Open Pit Barbecue
Char-Broiling
Hawaian Imu
Thai Satay

Greek Arni Kleftiko
Santa Maria Barbecue
Indian Tandoori
Campfire Cooking

Kentucky Barbecue
Korean Barbecue
South African Braai
St. Maarten Lolo
Southern Barbecue

Chinese Barbecue
Mongolian Barbecue
Carolina Barbecue
Asado & Currasco
Japanese Yakiniku

THERMOMETERS ARE VITAL

*"Bimetal coil thermometers are about as accurate as
a sniper scope on a Nerf gun."*

—ALTON BROWN

Cooking is the process of hot air warming the outside of the food and then the outside of the food warming the inside of the food. Hot air can't get in. That's why the single most important tools for cooking, indoors or out, are good digital thermometers. Cooking is all about temperature control. Thermometers are not just quality-control devices. They save lives by making sure you have killed pathogens. It is not nice to make your guests sick.

Almost all my recipes tell you what temperature to cook at and what temperature the food should be when done. You cannot tell if your steak is medium-rare by poking it. Neither can you tell if your steak is medium-rare by cutting into it. A medium-rare steak is 130° to 135°F in the center. Period. End of story. How long it takes depends on how hot the grill is and the thickness of the steak. Run away when you see a recipe that says something like "Cook the steak on medium-high heat for 6 minutes on one side, turn it and cook for another 4 minutes on the other side." Cook with a thermometer not a clock.

GET A HANDHELD RAPID-READ DIGITAL THERMOMETER. A good digital can give you a *precise* reading in 5 seconds or less so you

MYTH

The Meat Doneness Hand Test

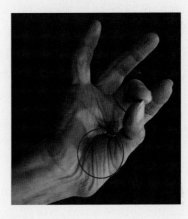

BUSTED. Surely you've seen it, the infographic of someone poking their hand. It's all over the internet. It could not be more wrong. First, everyone's hand is different. An overweight old man's hand is a lot different than a young female athlete's hand. But more important, flank steak is firmer than sirloin, which is firmer than ribeye, which is firmer than tenderloin. The only way to tell the doneness of meat is with a thermometer.

You can tell if your meat is done by cutting into it

BUSTED. Here is a flank steak I cooked hot and fast over grapevine cuttings (see Vigneron Method for Flank Steak Subs, page 231). It came off the grill at perfect medium-rare, 133°F. I carved it and served it. While it sat on the board, the myoglobin interacted with air and changed color, turning bright pink (A). Looks pretty rare, right? So I sliced off ¼ inch and beneath it was the true color (B). Looks medium, right? Moral of the story: You cannot tell doneness by cutting into a steak. Air changes its color. And the color is also influenced by the lighting. Incandescent bulbs have a yellowish hue, fluorescent have a greenish hue, and LEDs can be just about anything. And BTW, these were taken with window light and were not enhanced.

can check the food temperature in more than one location. A steak is not a balloon that will go pffffttttt and deflate from loss of juice. An 8-ounce filet mignon has 6 ounces of water. If a few drops of juice leak out, there is still plenty left. On large meats, like a chicken or turkey, be sure to test in multiple locations. Thermometers can also be used to measure fry oil. In the image at right are two excellent thermometers. At the top is the ThermoPop, about $30. It gives you a precise reading in about 5 seconds. Below it is the FireBoard Spark for about $150. It gives you a precise reading in about 1 second and it can be hooked up to a probe on a wire that can be left in the food or on the grill. It can also talk to your smartphone and computer and make charts so you can go indoors and watch the game and the food and smoker at the same time.

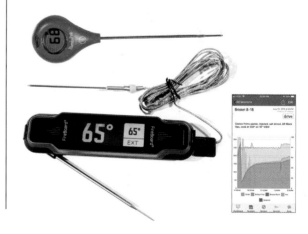

GET A PROBE TO READ THE TEMPERATURE THE FOOD FEELS. For total control, I also recommend a good digital continuous reading thermometer to measure air temperature in your grill, smoker, or oven. That cheap dial thermometer in the lid of your grill is not accurate and it is measuring the temperature high above the food, which can be vastly different from the temperature down on the cooking surface. Not very useful unless you plan to eat the lid. Especially worthless if you cook in a 2-zone setup because the thermometer sits right in the middle of the zones. Many digital thermometers also have a probe that can give continuous reads of the temperature of the interior of the food you are cooking. Some have as many as 6 probes.

The wireless Combustion model shown below has a probe with 8 sensors that locate the center and it is pretty good at predicting when the food will be done.

Photo courtesy of
Combustion Inc.

The pop-up will tell you when the turkey is done

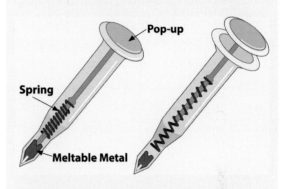

Pop-up

Spring

Meltable Metal

BUSTED. The plunger that pops up is anchored in metal that is supposed to melt at a set temperature, usually somewhere around 180°F, way past our target of 160°F. That 20°F is the difference between succulent and sucky. In November 2013, *Consumer Reports* tested pop-up thermometers and wrote that many popped too late but three popped up when meat was still unsafe, one as low as 139.5°F. Worse still, if you stuff your bird, it is not measuring the temp of the stuffing, which is several inches farther away from the heat than the tip of the pop-up, and stuffing has contaminated juices in it. That's one of the reasons I recommend you do not stuff the bird.

Some devices are attached to a blower that controls air to the charcoal so they can control a charcoal cooker's temperature.

BEWARE. New induction cooktops work by generating electromagnetic fields that vibrate the molecules of the pan, heat the metal, and they can interfere with some thermometers. The workaround is to momentarily turn off the cooktop when you take the temperature of something on it. This short interruption in the cooking process will not harm the food or significantly slow the outcome. Thermometer makers are fixing this.

HUNDREDS OF THERMOMETER RATINGS AND REVIEWS IN A SEARCHABLE DATABASE. Bill McGrath is an electrical engineer who works for Meathead's AmazingRibs.com testing thermometers with special equipment that measures accuracy and response time. He maintains a growing database of more than two hundred devices. You can search by price, type, and other variables.

FOOD TEMPERATURE AND SAFETY

Here's my handy Basic Food Safety Temperature Guide with common foods. It contains temperatures that professional chefs and I recommend for both safety and pleasure. It varies slightly from the US Department of Agriculture (USDA) recommended temperature guide, which is an admirable effort but is oversimplified to make it easier for the general public to remember, and it also errs on the side of caution. Worse, it has not been updated in years. For example, the USDA recommends 145°F for steaks. That's medium to medium well, barely pink. If steakhouses followed this guideline, they would go out of business in weeks. Steak lovers know steaks are most juicy and tender at medium-rare, 130° to 135°F.

Copy this and hang it from your fridge or order a magnetic version on Amazon. It has a QR code on it. Scan it and I'll send you a free e-book.

Here are some insights into this food temperature guide.

IT IS NOT JUST TEMPERATURE. What the USDA doesn't tell you is that making food safe is not just about temperature. It is about temperature, time, thickness, how much contamination there is, and the desired kill rate all taken together. According to the USDA you can make turkey safe when the internal temperature hits 165°F, but it is not well known that it is

Meathead's AmazingRibs.com
Basic Food Safety Temperature Guide

BEEF STEAKS, CHOPS, ROASTS 🐄 LAMB 🐑 VENISON 🦌 DUCK 🦆

Well Done	**155°F (68°C) or more**	Tan to brown, no pink, chewy, dry
Medium Well	**145 to 155°F (63-68°C)**	Tan, pinkish, firm, slightly juicy
Medium	**135 to 145°F (57-63°C)**	Pink, yielding, juicy
Best: Medium Rare	**130 to 135°F (54-57°C)**	**Bright red, tender, very juicy**
Rare	**120 to 130°F (49-54°C)**	Bright purple to red, stringy, juicy

PORK CHOPS, STEAKS, ROASTS 🐖 UNCOOKED HAMS 🍖 VEAL 🐄

Well Done	**155°F (68°C) or more**	Cream colored, tough,
Medium Well	**145 to 155°F (63-68°C)**	Cream colored, firm, slighly juicy
Best: Medium	**135 to 145°F (57-63°C)**	**Creamy pink, yielding, juicy**
Medium Rare	**130 to 135°F (54-57°C)**	Pink, tender, very juicy

Smoked Pork Ribs Pork Butt, Beef Ribs Beef Brisket
Cook at 225°F (107°C)

195-205°F (91-96°C)

Tuna Steaks
Sear hot & fast

120-125°F (49-52°C)

Lobster, Shrimp, Crab, Crawfish, Scallops, Oysters, Clams, Squid, Mussels, Octopus

Cook hot

131°F (55°C)

Raw Ground Meats, Burgers, Sausages, Meatloafs
Use 20 to 30% fat blend

Cook hot

160°F (71°C)

Chicken & Turkey
Whole or ground, with or without stuffing cook at 325-375°F (163-191°C)

160°F (71°C)

Most Other Fin Fish
Cook hot until slightly translucent, flaky, tender

125-130°F (52-54°C)

Casseroles & Leftovers

165°F (74°C)

Pre-Cooked Hams, Hot Dogs, Sausages

140°F (60°C)

Potatoes
210°F (99°C)

Ratings and reviews of 200+ digital thermometers and BBQ thermostats on https://AmazingRibs.com/thermometers

Free Cookbook

How long does it take to pasteurize?

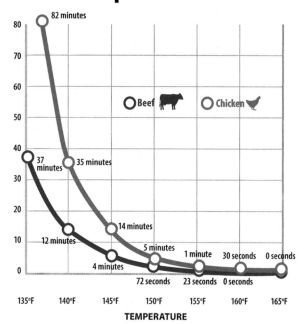

Pasteurizing to make food safe depends on both temperature and time. The higher the temperature the faster you kill bacteria. This chart, based on data from USDA Food Safety Inspection Service, shows the approximate time it takes to kill to a level called "7D." At about 131°F you start killing all types of pathogens very slowly. At 160° to 165°F you kill them almost instantly.

also safe if kept at 160°F for 27 seconds, and at 150°F for 5 minutes. So cooking below USDA recommended temps is safe, as long as you pay attention to time. Kill rate is also a factor. USDA wants something called 7D, but some other countries think 6D is plenty safe.

BEWARE OF CARRYOVER. Be aware that if you let hot meat rest, the heat built up in the surface will continue to cook, called carryover cooking, and it could be overcooked by the time you serve it. That's a major reason why I am not a fan of "resting" meat. I explain carryover in detail on page 54 and resting meat on page 55.

GROUND MEAT, BURGERS, AND SAUSAGE. The recommended temperature is 160°F, much higher than for whole muscle meats, and it should be adhered to closely. It is easy for bad bacteria to get on meat in the slaughterhouse. They are usually on the surface because they are too large to get into the interior. Cooking kills them almost instantly. But if you grind the meat, the contaminants are mixed throughout. So ground meat must be cooked to a higher interior temperature than whole muscle meat. Don't screw around. The risk is too high, especially for young, elderly, and immunocompromised people at your table. There are tricks for making safe burgers that are medium-rare or pink in the center and I describe them on page 289.

PORK RIBS, BEEF RIBS, PORK SHOULDERS, AND BEEF BRISKET. These cuts are safe at the same temps as other pork and beef cuts, but barbecue experts cook them up to 203°F to render fats and melt the connective tissues that are rife in these tough cuts. That may seem extreme, but I'm here to tell you that low, slow, long cooking these meats produces amazing results.

EGGS. Eggs pose a safety risk. *Salmonella enteritidis* is common among hens. In the ovaries, it infects eggs before the shells are formed. Salmonella growth is inhibited by refrigeration, so eggs should not be kept at room temp. If you like dishes made with raw eggs, such as pasta carbonara, eggnog, Caesar salad dressing, or Béarnaise and hollandaise sauces, it is strongly recommended that you use pasteurized eggs. To pasteurize whole eggs at home, put them in a strainer and sous vide them at 135°F for 90 minutes. The strainer keeps them from banging around and cracking. Don't go longer. Immediately, when they are done, dump the hot water and shock them with cold water. More on sous vide in Sous-Vide-Que (page 107).

VEGETABLES AND OTHER FOODS. The plain fact is that anything eaten raw has higher risk than anything cooked. Heat kills. That's why there have been more outbreaks from spinach and other greens than from meat. They grow close to the ground where rabbits, deer, and mice roam, and contaminants can come from birds, irrigation water, and human handlers.

ENERGY

Understanding how heat is produced and transferred, and its impact on food, is foundational. Max Good tests grills and smokers full time for us (and no, he doesn't need an assistant). He frequently gets asked, "How hot does that grill get?" It. Just. Doesn't. Matter.

Cooks often talk about *heat* and *temperature* interchangeably, but, although they are married by the kinetic theory, they are different. It is helpful to think of heat as one of many forms of energy and temperature as a measure of how hot or cold something is as a result of that energy. The more energy absorbed by something, the faster the atoms in that something move. The faster they move, the higher the temperature of that thing gets.

Do this experiment: Go set your indoor oven to 225°F. Stick your arm in there. You can actually hold your arm in 225°F air for a few moments. Now press your hand against the metal side of the oven. When you get back from the hospital you will understand that, although the air and metal were both a temperature of 225°F, the heat, the energy, in the two is different. The metal stores and transmits fifty times as much energy as air because the metal has more molecules per square inch than air.

Energy melts fats, liquefies collagen, browns proteins and sugars, thins liquids like honey, dries out surfaces, crisps poultry skins, transforms solids to liquids and liquids to gases, creates vapors that smell seductive, makes air and other things expand, and, perhaps most important, kills bacteria.

When you heat food, the water molecules on the surface get excited, heat up, and start shaking their booty. They get the molecules next to them excited and shaking, and they get the ones next to them excited, and slowly, the energy works its way to the center of the food. That's how the outside of the meat cooks the inside of the meat.

Energy inside the food can only move as fast as one atom can impact its neighbor, and then its neighbor's neighbor, so it can take time, perhaps even hours, to get the chemistry altered to the level we want it. We call that dinnertime. Since the process is slow, energy tends to build up in the surface. That's why so many steaks have a dark brown crust, a brown layer just below the surface, a tan layer below that, a pink layer below that, and finally, a layer of rosy medium-rare in the center. Half the steak is overcooked. That's because the energy moves slowly to the center, overcooking as it goes. I call this the rainbow effect, a problem that I will show you how to avoid later when we dive into the reverse sear.

INDIRECT CONVECTION ENERGY

Technically, convection energy is the transfer of energy by a fluid—and yes, air, oil, and water are all fluids. The warm air circulating inside a grill, smoker, or indoor oven is carrying convection energy. Convection air does not have a lot of energy. Boiling in water has more convection energy and frying in oil has more energy still.

CONDUCTION ENERGY

Conduction of energy from metal grill grates, a pan, or a griddle in contact with food is the most concentrated energy source on a grill and the fastest way to cook the surface. Conduction is how we get grill marks.

The metal grates transmit strong concentrated energy by conduction and produce grill marks. The glowing coals produce weaker but still very strong energy by infrared radiation and brown the meat between the grill marks on the side facing the coals. The top of the steak is warmed by gentle convection airflow trapped under the lid. Within the steak the molecules vibrate and generate conduction energy from excitation.

DIRECT INFRARED RADIANT ENERGY

When you are shopping for a grill, don't be swayed by a manufacturer who says, "Our grill can get up to 600°F." He's talking about hot air and he's talking with hot air. The temperature of the air is not that important. We need to know how much *energy* is produced. We love hot air, but if you want to sear a steak, you

The Electromagnetic Spectrum

Size Comparison of Wavelengths

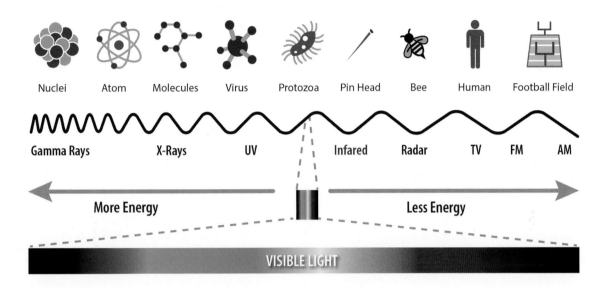

Nuclei · Atom · Molecules · Virus · Protozoa · Pin Head · Bee · Human · Football Field

Gamma Rays · X-Rays · UV · Infared · Radar · TV · FM · AM

More Energy ← → **Less Energy**

VISIBLE LIGHT

need direct infrared (IR) radiant energy or conduction energy, not hot air. A lot of grills just can't get it up. And if you love great steaks or burgers, you want a lot of IR.

IR is an intense form of light that glows red when it is visible to the human eye but some of it is invisible. You see it every morning from the coils inside your toaster. On a grill, IR is energy emitted by glowing coals or flames. This energy arrives at the speed of light and carries a real punch. That's why food browns and cooks much faster when exposed to IR. IR is a great way to get a sear on your steak.

Now please keep in mind that radiation takes many forms and not all are as scary as X-rays or gamma rays. Radio music arrives as a most welcome form of radiation. In the graphic of the electromagnetic spectrum (above), you can see that radiation comes in the forms of waves, and the distance between the waves, the wavelengths, can be as big as a football field and smaller than the nucleus of an atom. The closer together the waves, the more energy they pack.

TRY THIS: Turn your oven to 300°F and pop a steak in it. When it is done on the inside you'll notice it doesn't have a nice dark mahogany seared crust. Hot air can't do it. But you can broil it indoors because you are using direct IR from flame overhead. The broiler is like a grill, only the IR comes from above.

Because IR packs such a wallop, the surface of the food can burn before the interior cooks, so IR must be used judiciously, usually just to brown the surface. This is where a lot of newbies go wrong. They put the food directly over hot coals or gas burners and when the exteriors are really dark or starting to burn they take the meat off. But it is often raw in the center.

Searing seals in the juices

BUSTED. This is an Old Husbands' Tale that will not die. Meat fibers are not balloons that are tied off by high heat. Searing is highly desirable because it tastes good, not because it has any impact on juiciness. In fact, meats cooked at high heats actually lose more moisture than those cooked at low temps. But we don't mind because there is plenty of juice left and seared meat tastes sooooo goooood.

EXCITATION ENERGY

Microwave ovens work by excitation. Radio waves penetrate food and vibrate the water molecules inside it until the food gets hot without noticeably heating the air around it.

INDUCTION ENERGY

Some stovetops work by induction. A copper coil is placed under a smooth cooktop and an alternating current is sent through the coil creating a rapidly changing electromagnetic field. Electrons in steel or cast-iron pots on top of the cooktop are jostled by the rapidly changing magnetism. They get excited, but they resist and the resistance heats the pot. The pot then transfers the heat by conduction to the food without the cooktop or the air around it getting hot.

Induction is the most efficient way to cook, but it does not work with aluminum, glass, or copper pots. It has many advantages. It boils water rapidly, produces no combustion gases, and responds rapidly to control knobs.

THE INVERSE SQUARE LAW

The inverse square law says that radiation such as infrared, light, and sound dissipates rapidly as it moves away from the source. That's why people close to the camera in a flash photo are brightly lit and people in the background are dark.

Theoretically, if the food is 6 inches from the energy source (let's call that 1d) and you double the distance to 12 inches (2d), you don't cut the energy 50%, you cut it to 25%. The formula is **intensity = 1/(2d²)**. In other words, the energy intensity equals 2d squared (2 times 2) which is 4 which is also expressed as $^4\!/_1$. The inverse is ¼. So by doubling the distance you have cut the energy reaching the food by ¼ or 25%.

In practice, other factors such as reflection from the interior sides of the grill confound the calculation, but the lesson is simple: As you move away from the energy source, the energy drops rapidly. That's why I wish more charcoal grills had the ability to raise and lower the food or the charcoal bed like the Hasty Bake grills.

LOW AND SLOW

This we know: Heat causes muscle fibers to shrink and squeeze out juice. Heat also causes water to evaporate. It converts connective tissue to gelatin. It renders fat. It catalyzes chemical reactions like the Maillard reaction. It activates and deactivates enzymes. And it kills bacteria. So we have learned that cooking many things low and slow results in better outcomes. Throw a slab of pork spareribs on a hot grill and you'll get a tasty meal, but it will be as ornery as Clint Eastwood. But cook them at about 225°F for 5 to 6 hours and they will be tender as a baby's bottom and as juicy as an Agatha Christie mystery. Low and slow is the motto if you are cooking tough cuts, even if they are cooked to 203°F.

THE WARP SCALE

For searing we move the food directly above flame or glowing coals and bathe it in IR. *Because we are focused on searing the surface, we lift the lid.* We don't want hot convection air circulating around, warming the meat from above, and overcooking it. And we stand right there. We don't go to the bathroom or check the score in the game because you can go from a great dark Maillard sear to carbonized black in the blink of an eye.

You can't measure infrared from glowing coals or gas jets with a normal thermometer. For that you need a special device like a special thermocouple into the heat source or a spectrophotometer pointed at it. And because of the inverse square law, the energy dissipates rapidly over distance. And IR is often measured in watts per square meter, not degrees Fahrenheit or Celsius.

When I want you to cook with direct IR energy, I'll use a scale I made up rather than watts: The Warp Scale. Warp 10 is maximum IR on your grill, "Give 'er all she's got, Scottie." On occasion I may ask you to dial it back to Warp 7 or something like that when I am afraid Warp 10 is too harsh.

The Warp Scale

Cooking is all about temperature control. When talking about cooking in the indirect convection heat zone, I specify a temperature such as 225°F. That's easy to measure with a thermometer. But because infrared in the direct zone is measured in watts per square meter, I use a number scale I made up: The Warp Scale. Warp 10 is maximum IR, "Give 'er all she's got, Scottie."

THE IMPORTANCE OF THICKNESS

Repeat after me: Thickness determines cooking time not weight. This is a vital, foundational, essential core concept, and so many TV chefs

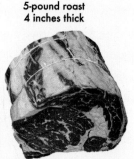
and cookbook authors just don't get it. So when you see a recipe that says to cook a roast at 325°F for X minutes per pound you're looking at a recipe for disaster. A 10-pound roast that is 4 inches thick and 12 inches long and a 5-pound roast that is 4 inches thick and 6 inches long will take pretty much the same time to cook because, even though one weighs twice as much, the thickness is the same.

Thicker foods take longer to cook and therefore the surface dehydrates more. And they are at greater risk of burning. That's why it's usually a good idea to cook at a low temperature. That's why it's better to cook two 12-pound turkeys than one 24-pounder. The thickest parts, the breasts, on the smaller birds are much thinner and will cook faster and they will be juicier than the big bird (not to mention that younger birds are more tender).

If you are cooking something thick on one end and thin on the other, the thin end will cook faster than the thick end. That's why it is important to inspect the meat you buy carefully. You want your steaks of even thickness. If the butcher has cut a steak poorly, it will cook poorly. That is why it makes sense to pound boneless chicken breasts flat and why you want to cut vegetables into chunks about the same size.

Let's say you've bought some pork chops and when you open the package you discover that one is thinner than the others. Solution: Put the thick one on first and the thin one after a few minutes.

THE HOCKEY STICK

Didn't Tammy Wynette sing the immortal song "Stand by Your Grill"? She must have known about the hockey stick, the curve that has spelled doom to many a griller. It is why you need to stand by your grill, especially toward the end of a cook.

When you throw cold meat on the grill it warms slowly and you get lulled into thinking you understand the pace at which it is cooking. So you head indoors to mix a cocktail, and when you get back outside the meat is overcooked. That's because the energy that cooks the meat builds up within the meat and speeds the cooking. You end up with a curve that looks something like a hockey stick. As you can see, the meat temperature rises slowly

at first and then it takes off, so the pace is not uniform and if you allow the first few minutes to set a mental clock, you are going to overcook your meat.

2-ZONE SETUP FOR ALMOST EVERYTHING

When cooking we are faced with a problem. We want high energy from conduction and radiation for the Maillard reaction and caramelization, but high temperature shrinks proteins and muscle fibers, squeezes out juices, and makes meats hard to chew, and high energy can burn things. At low temps you have more control and it is easier to hit your target temperature without overshooting it and meats are juicier and more tender. It is also easier to hit a slow-moving target than a fast-moving target.

So what to do? This is perhaps the most important method I can teach you: Set up your grill in two zones for almost everything.

On a charcoal grill, push all the coals to one side. If you wish, corral them with bricks. Now you have temperature control with a hot infrared zone and a cool convection zone. You can do just about everything this way. On the indirect side the warm air circulates around the food, gently cooking it from all sides. You shouldn't even have to flip the food over. If you want to smoke the food, all you need to do is throw wood right on the coals or flames. When it is time, you can move the food over to the direct, infrared zone and sear.

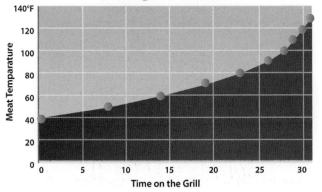

The Hockey Stick
Cooking a Thick Steak

Charcoal Grill 2-Zone Setup

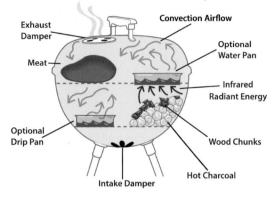

Gas Grill 2-Zone Setup

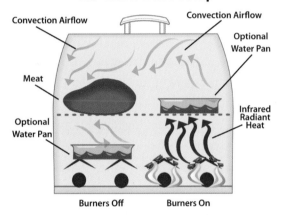

In these illustrations I have inserted water pans. They are useful for stabilizing the temperature and adding water vapor to the air. It condenses on the cold meat and smoke sticks to it. But water pans are optional.

On a gas grill, turn one or two burners on one side on and leave the others off. Wood chunks can go on top of the deflectors that cover the burners. Don't worry; burning wood cannot harm your metal burners or the covers. If you don't have chunks, use a disposable aluminum pan and poke some holes in the bottom. Measure how much you used so you have a benchmark for future cooks.

If you only have one burner, put a water pan on top of the cooking grates and steal a wire grate from the kitchen to put on top of the pan. With water in the pan you now have a nice cool indirect zone because the water can't get over 212°F. Try to keep the water from boiling, you don't want to steam the meat. When you need IR, simply remove the pan and turn up the gas.

Many grill manuals tell you to bank the coals on both sides like this. Don't do it. It leaves you little room in the indirect zone and it puts the edges of the food too close to IR.

If your grill has only one burner, try this setup.

3-ZONE COOKING

There is another configuration to play with: 3-zone for when you need flame. Sometimes I want the extreme heat of open flame to kiss a steak or burger. Set up a 2-zone configuration but place a log on top of the coals or a gas burner. When it ignites and starts flaming, move it to the side of the coals or burner. You now have an indirect zone on one side, an IR zone on one side, and a flame zone in the middle.

Charcoal Grill 3-Zone Setup

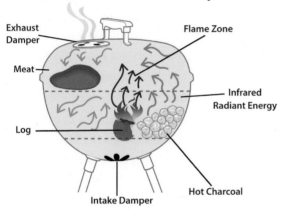

MASTER THREE TEMPERATURES

To keep things simple, almost all recipes in this book, in my last book, and on my website target three temperatures. Practice getting your cooker to these three numbers in all kinds of weather.

225°F. At this temperature in the indirect zone you are cooking low and slow, at a temperature that muscle fibers don't immediately seize up, shrink, and squeeze out all the juices. You can also render fat and melt collagen, perfect for tough cuts of meat like ribs and pork shoulder.

325°F. At this temperature you are cooking hot enough to render fat on poultry skins and get them crispy, but not too hot to squeeze a lot of juice out of meat.

WARP 10. This is the top end of your grill in direct infrared for searing. Pedal to the metal.

THE GENIUS OF THE REVERSE SEAR

Now here's where things get interesting. If you cook a thick steak over IR, you can get a gorgeous dark Maillard crust. As it continues to warm, the layer below turns brown, the layer below that turns tan, the layer below that turns pink, and finally, in the center you get a layer that is perfect medium-rare. Most of the meat is overcooked and the exterior may well be burned. I call it the rainbow effect.

Cooked hot and fast

Reverse sear

But if you put the steak in the indirect convection zone at about 225°F with the lid down, the energy transfer goes slowly and more evenly. At low temperatures energy has time to swim all the way to the center without overcooking the surface. The temperature can be close to the same bumper to bumper. Professor Blonder says, "It's like watering a potted plant. Sprinkle too fast and the pot overflows before the water can seep in. Sprinkle slowly and all the soil is moistened."

Called reverse sear because we sear at the end rather than the beginning, it works best on thick cuts, more than 1 inch thick. Start a steak on the indirect side at about 225°F with

Flip only once

BUSTED. For years we have been told to flip only once and many cookbooks still perpetuate the myth. They tell you to leave the meat alone. But this defies physics.

When searing a steak with infrared on a grill or conduction in a pan or griddle, the energy comes from a single source—below. Just as if you are lying on the beach in the sun, the side facing the sun collects energy, darkens, and cooks much faster than the other sides. So you need to flip or burn.

When you flip meat, the surface that faced the flame now faces up, but some of the energy stored in that surface continues to move gently down into the meat, and some of the energy dissipates into the relatively cooler air. Leave it alone and you get the rainbow effect. Flip often and you don't. Be the human rotisserie.

Flipping often also cooks the meat faster because heat is now entering the meat from more than one side at a time. Food scientist Harold McGee first explained the phenomenon in *Physics Today* in November 1999. J. Kenji López-Alt and I jumped on the bandwagon soon after, and Professor Blonder ran experiments that proved the theory in 2016. He proved the meat cooks 20 to 30% faster by flipping often. In March 2022 Jean-Luc Thiffeault, a professor of applied mathematics at the University of Wisconsin, published an equation-filled twelve-page paper in the peer-reviewed journal *Physica D* titled "The mathematics of burger flipping." His conclusion: Flip often.

How often is often? I say once a minute is too fast to build a dark crust. Every 2 to 3 minutes will do the trick. The result is a steak with more evenly cooked meat done sooner. A word of caution: Frequent flipping is not recommended for fish. It just falls apart.

the lid down so it warms gently and evenly from all sides. Then, when the center gets 10° to 15°F below your target temperature—let's say target is 130°F, so when it hits 115° to 120°F—take it out of the heat for about 10 minutes so the surface can cool a bit. Don't worry, the inside will stay warm. But this cooling allows you to sear a bit longer without pushing too much heat into the core. Then move it over direct scorching infrared at Warp 10 and lift the lid so all the energy is pounding only one side. Because the lid is up, the top side is cooler and energy isn't trying to overcook it from above. Then flip it and pummel the other side with IR and let the energy that has built up on the side that was facing the fire a minute ago bleed off into the cooler air rather than migrate to the center. And in defiance of tradition, we flip the steak every minute or two. Reverse sear. It is an important foundational method outdoors and indoors.

How many cookouts have you gone to where the chicken skin was charred and the meat was close to raw in the center? Because poultry skin has a lot of fat in it, direct IR can wreak havoc with it. Chickens are best cooked with indirect convection airflow for most of the process, and then, when they are almost done, you can move them over IR with the lid up for just a few minutes, watching carefully, to crisp the skin.

I use reverse sear for everything from huge prime rib roasts to baked potatoes (they are optimum at 208° to 210°F). The only time it doesn't work is on thin foods like skirt steak and many vegetables.

Reverse sear is perfect for pellet cookers. Because most are built like your indoor oven with a metal plate between the flame and the food, they rarely produce enough IR to get a good sear. So when I want that scrumptious smoked pork chop with a killer sear, I fire up my pellet cooker at 225°F and put a cast-iron pan in there. The meat doesn't go in the pan yet, it bathes in the warm convection air at first. When the meat hits 120°F on the interior I put a thin layer of oil in the pan (bacon fat anyone?), plop the meat in the pan, press it hard against the hot metal, let conduction sear it, flip, and pull it off in the 135° to 140°F range.

FOR THIN CUTS

I love *fajitas al carbon*, fajitas cooked over charcoal. They are usually made with skirt steak, a thin flap of tasty hardworking muscle from the underside of beeves. There are two cuts of skirt, inside and outside, and outside is a bit thicker and my preference. The problem is that they are both so thin that they can be well-done long before the exterior gets some nice Maillard flavor. Especially if you marinate it, because that makes the surface wet and the first few minutes of cooking are wasted evaporating the water, which cools the meat and retards the cooking. Yes, you can pat the meat dry, but then you are removing most of the marinade.

For thin cuts like skirt steak, skinny pork chops, or shrimp, reverse sear doesn't work well. There are four good solutions.

SEAR ONE SIDE ONLY. Cook it for only a minute on one side, just to take the rawness out. Then flip it and crank it to Warp 10 until it gets a dark sear. It will have a great crust on one side, be tan on the other, and remain rosy or pink in the center.

COOK FROZEN MEAT. Yes, you heard me. Get it frozen in the center and then sear it over Warp 10 on both sides. By the time the exterior is a beautiful dark brown, the interior has thawed and risen to optimum temperature. This also works if you marinate the meat. Skirt is a good cut to marinate because the striations in the grain have nice grooves for the marinade to nestle into, and because the meat is so thin, even if the penetration is shallow, it is still a large percentage of the thickness. But the problem with wet meat is that it must dry out before it can brown. If you marinate skirt for a few hours and then freeze it, then grill it over rip-snorting high heat, the surface can dry and brown before the frozen center turns gray.

THE AFTERBURNER. The afterburner is the name I've given to searing on top of a charcoal chimney as described on page 110.

TORCHING. You can get a decent shallow sear with a propane torch as described on page 230. It runs at about 2,000°F. The problem is that it can burn the meat in a hurry, and because you have to keep it moving, the sear tends to be shallow.

CARRYOVER COOKING

Even when meat comes out of the cooker, the heat built up in the outside layer of the meat continues to warm the interior. This is called carryover cooking.

As a rule of thumb, small pieces of meat like steaks, chops, and chicken will not continue to cook much after you take them off the heat, certainly no more than 5°F. But large thick roasts of beef, lamb, veal, pork loin, or turkey breasts can rise 10° to 15°F or more if they sit around before you start carving or serving.

Carryover can be a curse. You take your turkey off the grill at the perfect temperature and by the time it gets to table it is overcooked. That is one reason why I recommend you serve ASAP after cooking and don't let meat rest.

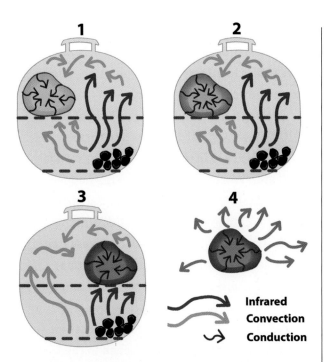

1 **2** **3** **4**

Infrared
Convection
Conduction

Here we see a reverse-seared roast. A charcoal grill is set up in 2 zones. (1) We start out with the roast in indirect convection heat until (2) the interior approaches the desired temperature. (3) We then move it to direct infrared radiation to sear the exterior and roll it so all sides sear and make a rich crust. (4) When it is almost finished, we remove it from the heat and some of the heat built up in the exterior of the meat bleeds off into the air while some continues to migrate to the center, cooking the roast further by carryover.

RESTING MEAT

"When it comes to all this talk about resting meat, I say give it a rest, stop crying over spilled juice, and clean your plate like Momma told you."

Practically every chef in the world says that when you take meat out of the heat it must rest at room temperature before carving or serving. Some say as little as 5 minutes, others as much as 30 minutes. Some even want you to "tent" it with foil. When it comes to normally grilled or roasted meats cooked to 160°F or below, it just doesn't make sense to me and I think it *may* be a mistake.

Notice the way I've hedged my bet. It *may* be a mistake for meats cooked below 160°F. I'm not sure, because there isn't any *real lab research by meat scientists*. Most advocates of resting say it is necessary to "redistribute" the juices and prevent juice from coming out of the meat when cut. In the journal *Meat Science*, in 2011 Pearce et al. surveyed the literature to present a summary of what is known of water in meat and they say, "Total water content of the meat and cooking loss cannot explain juiciness of the cooked meat product." Food scientist Chris Young, Professor Blonder, and Yours Truly have also done kitchen research and we think resting is a mistake.

That's because the juiciness sensation is dependent on not just water, but rendered fat, melted collagen, and even saliva.

SALIVA. When I put a sizzling steak in front of you, you salivate. The sizzling, the smell, all get your glands going. There is a lot of saliva in your mouth when you eat. Why rest meat, kill the sizzle, serve it cold, and stifle saliva?

FAT. Fat gets runny in the 130° to 140°F range. If you take it off the heat and let it sit around, it cools and the fat solidifies. Less juicy.

COLLAGEN. As meat cooks, collagen softens, producing a rich gelatinous texture. Let it rest and it solidifies.

WATER. Finally, there is water, or to be more precise, myowater, the juices made of water and tinted pink with myoglobin. Most kitchen research shows that if you cut into hot meat right off the grill or oven, *slightly more* juice comes out onto the plate than it does with meat that has rested for 5 minutes. But when you look closely at the experiments, the differences in the amount of juice spilled is small.

The theory is that meat fibers are like skinny balloons that shrink. If you cut into the meat when it is fresh off the heat, the theory says pressurized juices come gushing out. If you let meat rest and cool, the theory says water pressure drops, fibers relax, and fewer juices escape. A variation on the theory says that the juices run away from the heat from the side facing the flame.

Not so, said the late Dr. Antonio Mata, a meat scientist with whom I consulted often. "Water moves back and forth between compartments. It is not trapped in the fibers. Fibers are not balloons." The pressure equalizes quickly. Furthermore, water cannot be compressed under cooking conditions. In other words, this theory just doesn't hold water.

When you dry-age a steak it can lose 20% of its water during the aging process. But nobody complains that it is dry. Not to mention that during cooking another 10% or more is lost due to evaporation and drip loss. When you sous vide, a lot of water comes out in the bag but nobody complains it is dry.

Resting has other detrimental impacts. Resting meat softens the crust and turns crispy poultry skin to rubber, especially under a foil tent. Worse, carryover cooking happens. That can take your perfectly cooked steak or turkey to cardboard before you know it.

Furthermore, a lot of the kitchen research I have seen is flawed. Many experimenters slice up a whole steak to measure how much juice is spilled. But that's not how we eat a steak. First of all, it takes a few minutes for the steak to go from grill or kitchen to the dining room. Then we cut off a bite. One bite. Chew. Have a sip of wine. Eat some mashed potatoes. Then some green beans. Debate politics. Then back for another bite of meat. And if juices escape, what do we do? We mop them up with the meat on the fork. Nothing is lost. When the meal is done, the plate is clean. Have you ever seen a plate sent back to the kitchen with puddles of juice?

HOLDING MEAT

We need now to make a distinction between *resting meat* and *holding meat*. I think I have made a good case that meats cooked to 160°F or below should not be rested. But some meats, those cooked way beyond well-done, up beyond 190°F, like beef brisket or pork butt, are very different. Pitmasters have learned that when these meats are done cooking, wrapping them tightly in foil or butcher paper and *holding* them in an insulated box for an hour or more and letting them carryover cook and slowly cool benefits them. *I am not a fan of resting, but I am a fan of holding. There is a difference.*

These are very different meats than steaks, chickens, chops, etc. They are tough cuts with lots of fat and connective tissues that need to be cooked low and slow to a high temp to make them chewable. A steak cooked to 203°F would be inedible, but beef brisket is at its best at about that temp. Rendered fat and gelatin from connective tissues provide most of the juiciness. In fact, these cuts often can lose up to 30% of their weight during cooking to drip loss and evaporation of water.

During the holding period, very gentle, slow carryover cooking continues to cook the meat and tenderize it as it cools. These tenderizing, collagen-melting reactions take time. A few hours more in a hot smoker would dry out the meat, but a few hours in a warm beer cooler tenderizes without dehydrating.

Because these meats have been cooking for 8 to 12 hours, the surface has become dry, forming a jerky-like bark. Any water left is in the center. By wrapping it so no more water will evaporate, and cooling it slightly, that water can move back into the parched areas.

The problem is that if you let it go too long, it can soften the bark too much. It is a balancing act, and that is why top barbecue cooks are called pit*masters*.

Butter-basting steaks

BUSTED. Some chefs just hafta play with their food. Google "what's the best way to cook a steak" and you will find pages of videos of pan-seared steaks swimming in butter with chefs spooning the hot butter over the steak. Here's what that does: It slightly reduces evaporation, it keeps the top hot, and it adds a lot to carryover cooking. And that means that if you are foolish enough to rest your meat, it will almost certainly be overcooked. What it doesn't do is add flavor or speed cooking time. Chris Young, inventor of the amazing predictive cooking thermometer, Combustion, a wireless probe with 8 sensors, ran some tests and agrees with me. The only time basting added flavor was when herbs were in the butter and you can get that same effect with a rub.

Basting poultry

BUSTED. Painting your Thanksgiving turkey while it is cooking only cools the meat, slows cooking, and makes the skin soggy. I want crispy skin. No way a baste can penetrate the fatty skin and improve flavor.

Basting ribs, brisket, pork butt, etc.

Pitmasters often baste or spritz these low and slow cooked meats during cooking. It slows cooking and that is good because they are headed toward 203°F, and the longer they take, the more tender the meat. Alas, this does not add flavor. There is not enough flavor in the liquid. If you want to add flavor, head for the spice rack. The liquid may replenish some of the liquid lost to evaporation, but if you wet the surface during the last 30 minutes of cooking, the bark may be softer than you want.

UNDERSTANDING SMOKE

"Smoke is the soul of barbecue."

Campfires with logs were likely the first cooking fuel. Energy from combustion cooked the food, and smoke from the logs imparted a distinctive seductive scent. Compounds in smoke also helped preserve meats. Combustion, as it applies to barbecue and grilling, is a sequence of chemical reactions between oxygen and a fuel producing a change in the chemistry of both, creating heat, light, and a blend of water vapor, particles, and gases that we call smoke.

Cooking with logs became the norm for centuries of cooking both outdoors and in. But it is difficult to control energy and flavor when cooking with logs, so today, only a few expert pitmasters outfitted with special rigs cook with logs. Today, most grills and smokers use charcoal or gas to produce the energy, a few use electricity, and a rapidly growing number use wood pellets. They get flavor and aroma from wood in the form of wood chips, chunks, logs, sawdust, or compressed sawdust like pellets. Adventurous cooks can also get smoke from dried herbs, tea, and even hay. Think of smoke as another spice.

The best wood smoke comes from *deciduous* trees such as fruitwoods (apple, peach, cherry), nut woods (hickory, pecan, walnut, oak), and other hardwoods (maple, mesquite). Deciduous trees have compact cell structures. *Conifers*, evergreen softwood trees such as pine, fir, spruce, redwood, hemlock, and cypress, have more air and they burn fast. Their sap is loaded with highly flammable terpenes from which turpentine is made. They are not good for cooking.

Smoke can also come from meat drippings, which are laden with fat, protein, spices, and sugars. In the 1930s the Works Progress Administration (WPA), a federal government project, interviewed hundreds of former slaves and transcribed the recordings trying to capture the dialect. Wesley Jones of South Carolina, formerly a slave, remembered how he cooked hogs for special occasions and the impact of the drippings. The WPA transliterated his oral history into the colloquial accent. "Night befo' dem barbecues, I used to stay up all night a-cooking and basting de meats wid barbecue sass. It made of vinegar, black and red pepper, salt, butter, a little sage, coriander, basil, onion, and garlic. Some folks drop a little sugar in it. On a long pronged stick I wraps a soft rag or cotton fer a swab, and all de night long I swabs dat meat 'till it drip into de fire. Dem drippings change de smoke into seasoned fumes dat smoke de meat. We turn de meat over and swab it dat way all night long 'till it ooze seasoning and bake all through."

SMOKE is an aerosolized combination of tiny solid particles mixed with water vapor and a complex cocktail of gases. The exact mixture is crucial and it can add elegant vanilla and brown spice notes or coarse bitter or even ashtray taint. Too much can be as bad as too much salt. The right amount of smoke adds a dimension, a sexiness that you can't find on a spice rack.

Smoke contains as many as 100 compounds in the form of microscopic solids, including

How the Heat Source Influences Flavor

GASES
Carbon Monoxide
Carbon Dioxide
Nitric Oxide
Acids
Alcohols
Aromatic Polymers
More

SOLIDS
Char
Creosote
Ash
More

LIQUIDS
Water Vapor
Phenols
Creosote

Wood

GASES
Carbon Monoxide
Carbon Dioxide
Nitric Oxide
Polymers
More

SOLIDS
Char
Ash
More

LIQUIDS
Water Vapor
Phenols

Charcoal

GASES
Carbon Dioxide

LIQUIDS
Water Vapor

Gas

GASES
Air

Electric

char, creosote, and ash, as well as combustion gases that include carbon monoxide, carbon dioxide, nitric oxide, and liquids such as water vapor, phenols, and syringol.

Trace amounts of syringol are responsible for much of the smoke aroma we love, and trace amounts of guaiacol are responsible for much of the flavor of smoke. The composition of smoke depends a great deal on the composition of the wood, the temperature of combustion, and the amount of oxygen available.

Bill Karau, designer of my favorite smoker, the log-burning Karubecue (page 118), told me, "All smoke is created equal and equally nasty. Look closely at the smoke coming out of a fissure in a log in a fireplace. You'll see a plume of thick, yellowish gas. Stick a spoon in that gas for two seconds, let it cool for a few seconds, and then lick it. Revolting. You'll be tasting it two hours later.

"The key to great barbecue is what happens to the smoke after it's emitted. Crude smoke can be refined by burning it with flame. And that requires lots of oxygen. You don't get that when wood is deprived of oxygen and it smolders. With oxygen deprivation you get poor quality flavors."

HOT TIP

This is crucial: Smoldering wood produces less tasty smoke than wood burning with a flame. Fire is like you. Both need oxygen.

If the wood does not get enough oxygen, it can still undergo pyrolysis and gasification, but not combustion. It will not burst into flame, it will smolder, and *smoldering wood* produces lots of smoke and a different flavor than *burning wood*. Why would it not get enough oxygen? If the intake vents and flue are not open enough.

Wood also plays a role in the color of the meat and the formation of the crust on the meat. Here are two slabs of pork ribs with the same spice rub but no sauce. Both were smoked with applewood. The one on the left was cooked on a charcoal smoker and the one on the right was cooked on a gas smoker. You can see and taste the difference. They were both excellent, but different. The one on the left had a deeper fireplace aroma. The one on the right had a bacon undertone.

Whether you are cooking with logs or charcoal with wood chunks for flavor, you want to control the color and thus the flavor of your smoke.

BLUE SMOKE, CLEAN SMOKE. The most desirable smoke is almost invisible with a pale blue tint also called "clean smoke." Blue smoke is the holy grail of low and slow pitmasters, especially for long cooks. The color depends on the size of the particles and how they scatter and reflect light to our eyes. Blue smoke particles are the smallest, less than a micrometer in size. Just because you can't see it, doesn't mean it's not there. *You get blue smoke when the wood is burning with a visible flame, when impurities have been consumed, and when it has burned down to embers. That means you need lots of oxygen. Don't let the wood smolder.*

WHITE SMOKE, DIRTY SMOKE. This consists of larger particles, a few micrometers in size, and they scatter light in all directions. If wood doesn't get enough oxygen, it smolders and produces white smoke. Blue smoke may be best, but white smoke isn't always bad. It tastes fine and it can serve a purpose (and tell us when there is a new Pope). If you are cooking something thin, you want to cook hot and fast, and lots of white smoke works better than blue smoke when you are in a hurry. Some people prefer the flavor of dirty smoke.

YELLOW, GRAY, AND BLACK SMOKE. These contain particles large enough to absorb some of the light and colors. They happen when the fire is starving for oxygen, and they can make bitter, sooty food that tastes like an ashtray.

CREOSOTE. This is the Jekyll and Hyde of smoke cooking. On the Dr. Jekyll side, the right amount of creosote contributes positively to the flavor and color. Professor Blonder says,

MYTH

Soak your wood for more smoke

BUSTED. There's a reason they build boats out of wood. Wood doesn't absorb water. I've done the test. I've taken a bunch of wood chips and weighed them, soaked them overnight (not just an hour or so as many books tell you), patted the surface dry, and weighed them again. At most a 5% weight gain. And when I cut the chips in half and looked at them with a magnifying glass, I could see that all the water is on the uneven surface in cracks and pores and fuzz. Nothing on the inside. Chunks, because they have less surface area, hold less water.

So here's what happens when you throw wet wood on hot coals. The moisture cools your fire because boiling off the water consumes energy. And all that smoke? It's steam. That's because water boils at 212°F, but wood doesn't combust and make smoke until somewhere north of 500°F. So by wetting your wood you slow the formation of real smoke. Don't do it.

"Creosote is always present in wood smoke, and a few components of creosote (guaiacol, syringol, and phenols) are the largest contributors to smoke flavor. Without creosote the meat might as well have been boiled." On the Mr. Hyde side, "If the balance of the hundreds of chemicals in creosote shifts, it can taste bitter rather than smoky."

According to Professor Blonder, "when you smoke low and slow at temperatures like 225°F, many cooks control the fire by damping the oxygen supply, which moves it below the ideal combustion zone, creating creosote. Unfortunately, this is often the case with kamado and egg smokers. The best pitmasters cooking with logs burn small hot fires that combust at a high temperature to create the ideal flavor profile and direct a small fraction of the smoke across the meat." Although the fire may be hot, it is small enough that the cooking chamber does not get hot. When cooking with wood, to control creosote, practice good fire management methods and practice, practice, practice.

Cellulose and hemicelluloses are large molecules made of carbohydrates and sugars. Cellulose when burned is flavorless. Almost all the flavor comes from lignin. Lignin is another complex compound that gives wood strength, and it is found mostly in cell walls. The minerals in wood include potassium, sulfur, sodium, chlorine, carbon, and heavy metals. Although there are only trace amounts, these minerals can significantly impact the aroma and smoke flavor. Wood also contains pockets of air, making it such a good insulator that you can set one end of a wooden match on fire and hold the other end in your fingers.

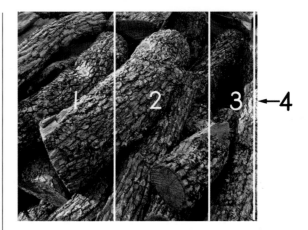

Not counting the water, here is what is in wood (actual numbers will vary depending on the species, subspecies, age, soil, and climate):

1. 40% of wood is cellulose
2. 40% hemicelluloses
3. 19% lignin
4. 1% minerals

CHUNKS, CHIPS, PELLETS, BRICKS, ETC.

You don't have to cook with logs to get the seductive flavor of smoke. You can generate heat with charcoal, gas, wood pellets, even electricity, and use pieces of wood for smoke flavor. Smoke wood comes in a variety of other compositions, shapes, and sizes.

CHUNKS. Wood chunks from egg to fist size are easy to find in hardware stores. Chunks burn more slowly than chips, and often a chunk or two is all that is necessary for a load of food. Because they burn slowly and steadily, they are in many ways the most desirable. When you use chunks, you can add one or two at the start of the cooking cycle and you don't need to keep opening the unit and messing with the equilibrium in the cooking chamber's atmosphere. You can make your own chunks by buying hardwood logs or lumber and cutting them.

CHIPS. About the size of fingernails, chips are easy to find in stores. Throw them on a flame and they burn quickly. You may find that you need to add them more than once during the cooking cycle. On electric smokers they smolder. Chips are fine for short cooks, but for long cooks, chunks are better.

PELLETS. Pellets are made by compressing wet sawdust and extruding it in long pencil-thick rods. They are broken into small bits about ½ inch long. They usually come in 10- to 40-pound bags. Food-grade pellets contain no binders, glue, or adhesives, and when they get wet they revert to sawdust. Many smokers use pellets as the main fuel for both flavor and

heat. You can also toss pellets on the flame for smoke on a gas or charcoal cooker.

They burn hot and clean. I love using these products because they are easy to measure and control. They are usually a blend of oak and other wood types. They use oak as a base because it feeds and burns more easily. Some brands even have flavoring added.

BRICKS. Another form of compressed sawdust is the brick, the most notable being made by Mojobricks. These are typically the size of a pack of cigarettes. They are used for flavor, not for fuel/heat. I have had good luck with them on a variety of smokers.

BISQUETTES. Bisquettes are another variation on the compressed sawdust idea made for the Bradley Smoker. They look like small brown hockey pucks.

SAWDUST. Sawdust can also be used for flavor, but it burns quickly. It is popular with restaurants. They scatter the sawdust in the bottom of pans, place a grate above the sawdust, put the food on the grate, cover the pan, and put it on the range over a burner. The sawdust smolders and the powerful overhead exhaust system whisks away any smoke that leaks out. Sawdust can be used effectively on thin, fast-cooking foods like shrimp and fish fillets. There are even a few small dedicated smoker pans that use smoldering sawdust. A popular one is by Camerons.

SMOKE GENERATORS

When I am smoking on gas grills, I use a smoke generator with wood chips, pellets, or sawdust. Sometimes I make sawdust by putting some pellets in a bowl and covering them with water. After a few minutes they dissolve into sawdust. Then I just dump it into a strainer, spread the wet sawdust on a pan, and it dries beautifully. I light them with a torch, butane lighter, or candle lighter.

FOIL PACKET. Use heavy-duty foil or two or three layers or regular foil and fill it with pellets, chips, or sawdust. Poke holes in the top so oxygen can get in and smoke can escape. Place the pouch as close to the heat as possible. It will smolder and make white smoke. If you are having problems getting the wood in a pouch to smoke on a gas grill, before you put the meat on, turn the burner on high, put the foil packet on, and wait for the chips to begin smoking. Then dial the burner down so you can get the oven to 225°F.

TIME BOMBS. Get two disposable aluminum loaf pans. Add dry wood to both. Pour enough water in one to cover the wood. The dry pan will start to smoke quickly. After it is all consumed, the other pan will have dried out and begun smoking. This method is ideal for both gas and charcoal cookers when you have a long cook and getting under the grate will be tricky, like when there's a whole brisket on board.

TUBE OR POUCH. Another way to add smoke flavor to something that cooks fast like a fish fillet or to cold-smoke safe foods like cheese is with a smoking tube or pouch. Tubes are

typically perforated stainless steel that you fill with wood pellets or chips. You light one end with a propane torch, and they smolder up a cloud of white smoke for hours. I like the Smokist Smoking Pouch, an envelope of fine-mesh stainless steel that holds wood chips or pellets.

SMOKE DADDY SMOKE GENERATOR. Smoke Daddy makes a small tank that holds pellets or chips, and when ignited, a blower pushes the smoke into your grill. Great for cold-smoking cheese and chocolate.

SMOKING GUN AND CLOCHE. The Smoking Gun is the original of many brands that have a thimble-sized cup in which you place

sawdust and light it. A motor sucks air through the smoldering wood and blows the smoke out a tube. The tube can be placed in a cocktail glass, in a pot with a lid, or as we see here, under a cloche at Smith & Wollensky steakhouse in Chicago.

STOP OBSESSING OVER WHICH SPECIES OF WOOD TO USE

I get a lot of emails like this: "I'm smoking a turkey, which wood should I use?" The answer is "Yes."

Far more important than the species of wood is how the wood burns. I cannot overemphasize this point. Controlling combustion is by far the most important aspect of the flavor of smoke, far more important than the species of wood. Sterling Ball is a champion pitmaster whose trophies include the big one, the Kansas City Royal Invitational. He also owns the famous Ernie Ball guitar business. He describes the art of making tasty smoke as similar to tuning a guitar: "You need control of your instruments, the pit, fuel, oxygen, fire, and heat. And you need to practice."

Although I can often smell differences in the woods I use, it is hard to taste them in the food. That's because if you use wood properly,

Why Smoke Gets in Your Eyes

It seems to me that no matter where I stand, the smoke blows in my eyes. Professor Blonder saw the cartoon I commissioned and, sonofagun, he proved it! He actually wrote a paper on the fluid dynamics of grill smoke complete with computer animations explaining that, as hot air rises, cold air rushes in creating a radial inward air current around a fire. Your body partially blocks this current. Like water coursing around a piling, whirlpools form downstream. These vortices sweep around your head, over the fire, and eventually circulate smoke back to your face.

it is just one instrument in the orchestra: The meat, salt, rub, cooking temperature, smoke, sauce, side dishes, and drinks all influence the overall impression. To be sure, some woods (I'm looking at you, mesquite and hickory) are stronger than others, but anybody who attempts to describe wood flavors in florid terms like wine flavors, as I see on many websites, is on a fool's errand.

The differences between hickory grown in Arkansas and hickory grown in Ohio may be greater than the differences between hickory and oak grown side by side. And there's really no such thing as "hickory." There's shagbark hickory, shellbark hickory, scrub hickory, pignut hickory, mockernut hickory, pecan (technically it is a species of hickory), and more. There's post oak, white oak, black oak, live oak, pin oak, and more. The climate and soil the tree is grown in can make a difference. Hot climate drought-stressed Texas oak grown in sand is different from cool-climate Michigan oak grown in river silt. Moisture content is crucial. Is the wood still fresh and green? Air-dried? Kiln-dried? Bark on or off? Are you using logs, chunks, chips, or pellets?

To make matters worse, there is no guarantee that the wood in the bag is the wood on the label. How do we know that the bag of cherrywood is really cherry? Coffee, olive oil, and fish markets are regularly rocked by scandals of fraudulent labeling. When Big Box Hardware calls and orders ten thousand bags of hickory and the rickyard only has enough for seven thousand, do you think they say,

"Gosh, we don't have it, why don't you call somebody else"? You think that maybe they might mix in a little oak to complete the order?

Do you really think anyone can taste a smoked rib with a rub and sauce and tell me what wood was used? Get real.

> *"If the sign out front says 'Barbecue' and there's no wood pile out back, keep driving."*

AVOID SOFTWOODS. Never use wood from conifers.

NEVER USE LUMBER SCRAPS OR PAINTED WOOD. Some lumber is treated with chemicals that are poisonous. Never use wood that has been painted.

NEVER USE WOOD THAT IS MOLDY. Some molds contain toxins.

USE DRYISH WOOD. If you buy logs, ask for something around 20% moisture.

BARK OR NO BARK? Some wood has more bark than others. This is a point of contention. I have not heard a good reason to either take it off or leave it on.

WHAT DO I PREFER? If I was on a desert island, I would want a bag of applewood chunks and a bag of small apple chips or pellets. I would use the chunks for steady slow-release smoke, and the chips or pellets for quick smoke. Applewood is mild and rarely tries to take center stage.

WHERE TO GET IT? Many hardware stores carry only hickory, but a few carry a selection of woods. We have links to many wood suppliers on the website. Another option is to go to an orchard and ask if you can have some dead trees or limbs, but be careful: They could be laden with pesticides or other sprays.

SMOKE AND FOOD

In a smoker or grill, after combustion, the smoke rises and flows from the burn area into the cooking area. Most goes right past the food and up the chimney. Professor Blonder explains why: "Around every object is a stagnant halo of air called the boundary layer, especially around cold meat. Its thickness depends on airflow, surface roughness, and meat temperature. When smoke particles approach the meat's surface, they follow that boundary layer around the food. Very few ever touch down. We've all cursed a form of this piece of physics while driving: Gnats follow the airstream over the windshield, while larger insects make green sticky splats."

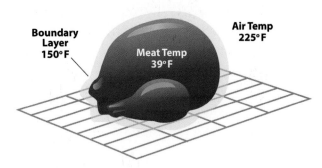

Boundary Layer

Boundary Layer 150°F

Air Temp 225°F

Meat Temp 39°F

The boundary layer also impacts cooking in your indoor oven because it can be a lot cooler than the temperature to which you set the oven. The cool meat cools the air around it and steam evaporating also cools the air. The boundary layer can be as much as 2 inches thick, and the air in contact with the food can be as much as 200°F below the oven temperature. If you are using a thermometer to measure the oven/grill/smoker ambient air temperature, make sure to get it at least 2 inches away from the food. That's the magic of convection ovens and air fryers: There is a fan that blows the cold boundary layer away.

Some smoke particles and gases do get through and contact the surfaces of foods. They may dissolve and penetrate as much as ⅛ inch, but not much more. Meats, especially, are hard to penetrate. Here's how to prove it to yourself. Get a 4-pound section of pork loin about 3 inches in diameter (not tenderloin, it's too thin). Cut it in half. Do not use salt or any rub. Smoke the heck out of one half at 225°F until it is about 180°F internal temp. Roast the other half in your indoor oven at 225°F until it is about 180°F internal temp. That is way past well-done, but we want to expose one to a lot of smoke. Put them on separate plates. Let them cool a bit and cut a core sample out of the center of each. Have a friend serve these core samples to you with your eyes closed. Can you taste the smoke? Didn't think so.

Ice Water   Empty   Control

I showed this photo in my last book and I have to show it again. I painted three beer cans white. One was filled with ice water and two were empty. One empty can sat on my desk as a control and the other went into the smoker along with the can with ice water. The empty one, center, got a nice even amber coat of smoke. The one with the ice water clearly attracted more smoke. Professor Blonder explained that this is a phenomenon called *thermophoresis*, a force that moves particles from a warm surface to a cold surface. That's why your bathroom mirror fogs up when you shower.

Thermophoresis happens to meat when you put it in the smoker. It will start at refrigerator temp, let's say 36°F, and rise to perhaps 205°F. In theory the surface could dry out completely, forming a very hard bark, and it might get slightly hotter than 212°F, but that is never the case. In fact, as the interior temperature rises to about 200°F, it might even be hotter than the surface that is still being cooled by evaporation. So if you are cooking at 225°F, the meat will always be cooler than the air.

One of the biggest mistakes we make is using too much smoke. Too much smoke can make your meat bitter or taste like an ashtray. I cannot give you a precise amount because each cooker is

At about 150°F, meat stops taking on smoke

BUSTED. Sorry, but meat does not have doors that it shuts at some time during a cook. There is a lot of smoke moving through the cooking chamber, although sometimes it is not very visible. If the surface is cold, or wet, or rough textured, more of it sticks. Usually, late in the cook, the bark gets warm and dry, and by then the coals are not producing a lot of smoke. Smoke bounces off warm dry surfaces so we are fooled into thinking the meat is somehow saturated with smoke. Throw on a log and spritz or baste the meat and it will start taking on smoke again. Just don't baste so often that you wash off the smoke and rub.

Let meat come to room temperature

BUSTED. So many recipes say to start by taking the meat out of the fridge and let it come to room temperature, called tempering. There are some benefits. If you have a steak and the target temperature is 131°F, going from fridge temperature of 38°F to 131°F means a trip of 93°F. But if it comes to room temperature of 70°F, the trip is shortened to 61°F, so the steak will cook faster and lose less weight.

The problem is that it can take 2 hours for a 1½-inch-thick steak to come to room temperature, during which time bacteria have a rave. A 3½-inch-thick roast takes 10 hours! Besides, cold wet meats are smoke magnets. So it is a best practice to go straight from the fridge to the grill or smoker for more grill or smoke flavor.

different and the amount of wood to get the right flavor depends on the volume of the cooking chamber, the airflow, leaks, how often you peek, the kind of wood you use, basting, humidity, the weather, and, of course, your preferences. You will need to experiment, but a good rule of thumb is to start experimenting with about 4 ounces of wood, regardless of the cut or weight of the meat. If the results are not smoky enough, you can add more wood on your next cook. Keep records of your experiments on a cooking log (see page 144).

THE SMOKE RING

Smoked meats often have a pink band directly below the surface called the smoke ring. Backyarders know they have arrived when they make their first smoke ring.

Alas, as with so many things barbecue, there is a lot of misinformation about the smoke ring, and anybody who tells you that good barbecue needs a smoke ring is blowing

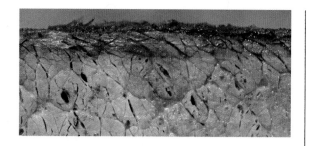

smoke. Professor Blonder did some great research for us and has pinned down the facts, among them, this: You don't even need smoke to make a smoke ring!

IT'S ALL ABOUT THE MYOGLOBIN. Professor Blonder has proven that the smoke ring is an interaction between myoglobin, which is pink, and the gases nitric oxide (NO) and carbon monoxide (CO). NO and CO are made during the combustion of wood or charcoal. Like many proteins, myoglobin changes color when it breaks down after exposure to heat. That's why beef goes from purple or red to tan.

NO and CO willingly mix it up with meat juices and basting liquids. These dissolved gases react with iron in the myoglobin and prevent it from changing color. But they do not change the flavor. That's right, smoke rings have no flavor. If you want a smoke ring, here's how to get it.

> *"A cook knows how, a chef knows why."*
>
> **—ANONYMOUS**

BURN WOOD. Wood produces more NO and CO than charcoal or gas.

DON'T USE ELECTRIC SMOKERS. Electric smokers use a heating coil like old-fashioned electric ranges and produce smoke by smoldering wood chips in a pan above the coil. The wood smolders at a low temp and combustion at high temps are required to create NO and CO. Experts at cooking in electric smokers sometimes add a charcoal briquet to the wood pan because some briquets contain powdered sodium nitrates, which enhance ring formation.

REMOVE THE FAT. NO and CO can penetrate fat, but if the fat is too thick or if the underlying meat changes to brown before the gases get there, you won't have a smoke ring. There are many other good reasons to get rid of the fat cap discussed in Trimming Fat (page 20).

MOISTURE IS A FACTOR. If you want a smoke ring, keep the surface moist by basting or spritzing it with a thin water-based mop during the first hour of the cook. Putting a pan of water in a smoker helps because the evaporating water condenses on the cool meat. But keep in mind, wetting the meat cools it and slows cooking.

TEMPERATURE OF THE MEAT MATTERS. Start with cold meat and keep the temp down at first. Then basting or spritzing the surface cools it and allows the ring to grow. Yet another reason to cook low and slow.

Professor Blonder, a scientist with a little poetry in him, observes that "Myoglobin, a molecule essential for life, is as fragile as life itself. It can breathe in gases and react by changing color. It can die when it overheats. And like every individual, it has its own, unique personality. Understanding myoglobin is a key to understanding how meat cooks, looks, and tastes."

BUYING A GRILL OR SMOKER

The law of diminishing returns applies. A $2,000 gas grill is probably better than a $1,000 model, but rarely twice as good. But more expensive units tend to be better built and will last longer. Be sure to inspect the welds and moving parts.

I am not a fan of stainless steel. There are different grades and some can rust. They look pretty until they don't, and I just don't care what the outside of my grill looks like. The inside is another story altogether. It is easy to become obsessed with making stainless shine.

In the fall, hardware stores are making room for Halloween and Christmas displays and that's the time they decide to clear out unsold grills and smokers and display models. The selection may be limited, but there are usually deep discounts. Just check carefully for broken knobs and missing parts. Find the store manager and make them an offer. Start at half off. I once got a $2,000 gas grill for $600 because some knobs were missing. I ordered them from the manufacturer for under $50. And be sure to check the detailed reviews by Max Good on AmazingRibs.com.

COOKING WITH LOGS

"Wood is not a fuel. It is a fuel source."
—BILL KARAU, INVENTOR OF THE KARUBECUE SMOKER

Without a doubt, the best-tasting steaks I have ever made were cooked with direct heat over split logs. Without a doubt the best ribs I have ever tasted were smoked with indirect heat from split logs. Cooking with logs makes special food.

If you want to try it at home, you can cook with logs in just about any charcoal grill. Fireplace logs are usually too large for grilling. They should be about the thickness and length of a baby's arm, perhaps 2 inches thick and 10 inches long. You can cut the wood with a saw or an axe. I use a splitter called the Kindling Cracker.

Perhaps you have noticed how many restaurants now grill with wood. The restaurant Birch in Milwaukee cooks almost everything over wood. And the food is wonderful. Notice how the chef is burning logs down to embers on the left of the hearth? Embers fall through the grate and are shoveled to the right where they cook customer meals. High above the flames he is searing steaks. And yes, those are greens on the grill. Notice the split logs drying underneath the heat? See the bricks wrapped in foil on the right? They can be heated to cook Chick Under a Brick (see recipe, page 272). They have sheet pans warming high above. Notice how far back the chef is standing? Poor devil was soaked in sweat and he had no hair on either of his arms. But, bro, that was one tasty meal.

When wood burns, it decomposes into water vapor, charcoal, and smoke. It goes through four stages. In a typical log-burning cooker, all four stages happen at once.

STAGE 1: DEHYDRATION (212° TO ABOUT 500°F).
For grilling with wood, build a small log house with splits. In this stage the splits must be heated from an external source. Ignite them with a pile of twigs in the center or some charcoal. If you try newspaper, splash some used frying oil on the logs first. A lot of energy is consumed in evaporating the water in the wood and that keeps the fire cool at first. Steam and some gases like carbon dioxide are given off, but there is not much flame or heat produced. Don't put the food on yet. Too many nasty gases.

STAGE 2: PYROLYSIS (ABOUT 500° TO 700°F).
This is when wood decomposes and gasifies. Cellulose and lignin break down and boil off in a gaseous cornucopia of volatile organic compounds and particulates. You can see this in a log in a fireplace or campfire: As gases shoot out through cracks, they ignite and burn orange. Resist the temptation to start cooking. Too many bad-tasting impurities in the smoke.

STAGE 3: COMBUSTION (ABOUT 700° TO 1,000°F).
Escaping gases burst into flame, among them nitric oxide (NO) if there is sufficient oxygen. Nitric oxide is essential for formation of the smoke ring (page 69) in meat. In the sweet spot of about 650° to 750°F, the best aromatic compounds vaporize, among

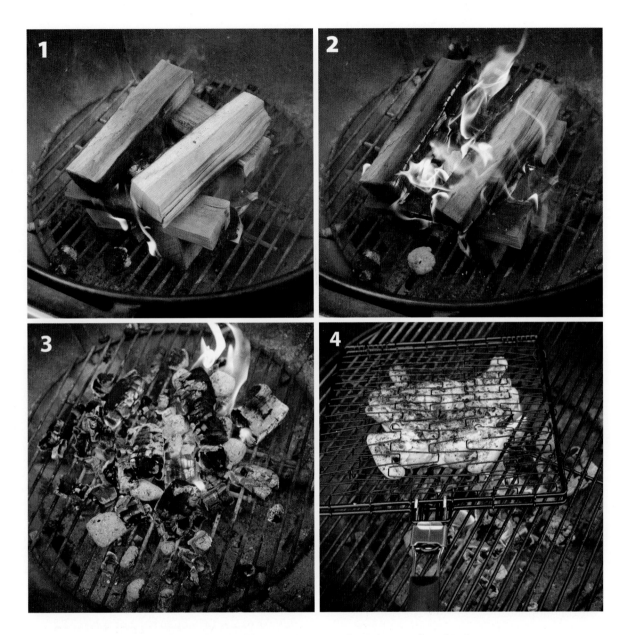

them guaiacol and syringol. Some are ethereal and dissipate, and that's why smoked meats don't taste the same after they have been reheated. As the temperature rises above 750°F, acrid, bitter, and possibly hazardous compounds are formed. Still too many flames at first, but as they die down you can start cooking.

STAGE 4: CHARCOAL FORMATION (ABOVE ABOUT 1,000°F). Most of the organic compounds have burned off, leaving behind pure carbon, or char, which burns as red embers with little smoke, odor, and flavor, just some tasty smoke from wood that hasn't carbonized yet. If you want more smoke flavor, you can throw a chunk of wood on. When grilling with wood, stand near the fire until you have mastered the methods. Dripping fat can start an inferno in a hurry.

SMOKING WITH LOGS

LET THE AIR FLOW. Grills and smokers must have a way for oxygen to enter, an intake, and a way for heat and smoke to exit, an exhaust, also called a flue or chimney.

THE INTAKE DAMPER. This is near the fuel and its job is to provide it with oxygen. The intake damper is the engine that drives the system. Close it off and you starve the fire and make dirty smoke. Eventually the fire burns out even if the flue is open. Open the intake all the way and the temperature rises.

THE FLUE. As heat and smoke exit the flue, they pull oxygen in through the intake damper. The flue needs to be at least partially open at all times, otherwise combustion gases will smother the fire. Don't be tempted to control temperature by closing the flue. The fuel will use only the air it needs. The best way to control temperature and smoke quality is with the amount of fuel, not oxygen. That said, until you become expert, you may have to throttle back a little on oxygen if you start running too hot.

LET YOUR WOOD BURN. A lot of cooks using gas or charcoal fret when their wood catches fire. They want to see it belch smoke. But you want it to burn with a bright flame. You'll use more wood but you'll get better flavor. If the temperature runs up a bit when you add wood, it's not a big problem. The meat is cold and it can take a little extra heat. Then, as the wood burns down, stabilize the temp. Don't let embers sit in ash, which can smother them. Keep them on a grate above the bottom of the firebox. Knock ash off occasionally and, if necessary, remove it.

KEEP YOUR COOKER CLEAN. Grease on your cooking grates and the walls of your smoker can create dirty foul-tasting smoke. A thin layer of neutral carbon is harmless, but black sticky goo is not. Many competition pitmasters power-wash after a cook-off.

SIZE MATTERS. For long cooks on charcoal, such as ribs and roasts, chunks of wood from golf ball to baseball sized work best. For short cooks, like a steak, chicken, or fish, smaller pieces like chips and pellets work best because they produce more smoke in a short burst. If you are burning only logs, shorter lengths of logs are easier to keep in the hot zone.

START CHARCOAL AND LOGS ON THE SIDE. Start your logs or charcoal outside your cooker so you can cook with only hot embers. I start them in a wheelbarrow. Remember, when it is burning properly, charcoal doesn't produce much smoke.

PREHEAT. Start the fire well before the food goes on. Warm the walls of the cooker. When one of my recipes says "fire up" or "preheat" your cooker, do it. Adjust your airflow and get the temp, fire, and smoke stabilized. Be aware of the weather. It is harder to get blue smoke in cold, rainy, or windy weather.

COOK INDIRECT. If the meat drips on the fire, sometimes it produces good flavors, but fat can burn and produce dirty smoke and soot. This is not a problem when you cook indirect.

USE DRY WOOD. Fresh-cut *wet* or *green* hardwood has a lot of water in it, up to 80% by weight. It produces a lot of steam, sparks, and off flavors during combustion, and it takes a lot of energy to dry it out when it is burning. Dehydrated woods are a lot lighter and they burn hotter and cleaner than wet wood because energy isn't wasted boiling off the water. That's why pitmasters prefer hardwood that has been dehydrated to 20 to 30% moisture. A 5-pound log that is 25% moisture holds a pint of water.

Dead trees can still have as much as 50% moisture from sitting in the rain, so they are too wet for cooking. When a tree is cut, the moisture moves out through the ends, the same way it moves through a living tree. Splitting it and cutting it to shorter lengths and stacking it in a covered space with airflow speeds the process. This is called *seasoning* or *curing* and can take 6 months to a year depending on the weather. Put it under an overhang or cover it with a tarp and raise it off the ground because after many months it can absorb some water

from the ground, rain, and snow. Do not bring it indoors, because it almost always contains insects. Seasoned wood often settles in at about 20% moisture but it can get drier. Drier wood often has cracks and the bark is loose.

Sometimes wood is dehydrated quickly in a heated kiln or dehumidifier. When done it can contain as little as 5% water and it is expensive. This process is often used for construction or woodworking projects where the wood needs to be dried quickly to prevent warping, cracking, or other damage. Kiln-dried logs used to be rare, but with more and more restaurants using wood for grilling, health departments demand it because kiln-drying kills insects, molds, and bacteria.

Some pitmasters even buy small handheld moisture meters to test logs. They measure electrical resistance since water is a good conductor and wood is not. Aaron "Huskee" Lyons is the "Pitboss" of our Pitmaster Club. He uses his log-burning smoker as a kiln. "I put the wood in the food chamber and cook at 225°F for as long as a day. If it is still above 25% I let it go for another day or so. I have a lot of pine near my home so I burn pine when I do this so as not to waste good smoking wood while trying to make more."

USE YOUR SENSES. It's hard to see the color of the smoke at night, but the smell should be sweet, with meat and spice fragrances dominating. The smoke aromas should be faint and seductive, perhaps like vanilla, not like a bonfire smell.

PRACTICE. Practice without food until you can anticipate when more fuel is needed, how to adjust the airflow, and how to react when the smoke starts going bad. Fire up your cooker and learn how to dial in 225°F and 325°F in the indirect convection zone, and how to get to Warp 10 in the direct infrared zone. Practice in hot weather, cold weather, and rainy weather. Master your tools.

COOKING WITH CHARCOAL

"Charcoal is for heat, not smoke. Wood is for smoke."

When it comes to fuels, charcoal can do it all. Use it close to the food and it can sear. Use it farther away and it can cook low and slow. Add wood and it can smoke. The only downsides are that it can take 15 minutes to fully ignite, handling it is dirty, and there is ash to clean up after. All easily overcome. Probably the best reason to go charcoal over gas is that it can produce a lot of powerful IR.

Charcoal is made by an ancient process of burning hardwood in a low oxygen container until it carbonizes forming "char." Char packs more potential energy per ounce than wood. The biggest problem with charcoal is that when it is just igniting it produces a lot of smoke and gases that can impart lesser flavors. You should wait until the coals are fully lit, covered in white ash, and producing little visible smoke.

LUMP CHARCOAL. This is made by chopping down hardwood trees, cutting them into chunks, and putting them into a large low-oxygen tank or pit and setting it on fire. The

> **HOT TIP**
>
> Most grilling problems occur because you have wandered off and are not paying attention. Be vigilant or you can go from undercooked to carbon in minutes. Keep the beer and bathroom runs to a minimum.

wood starts to burn, the container is sealed, and the oxygen in the wood and container are consumed rapidly. The wood smolders and volatile compounds, such as water, methane, hydrogen, and tar, are consumed and what is left is char that has retained its original shape but is only about 25% of the original weight. Lump is fashionable for the same reasons that organic food is fashionable. It has this aura of being more natural.

Part of a computer cable found in a bag of lump charcoal.

One of my many problems with lump is that the hunks range in size from fingernail to softball. That means that during production some large pieces are not fully carbonized so there can still be a lot of cellulose and lignin in there to spark and smoke. But smoke from pure wood is cleaner, better smelling, and better tasting than charcoal smoke.

Lump is more expensive than briquets, burns out more quickly, and varies in heat output per pound. Bags of lump often contain a lot of useless carbon dust from improper filtering in the factory and rough handling in stores. Some lumps can be quite large and don't fit neatly in charcoal chimneys, the best way to light charcoal (see page 79).

As a result, one never knows exactly how much fuel you have in a chimney. In addition, there are often carbonized lumber scraps in there. Lumber can be treated with chemicals to fight insects. It is not uncommon to find rocks, metal pieces, plastic pieces, and other foreign objects in the bag. I use lump only on my Big Green Egg, because lump makes less ash and the Egg's design is such that an accumulation of ash can block airflow.

BRIQUETS. These are made from chips, chunks, scrap, and mostly sawdust. The process is the same as for lump: Wood is converted to char in a sealed vessel, but this time the output is mostly powder and sawdust size. Some brands, such as Kingsford "Blue Bag," are then mixed with binders, mostly cornstarch, and pressed into uniform pillow-shaped chunks. A few contain other compounds to control burn rate.

If you are concerned about the additives, here are two brands that claim to use nothing but hardwood: Kingsford 100% Natural

Briquets covered in ash are fully engaged at maximum heat and emit little smoke and impurities.

Hardwood Briquets and B&B Competition Oak Briquets. You might also try Royal Oak 100% All Natural Hardwood Charcoal Briquets. They say it has only vegetable starch, presumably cornstarch, as a binder.

Each pillow is a finite hunk of energy. Twenty briquets produce about twice the energy of ten. Because I use a charcoal chimney (see Starting Charcoal, page 79) to start my briqs, I can regulate energy by filling the chimney all the way, halfway, one-third, etc. Because cooking is all about energy control, and because I am a control freak, this appeals to me as much as puppy breath.

Briquets with binders typically produce more ash than lump. Some folks say they can taste the additives in their food. I can't if the coals are fully lit. Perhaps those folks are not waiting long enough for the coals to be lit?

Self-igniting Match-Light charcoal briquets, which has an accelerant added to promote ignition, is a different story. It stinks and I can smell it a block away. Kingsford and government regulators say it is safe if you follow instructions, but I suspect it can taint the food. I don't use the stuff and I don't recommend it.

BINCHŌTAN. Japanese binchōtan is the charcoal used in yakitori and robata restaurants. In the US, binchōtan is expensive

but highly regarded. It is usually made from oak branches, and after they are carbonized, they sound like wind chimes when clanked together. They burn hot and smoke-free.

OTHER TYPES OF CHARCOAL. There is a trend to varietal woods such as coconut shells and coffee wood. Some produce interesting aromas.

CHARCOAL AND THE ENVIRONMENT. From an environmental standpoint, Tris West, a researcher with the Department of Energy's Oak Ridge National Laboratory, said in 2007 that "there is going to be twice as much carbon released from your charcoal grill as there is from your propane grill. When we consider the total carbon cycle and that charcoal is a renewable energy source because it's from wood, the story is completely flipped and you have more emissions from natural gas because emissions from charcoal are net zero."

BOTTOM LINE. Harry Soo of Slap Yo' Daddy BBQ, one of the top competitors and a beloved teacher, said, "I buy whatever is on sale." Mike Wozniak of Quau, winner of the Kansas City Barbeque Society Team of the Year award, told me, "I cook on whatever brand the competition sponsor is giving away for free." Remember: Charcoal is for heat. Wood is for flavor.

STARTING CHARCOAL

The best way to fire up charcoal is with a charcoal chimney. I never use liquid charcoal starter. It is usually an accelerant like mineral

spirits (aka paint thinner), and I can smell it a mile away.

There are about 16 Kingsford briquets in 1 quart. A Weber chimney holds about 5 quarts, or about 80 briquets. That's a consistent measured quantity of energy. There are too many variables in outdoor cooking and having a reliable steady heat source is crucial. There's a lot to be said for a fuel source that is rock-solid consistent from bag to bag.

You light a chimney with a wad of two sheets of newspaper from below (you do remember newspapers, don't you?). If you don't have newspaper, printer paper will work fine, perhaps with a drizzle of leftover frying oil.

Weber sells paraffin cubes that work great for starting charcoal chimneys. I have also been known to wad up half a sheet of newspaper and dip it in melted candle stubs. When it hardens it burns hot and long enough to light the coals. I have heard of people dipping dryer lint into paraffin. There are other starters that work well, such as one called Tumbleweeds—about the size of a wine cork, they are bundles of wood shavings.

Give the coals time. Once the coals are fully engaged and ashed over, after about 15 minutes, you can add them to your cooker. After you've added the coals to your cooker, open all the vents, close the lid, and wait for the metal to heat up. You will need more coals and more time in cold, rainy, snowy, or windy weather because the combustion air is cold and the cold air cools the metal of your cooker. Rain and wind are your enemies, so place your cooker behind a wall or build a simple windbreak. Do not put it in your garage. Smoke and carbon monoxide can build up in there.

Never close the top dampers all the way; you want gases to escape or they will snuff out the coals. Control temperature with the bottom dampers, but try not to close them all the way either.

A charcoal fire and its heat slowly decline in temperature as the embers are consumed. If you are doing a long cook that needs more fuel, it is best to start another chimney on the side rather than add unlit coals.

The Minion Method (below left), named for its inventor, Jim Minion, is a clever workaround for long cooks. Fully engaged coals from a chimney are poured on top of fresh coals, and as the lit coals burn out, they slowly ignite the fresh ones around them at about the same rate, and the temperature remains pretty steady. This technique can keep a smoker going at an even temperature for 8 hours or more.

Like the Minion Method, the fuse method (also called the snake method, below right) can keep an even temp for hours. The problem with the Minion or fuse method is that you are burning cold coals, so the smoke and gases are not ideal (but I can't taste the difference).

COOKING WITH GAS

"As a rule of thumb, a $400 grill will last three times as long as a $200 grill. Buy quality."

Gas grills are by far the most popular grills for good reasons: Convenience and control. Easy to fire up, ready to go in 10 minutes, increase and decrease temperature with a quick twist of a knob, steady temperature, easy to configure in two zones, quick to cool down, minimal flare-ups, minimal cleanup, good flavor.

HOT TIP

Always make sure you have a backup tank of fuel before you start to cook. Nothing is worse than running out of gas before you are done cooking.

When propane or natural gas combines with oxygen and they are ignited, they produce heat, light, water vapor, carbon dioxide, carbon monoxide, and not much else. Wood and drippings are needed for flavor.

More burners give you more control. I recommend three burners at the minimum. Gas grills have a venturi for each burner behind the knobs, a valve that blends the gas and oxygen like a carburetor. When properly blended, all the fuel is completely burned and the flame is nearly blue. When there is

HOT TIP

You need enough surface area on a grill to handle a party. A typical hamburger is 4 inches wide and you need ½ inch between burgers. That means you need 4¼ inches for each burger. That's 18 square inches. Measure the cooking surface, multiply width times depth to get square inches, and divide by 18 to see how many burgers it will handle in 1-zone cooking. Divide by 2 to see how many you can handle in a 2-zone setup.

A chicken cut into parts takes up about 12 × 12 inches or 144 square inches; then you need 288 to 300 square inches for one chicken in a 2-zone setup—and you *must* cook chicken in a 2-zone setup. If you will be cooking veggies or other sides, then you need more space.

Rule of thumb: Allow a minimum of 80 square inches total per person, about 9 × 9 inches, about the size of a dinner plate.

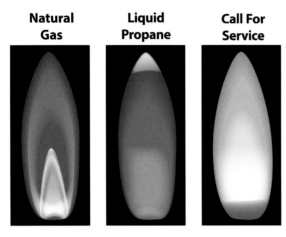

Natural Gas   **Liquid Propane**   **Call For Service**

Thanks to Appliance411.com for the idea for this illustration.

too little oxygen, the flame glows yellow and orange because some of the fuel is unburned.

Always remember, gas is explosive. You must leave the lid open when you fire up. If your grill goes out due to wind, a common occurrence, remember that the jets are still open and gas is filling up the interior. If you hit the ignition, even if the hood is up, there can be a violent explosion. If you flame out, open the lid and let it gas out before igniting.

Gas grills can suffer from carbon and grease buildups that need to be scraped or pressure-washed every few months. Grease can catch fire and it is difficult to put it out without a class B fire extinguisher. There are also gas jets that can clog. Spiders seem to like to hide down in the tubes and venturis, and I have even had to dig a wasp's nest out of one. Professor Blonder lives among some ravenous squirrels that like to gnaw on his gas hoses, so he wraps them in foil.

Assembly of new gas grills tends to be more complex than charcoal grills, and there are more parts to break and be replaced. As for fuel cost, it is hard to compare the two. Charcoal is often on sale, especially in spring, and propane fluctuates with petroleum prices. Either fuel is inexpensive compared to the food.

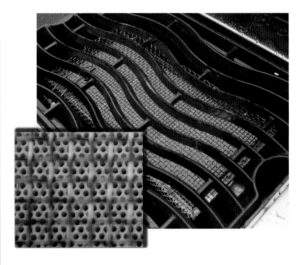

Because I cook outdoors so often, I frequently use my one and only gasser, a Hestan GMBR36-LP that the manufacturer sent me to review. I promptly fell in love because of the two infrared burners. They are similar to the burners used in high-end steak houses. They are ceramic with tiny holes and they produce every bit as much energy as my charcoal grills. Professor Blonder explains that "the ceramic plate is heated by scores of gas flames that come through the many tiny holes. The energy accumulates within the ceramic plate and it heats to a glowing red loaded with IR energy.

The Hestan is very expensive but Napoleon makes a gasser with a ceramic IR burner at a much lower price. Most other gas grills just can't generate enough IR to properly sear a steak, and many of those with so-called sear burners aren't up to the job. Worse, most sear burners are narrow and can only sear one or two steaks at a time—fine if you're an empty nester, but if you're hosting the July 4 party, you will want more real estate. I recommend you use a skillet or griddle to get your sear. It's not cheating and nobody will turn you into the grill police.

If you have a new gas grill, fire all burners up to Warp 10 to burn off any dust and cardboard and grease from the manufacturing and shipping processes. Whenever you cook, begin by checking your fuel supply. Always preheat your grill on high so the metal is hot. You can then scrape and brush the grates. I often fold a paper towel and push it over the surface with a brush to get up any residual grease or carbon. You want bare metal in contact with the food. Then I back it down to a 2-zone setup and cooking temperature.

> ### HOT TIP
> **Every few months clean the flavor bars or deflectors that cover the burner tubes. If they get coated in burned grease or carbon, they will not radiate heat very well.**

PROPANE GRILLS. Gas grills use either liquid propane (LP) from a tank or natural gas (NG) through a pipe from the house. Here's how propane grills work. A tank is filled with propane gas that has been liquefied under pressure. Some of the liquid turns to vapor and fills the headspace in the tank. As the valve to the tank is opened, this gaseous propane fills the tubes in the grill, and when you click the ignition, the gas ignites.

> ### HOT TIP
> **To keep food from sticking, it is better to oil the food than the grates. Oil applied to hot grates burns and creates acrid smoke. Oil on cold food sizzles on the grates and usually doesn't burn.**

There are several aftermarket liquid propane gauges. Problem is they don't work very well because they measure vapor pressure, not the liquid in the tank. As long as there is some liquid in the tank the vapor pressure stays pretty constant. Not very helpful.

Most standard grilling propane tanks have "TW" and a number stamped on the handle. TW is Tare Weight, the weight of the empty tank. Most standard tanks weigh about 17 pounds empty (TW) and hold about 20 pounds of liquid propane, a total of 37 pounds. To determine how many pounds of propane are left, weigh the tank and subtract the TW. So if a tank weighs 25 pounds and has a TW of 17 pounds, there's about 8 pounds of liquid left. Less than half a tank. You can weigh the tank on a bathroom scale or on a propane scale that looks like a luggage or fishing scale.

NATURAL GAS GRILLS. NG must be delivered to the grill by a pipe from inside your house so a certified heating or plumbing contractor will be needed to do the installation. Because the NG line is a hard line, that means you cannot roll your grill around once it is connected.

NG is mostly methane and is cheaper than propane, and you never run out (unless you don't pay your gas bill). Propane grills cannot be hooked up to natural gas without an adapter kit and the regulator may need to be changed. Some grills come with adapter kits, some are options, and some just cannot be adapted. If you want to use NG, make sure your choice of grill can be connected before you commit.

FLUX. Gas grill manufacturers brag on the number of British Thermal Units (BTU) their grills can produce. But the number is relatively meaningless. One BTU is the quantity of heat required to raise 1 pound of water by 1°F at about 39°F.

Energy delivery to the food is what we need to know and that must be calculated by BTU *per square inch* of cooking surface, called "flux," something they never tell you. In each gas grill review on our site, we calculate flux for you. But you can calculate it yourself. The formula is easy. Divide the BTU by the area of the cooking surface. The lower the flux the less energy delivered to the food. Here's the math:

1. If the burners produce 40,000 BTU and the cooking area is 36 × 24 inches (864 square inches), the flux is: 40,000 / 864 = 46.3 flux.

2. If the burners produce 60,000 BTU but have twice as large a cooking surface of 36 × 48 inches (1,728 square inches), the flux is: 60,000 / 1,728 = 34.7 flux.

Even though the second one has a much larger cooking surface and more BTUs, it is less efficient. Another problem, BTUs are measured at the burner level not at the food level. Remember the inverse square law tells us that heat dissipates and spreads out rapidly on the way from the burner to the food level, so the location of the burners can make the BTU measurement deceptive.

SEARING ON WIMPY GAS GRILLS. Most gas grills just don't produce enough concentrated infrared energy to produce a dark flavorful sear. The solution? Conduction. Let's do a 1½-inch steak on a cheap gas grill. We start with a 2-zone setup and the meat goes in the indirect zone. Throw a chunk of hardwood on the direct-side burners and let it catch fire. Put a black cast-iron pan or griddle over the flame on the direct side. Cast iron is a good choice because it stores a lot of energy and its dark color transmits a lot of energy. Close the lid and let the meat warm slowly and gently until it hits 115°F. Then lift the lid, pour a thin layer of oil in the pan, plop the meat in, and leave it there for 2 minutes. Flip, and keep flipping until you have a dark sear. Don't let the interior temperature go past 130° to 135°F. Shouldn't take more than 5 to 6 minutes. You have broken no BBQ laws and you have a killer steak.

SMOKING ON GAS GRILLS. Getting smoke on gas grills is sometimes tricky but it is doable on most units. One problem is that law requires them to be very well ventilated, which means a lot of smoke escapes. So you need to use more wood than on a charcoal grill. You need to experiment. Here are some things to try.

Get dry chunks of wood right on the burners or the flavor bars that cover the burners. Sometimes wood just won't burn. A reader, Nei Ng, found a solution: Make a small pile of 2 or 3 charcoal briquets on the flavor bars. When they start to glow place a wood chunk on them. It is possible that your wood might just catch on fire and not smolder. Don't worry! Remember, small hot fires produce the best-tasting smoke. You will use more wood than if you let it smolder, but you may just like the flavor better. Another option is to use a smoke generator. Try one of the ones I recommend on pages 64–65.

COOKING WITH WOOD PELLETS

Pellet cookers are incredible. I love them. They are the future of smoking at home and in restaurants. Precision temperature control. Probably better than your indoor oven. They hold temperature rock solid in all weather and recover rapidly after you open the lid. No more sitting by the smoker all day monitoring the temperature while adjusting the dampers and adding logs for that 12-hour brisket cook. No moldy bug-riddled logs. Set it. Forget it. But there are some issues.

Let's start with price. They are expensive, especially those with Wi-Fi. Most are more than $1,000, some more than $2,000. But prices are coming down.

The fuel is pure wood pellets made from compressed sawdust. Plug in the machine and dump 20 pounds into a hopper. Set the temperature you want on the digital controller. An auger turns and feeds the pellets into a metal cup about the size of a beer mug. A glowing rod ignites them, and a fan controls how many burn and how hot they get. The system is highly efficient, producing very clean smoke and little ash.

Before you buy, ask yourself, Do I need all the bells and whistles that come with the app-driven models? Do I need to be able to program a cook on my phone to start at 400°F and after an hour drop back to 225°F and then when the internal temperature hits 203°F drop to holding it at 160°F? Or can I just get off the couch during time-outs and check the meat with my eyes, nose, fingers, and a handheld thermometer? I don't rely heavily on the app. I rely on my senses.

Pellet smoker behavior is sometimes counterintuitive. The hotter they run, the less smoke they produce, and at their top settings, they don't produce much smoke at all. But down at 225°F, they produce that nice mild, elegant smoke. And even though the fuel is

HOT TIP

The smoke flavor from pellet smokers is delicate, elegant. Texans think it is wimpy. If you think your pellet smoker is too wimpy, use strong-flavored woods like hickory or mesquite, keep the temperature low because low temps make more smoke, and place a small log or wood chunk on top of the burn cup. I have also found that putting meat on the upper shelf gets more smoky air around it than the lower shelf. Huskee Lyons, Pit-boss of our Pitmaster Club, says, "I like to place the meat in a cold cooker, then turn it on. There is a thick cloud of smoke at startup and that's valuable, especially so on thinner quicker-cooking meats like a grocery store steak or fish fillet."

wood, it is hard to oversmoke with a pellet smoker. Burning wood on a charcoal grill produces much more intense smoke flavor.

There are no petroleum products in pellets, no fillers, chemicals, or binders. There is very little ash: 10 pounds of pellets will produce about ½ cup of ash. All the rest is converted to energy and combustion gases.

The average unit uses about 1 pound of pellets per hour at 225°F. At temperatures over 350°F they burn 2 to 3 pounds per hour. The burn rate will vary depending on the outside air temp and the size of the cooking chamber.

At press time, cooking pellets run about $1 per pound depending on the wood flavor and brand, if you get them on sale, and if you have to pay shipping. And remember to top off the pellet hopper before you start cooking.

Pellets are made from different woods, each of which creates a distinctive aroma. Hickory, oak, maple, alder, apple, cherry, hazelnut, peach, and mesquite are among the types available. I can smell the differences, but it is hard to taste the differences, especially if you use rubs and sauce.

It is important to note that there are home furnaces that burn pellets. You should not burn furnace pellets in a cooker. Furnace pellets often contain sappy softwoods such as pine and they can have treated lumber and other chemical contaminants in them. The smoke they put out is not good for food.

The biggest problem with pellet smokers is that the design of the devices almost always places the cup in which the pellets burn beneath a thick metal diffusion plate, a configuration much like an indoor gas oven. The food sees no flame, no intense infrared. So, like almost all other smokers, they cook with indirect convection airflow. Although many manufacturers call them grills, most just can't sear a steak properly or set up in two zones, cornerstone functions of a good grill. They just don't do it for me. I can sear a steak better on a $30 hibachi than most pellet smokers.

There is a workaround to the searing problem: Start the food indirect, low and slow in the smoker. Put a griddle or a skillet in the smoker. Bring the meat up to about 15°F below your target temperature. Take the meat out and crank the smoker to max temp. Oil the meat and sear by conduction in the metal pan.

There are a few pellet smokers that can sear and surely more are coming. The Weber models are designed so you can expose food to flame, and they do a nice job of searing. Other companies have gas burners on the side that you can sear with.

If you can, devote a 20-amp circuit breaker to your pellet smoker and use a GFI outlet. If the built-in cord is not long enough, get at least a 10-amp, 12-gauge, three-prong cord (oddly, 12-gauge is higher capacity than 14-gauge). That's more than you need, but better safe than sorry.

Humidity and rain can make pellets dissolve. They then harden into a solid concrete-like blockage that prevents the auger from turning. Getting them out is a real knuckle-buster. Protect the pellet hopper and the auger from moisture at all costs. Huskee says, "Empty your pellet hopper periodically and clear sawdust and broken pellets out of the auger tube by running it until the fire dies."

Cleanup can also be a pain. Grease buckets must be emptied often. Some have buckets to collect grease on the outside of the cooker where bees, raccoons, and dogs can feast. I had one that allowed water into the grease bucket after a rain. That overflowed the bucket and made a greasy mess of my deck. But my dog sure loved it.

After you go through 20 pounds of pellets you need to remove the combustion cup, dump the ash, and vacuum out ash that has scattered around the lower part of the grill body. The biggest cleanup issue is the buildup of grease and carbon on the heat deflector under the cooking grates. You can often reduce the mess by covering it in foil. On most models, a thorough cleanup means a 30-minute process of taking out the greasy grates and the gooey deflector plate, scraping them and washing them. You'll need good gloves and an apron. Resist the temptation to put these parts in the dishwasher. That grease is like tar and it could get all over the insides of the dishwasher and hang on for dear life. Then it's off to divorce court. I use an inexpensive handheld steamer to remove gunk.

ELECTRIC GRILLS AND SMOKERS

Electric grills use a glowing metal heating element for energy, like an old-fashioned electric stove, so there is no flame, no combustion, and no smoke. Most electric grills are not much different from a George Forman indoor electric grill, but a few have some nifty bells and whistles. Temperature control can be fairly accurate. They are popular in condos and apartment high-rises where gas and charcoal devices are prohibited.

Most electric grills are just frying pans with ridges. Drippings do fall into the valleys and can vaporize up to the meat and the ridges can make grill marks, but the flavor of the results bear no resemblance to a grill that has combustion.

Electric smokers have a metal pan above the lower heating element. You fill it with sawdust, wood chips, or pellets. They smolder and make decent smoke flavor, but again, not as good as charcoal, gas, or pellet smokers where there are flames and combustion gases. They don't produce the gases that make a smoke ring without a hack: Put a charcoal briquet above the heating element and it might produce a smoke ring.

Some units come with heating elements below and above, a griddle attachment, a smoker box for wood chips, and a steamer that can do double duty as a chafing dish. A very handy device.

Because there is no combustion, there is much less airflow through the cooking chamber, so there is much less moisture loss from the food. In fact, they can be quite humid. Meats are more juicy, but flavors are less concentrated and crusts are less crusty.

Alas, there is no substitute for the flavors from live fire. As I said in my last book, a VW Beetle can be a delightful drive until you sit behind the wheel of a Porsche. Still, if you live in a condo and they don't allow other grills, an electric is better than nothing. Weber's Lumin is my fave.

GRIDDLING

Most restaurants have a large griddle, a flattop as they call it, or a plancha as it is called in Europe. It is the backbone of many diners and taverns like Moe's from *The Simpsons* where short-order cooking is essential.

The great advantage to griddling is that cooking is fast and you get an unmatched dark sear because you are cooking by conduction. And no flare-ups.

An easy way to start is to put a flattop on top of your grill. Gas grills are especially well suited for cooking surfaces like cast iron, stainless steel, tempered steel, cast aluminum, stone slabs, even salt slabs.

Photo courtesy of Lodge Cast Iron

I use the Lodge Reversible Cast Iron Griddle that sells for about $50. One side is flat and one side is corrugated to make grill marks. I use only the flat side. It has a lip to keep the oil from dripping into the flame and creating a firestorm. It also works on my indoor stovetop. Some griddles come with a little moat around the edge. I don't like them. The oil runs into the moat and I want it on the griddle surface. There are larger and more elaborate griddles designed to go on top of your gas grill, but many are thin metal and warp.

There are several brands of stand-alone outdoor griddles with multiple gas burners similar in size to a gas grill. Cooking on them is like cooking on a giant frying pan. You can sear steaks or burgers beautifully, fry bacon, and make sunny-side up eggs, pancakes, hash browns, foie gras, and I've even done sweetbreads. One can make a strong argument that smash burgers with crispy edges cooked on a griddle are every bit as droolworthy as a steakhouse burger from a grill. Check out my recipe for the amazing Oklahoma Onion Burger (page 291).

**Photo courtesy
of Pit Boss**

Cooking surfaces range from black cast iron to shiny stainless steel and a few even have relatively nonstick ceramic coat. Most are rolled steel or carbon steel. Restaurants favor thick stainless. You want one with a lid so you can melt cheese. An easy-to-empty grease trap is a must. There should be a gap under the cooking surface so you can see if the burners are on. And the best ones have a way to level the cooking surface.

Most of the cast-iron and steel surfaces need to be seasoned as you would a cast-iron skillet. That means a thin layer of oil needs to be baked onto the surface to form a rustproof and relatively nonstick surface. And that's the rub. When you are done cooking, if you spend 5 minutes deglazing the surface with water, scraping off the residue and grease, wiping it down, coating it with oil, by then your burgers are cold. Or you can turn off the gas

and serve hot food, and then after dinner go back out, fire it up and clean and season it if you haven't drunk too much wine with dinner and watched Netflix. Or you can leave it overnight and hope the raccoons and wasps don't discover your sloth.

If you use a griddle that you put on your gas grill, you can bring it indoors and toss it in the sink.

Then there are the special tools you need: Two stiff, solid spatulas for smashing burgers and flipping pancakes and stirring stir-fries. They should have a square sharp leading edge so you can get under a smash burger to flip it without leaving the marvelous Maillard behind. You need two squirt bottles, one for oil and another for water. I recommend an infrared gun thermometer so you can dial in temperatures. I do most of my griddling in the 350° to 400°F range, although they can get much hotter. And just like my grills, I believe in two zones minimum. I have a 3-burner and I leave one off, one on medium, and one on high for most cooks—three zones.

Scrubby sponges are helpful for removing scorched residue. Don't use stainless steel or pumice blocks. Get a window washer's squeegee for sweeping grease to the trap. I buy disposable aluminum pans for the grease trap and keep a roll of paper towels and a flip-top waste can next to my griddle.

A dome or upside-down baking pan is handy for placing over cheeseburgers to rapidly melt cheese. Cast-iron sandwich weights are helpful to keep buttered bun surfaces in contact with the hot metal and for grilled cheese and paninis. I use clarified butter a lot.

At press time, my favorite griddles are the Pit Boss models (see photo, page 90) because they are the only ones that have a ceramic-coated cooking surface similar to ceramic frying pans. This surface is rustproof and very slippery, requiring much less cleanup. I suspect there will be more with ceramic coatings soon. The whole cooking unit can be lifted off the cart and thrown in my trunk for tailgaters.

WOK COOKING ON A CHARCOAL CHIMNEY

I know you love Chinese food, and probably you've attempted to stir-fry indoors. And the results were good, but not as good as the neighborhood Chinese restaurant. The reason is simple: Even if you used a good carbon-steel wok, your stove isn't hot enough. For great stir-fry you need food cooked very hot and fast so it sears almost instantly. If you don't have enough heat, the juices run out of the meats, you end up with a puddle of liquid, and the ingredients steam or poach and go limp. It can also smoke up the kitchen, set off smoke alarms, and spatter all over your stove and counters.

But you can make your favorite Chinese dishes in your backyard. A charcoal chimney can generate enormous concentrated heat. Most charcoal chimneys have holes near the bottom to supply the coals with oxygen. But

a wok on top of a chimney can reduce the heat, because the combustion gases have no way to escape, and they can choke out the coals. So here's the solution: Drill holes near the top lip of your chimney or use tin snips to cut vents, as shown on page 91.

The carbon-steel wok is a tremendously versatile pan and its design hasn't changed in centuries. It can be used for stir-frying, deep-frying, braising, steaming, simmering, boiling, poaching, sauce making, and even smoking. In a pinch you can stir-fry in a large saucier or curved-sided pan or even in a large skillet, but part of the trick to stir-frying is controlling the heat by pulling ingredients up the side of the wok where it is cooler.

Do not under any circumstances use a Teflon pan. Teflon and similar coatings can volatilize under high heat and make noxious gases that find their way into your food. And I am not a fan of cast-iron woks because they are too heavy to toss the food with a flick of the wrist. Also, they hold heat all the way up the sides.

You can find excellent woks in most Asian stores and they are incredibly cheap there. They are also available online. I recommend the one sold by Milk Street. Get a 14- to 16-gauge carbon-steel wok with a flat bottom that will be stable sitting on a flat surface. Woks come in a wide range of sizes but one that is 13 to 14 inches across the top is enough for 4 servings. Larger woks are harder to handle. There is a cult around hand-hammered woks, but machine-hammered or spun woks work just fine. Get one with a long

wooden handle that makes flipping with a wrist flick easy. Look for one with a small "ear" handle on the opposite side of the long

MYTH

Never use soap or metal utensils on carbon steel or cast iron

BUSTED. With use, carbon steel, like cast iron, captures a bit of the cooking oil in microscopic pores and seasons the surface so it becomes relatively non-stick. Conventional wisdom says not to use soap on them. Conventional wisdom is not so wise. Modern mild dish soap is good at removing oil, but the seasoning is a new compound made *from* oil but not made *of* oil.

Modern mild dish soaps are repelled by the seasoning and they will not seep in and make your food taste soapy. Old-fashioned lye-based soap and scouring materials like Comet can damage the seasoning, and that is no doubt how the prohibition against using soap began. If washing reveals bare metal, don't fret, your seasoning was not thick enough. Just reseason.

You can even use a metal spatula on carbon steel or cast iron. They do in restaurants. The seasoning is held on by some strong chemical bonds.

handle making it easier to carry. Make sure the handles are well riveted. If you plan to use it indoors, buy a wok ring that holds the wok steady above a stove burner. Get one with a lid and a little grate that goes inside for steaming.

Carbon steel can rust, so new woks often come coated with oil to prevent rust. When you get it, you must fill it within an inch or two of the top with water and bring it to a boil for 15 minutes, dump the water, and then scrub with an abrasive cleanser, steel wool, hot water, and elbow grease. Rinse off all the soap, towel-dry, and then warm it on the stove to drive off any water in the metal pores. Then give it a thin coat of vegetable oil. Reviews of excellent woks on Amazon are burdened with 1-star rankings because the buyers didn't follow directions for breaking them in.

Hardened Oil

Cast Iron and Carbon Steel

The seasoning process is like seasoning cast-iron pans and steel griddles. It can be smelly, so I do it outdoors on my gas grill. The goal is to fill the microscopic irregularities in the surface with oil and then harden it into a relatively slippery surface with heat. If your wok has wooden handles, remove them. Warm the wok over medium heat to remove any moisture and open the pores of the metal. Lightly coat the inside with vegetable oil. With a lint-free cloth or Bounty paper towels, wipe off the excess. The oil starts to fill the

The seasoning on a carbon-steel or cast-iron pan is a polymer

BUSTED. According to Professor Blonder, it is actually a glass. "A polymer is a chemical with repeating molecular elements strung out like soldiers marching in formation. A glass is a randomly bonded mesh like a Rice Krispies Treat. It is neither as uniform nor as effective as a Teflon coating, which is a polymer. Oil seasoning is impressive, but it ain't Teflon and it ain't a polymer."

valleys in the metal. Don't let any oil pool in the center. In order to form the glassy coat, you need to close the grill's lid and heat the oil to 400° to 450°F, but you don't want to overheat it, or it could blister and flake off during cooking. An infrared gun thermometer is helpful. For new pans, two or three treatments is enough. When you are done, the oil should have formed a smooth, dark nonstick surface.

Going forward, every time you wok, the surface will darken. Then, when you are done wokking, clean it with a soft brush or scrubby sponge under warm running water with mild dish soap. But don't use steel wool or Comet. Then wipe it dry and give it a very thin coating of oil to prevent rusting. Here are some tips on wok technique:

GET A GOOD BOOK. I have lots of Asian cookbooks, but *The Wok* by J. Kenji López-Alt is my favorite. Full of great tips and recipes.

DO ALL YOUR PREP FIRST. Things move very quickly when you are wok cooking, so make sure you have prepared all your ingredients and tools and have them handy, right by your side. Cut pieces so they are similar in size so, if they are to be cooked together, they will cook evenly. Make the sauce, and to thicken it thoroughly mix 2 teaspoons cornstarch in 1 tablespoon cold water to thicken ½ cup sauce.

USE DAY-OLD RICE. Freshly made rice is too wet and sticky for most stir-fries. Use day-old rice or leftovers from your last take-out order.

TOOLS. Most chefs recommend a special wok spatula that is like a long-handled spatula with a curved edge so it can scrape along the curved sides. They also recommend a spider, which is like a sieve with a long handle. The slotted spatula shown here can do the work of both a spatula and a spider and it has a sharp edge, for scraping the stuff stuck to the wok. When I am wokking outdoors, I keep a pitcher of water nearby in case I need to scrub out the wok, a bucket to dump it in, a roll of paper towels, and an infrared gun thermometer.

GET THE WOK HOT. Fill the chimney only halfway with briquets. Get the coals as hot as possible, which means wait until they are covered with white ash. Then put the wok on and get it hot. I use an infrared gun

thermometer and shoot for a minimum of 500°F.

COOK THE INGREDIENTS IN FLIGHTS. Start with no more than 2 tablespoons oil and get it good and hot. Swirl the wok so the oil climbs the sides. Add the things that take the longest, usually thick things like carrot coins, cauliflower, broccoli, etc. Don't crowd the wok. Each piece needs a chance to make contact with the metal at all times. If you pile things up, they will steam not fry. Cook in batches if necessary. Stir or flip every 10 seconds or so, keeping the pieces in contact with the hot part of the wok until browned, then move them to a bowl. A well-seasoned wok should come clean with a curved metal spatula. If not, a splash of water and a scrubby sponge will do the job, then let the heat dry the wok. Add more oil, swirl to coat, add the next batch of food, stirring frequently, and move it to a bowl when done. During the last batch I add the aromatics like garlic and ginger because they cook fast.

KEEP THINGS MOVING. Surprising as it seems, keeping the food moving cooks it better and faster than leaving it sit. The motion liberates steam so you sear better. Just be careful not to knock over the tower of power.

FINISHING. When all is done, combine everything in the wok, add quick-cooking ingredients, like peapods and greens like spinach or scallions, last. Then drizzle the sauce down the sides so it heats rapidly rather than pour it on top of the relatively cool food.

Stay right there and don't go to the bathroom. When the sauce thickens, it's time to eat.

CLEAN. When you are done cooking and you've cleaned your wok, you should coat it with a thin layer of oil to prevent rust. But the oil can go rancid. So before you start again, rinse the wok with hot water and a sponge. Then dry it thoroughly and give it another light coat of high-smoke-point oil.

CLEANLINESS IS NEXT TO DELICIOUSNESS

Clean your damn grill, willya?

I don't care what you've heard, a layer of grease on the inside does not make the food taste better. It makes it worse. A layer of carbon will reduce reflectivity of shiny metal and that can be either good or bad, but grease will produce stinky smoke and a buildup of carbon that can flake off onto your food. All you need to do with a new cooker is run it on high for 15 to 30 minutes to burn off any grease from manufacturing and packing materials from shipping.

Before you start to cook, get the interior good and hot to burn off any grease and food residue. Use a sharp-edged scraper and a brush on both sides of the grates. Never put the grates in your dishwasher or you'll be sleeping on the couch for a month.

Metal-wire bristle brushes are a fine way to clean grates, but after using a brush always check for bits of wire on the grates. A tiny bristle can stick to food and lodge in the throat or stomach of someone you love. Before you buy a brush, inspect the construction. In general, anchoring bristles in wood is riskier than in metal or plastic.

If you forgot your brush, fire the grill up to Warp 10, get a foot or two of aluminum foil, scrunch it up, grab it with your tongs, and scrub. Remarkably effective. If you have a charcoal grill, scoop the ash out. Ash just absorbs energy and if there is wind, it can fly up onto your food. It can also block airflow and if it gets wet, it can turn to concrete.

I have practically every grill brush ever invented. Grill Rescue is the grate cleaner I grab first. It has a metal scraper for stuck-on food bits, and a pad made of polyurethane foam wrapped in tough aramid fiber (the same stuff in firefighter gear). I dip it in a loaf pan filled with water and when I wipe it across hot grates, the water boils and steams off most of the grease. The pad gets black quickly, but it can be washed many times and it still works like new. The pad can be replaced.

On many grills and smokers, there is a grease collector. Empty it regularly. It is a grease fire waiting to happen.

Periodically, if you have a pressure washer, pull the grates and other parts out and go at them. Or, if you have a pickup truck or a trailer, take your grill to a self-service car wash and use the pressure washer there. Just be careful to protect electrical contacts.

Occasionally, if you haven't used your cooker for a while, you might get a nice fluffy coating of mold on the grates and other parts. Fire up as hot as possible for 30 minutes. Then power wash and put on your N95 mask and scrape and brush everything in sight. Then spritz all surfaces with 1 part chlorine to 4 parts of water. Let it stand for 30 minutes. Then fire up again and burn off any residual chlorine. Let it dry in the sun. If you have a ceramic cooker, do not use chlorine or a power washer. Heat is your only tool. Fire it up as hot as you can get it and burn it off. If your lava rocks are moldy, buy new ones.

Clean the flakes of carbon scale from the inside of the lid. They can drop onto the food, especially if you slam the lid. I use a plastic plaster spreader.

Gas grills often have bars that sit above the burners. They serve three purposes, protect the burners, distribute the heat evenly, and incinerate drippings so they vaporize and land on the food adding flavor. Food that drips on them can carbonize and cake. Clean them.

Wait until dark so you can see the flame better, remove the protective covers, and fire up your gas grill with the burners exposed. Some of the gas jets may be plugged. You can clean them out with a straightened paperclip. Then check the venturi, the gap just behind the knobs where the gas and air mix. Look for spiderwebs. The flame should be blue with a tiny bit of orange at the tip. Venturis usually have a set screw. Loosen the screw and rotate the venturi until the flame is right.

If you want to see your reflection in your grill, oxalic acid, the active ingredient in Bar

Keepers Friend, is very good at restoring the luster to stainless. Use a sponge and warm water. Work on a cool grill and follow the grain. Never use steel wool or metal brushes. Don't let it air-dry or it will leave water spots. Polish it with a soft cloth while it is wet.

WHY FOODS STICK

Metal surfaces, no matter how smooth they feel or look to the eye, are not smooth, even new ones. There are microscopic scratches, pores, pockmarks, hills, and valleys. Chemical bonds form between food and metal. As proteins brown, their chemistry changes, their surfaces dry, and they usually let go. Sometimes they don't and they permanently bond. The problem is exacerbated by the condition of the metal.

The way to prevent sticking is to put something between the food and the metal that smooths the surface and prevents the bonding by covering over the metal.

Oil is a fluid and it easily fills the hills and valleys, and its slippery chemistry helps prevent the bonding. But if the food sits in one place for a while, it can push the oil away and hook up with its metallic mate. I prefer to oil the food, which is cool. Oiling hot metal usually cracks the oil, it smokes, and it tastes bad.

SMOKING WITH TEA, HERBS, SPICES, HAY, PINE NEEDLES, AND ORANGE PEELS

You can burn tea, herbs, spices, sawdust, pine needles, even orange peels for a quick burst of mind-blowing flavors. They are great for quick-cooking foods like fish fillets and shellfish, and the flavor is significantly different than wood. My recipe for Tea-Smoked Whitefish is on page 310.

For fast puffs of smoke, just throw dried herbs on the fire, put the food on the indirect side, put a pan or metal bowl over the food to trap smoke, a method I call the "smoke catcher," and keep the lid closed.

I burn year-old spices and herbs and replace them with fresh purchases. We have a small herb garden, and at the end of the season there are always a few unpicked oregano, basil, and other herb bushes. I cut them above the roots and stick them in paper bags to dry. Then I crumble them and throw them on the grill after the meat is on. To make them smolder for longer cooks, stuff them into a foil packet and poke a bunch of holes in the top to let the air in and smoke out. They can also be scattered between the rails of GrillGrate® (see page 121).

Hardwood sawdust is also a good smoke source for short cooks. The best source of sawdust is to take a handful of wood pellets, get them wet, and let them air-dry. You can also get quality sawdust from a local cabinet maker or lumberyard, but be careful: It is often mixed with pine and other undesirable resiny woods.

TANDOORS AND MAKING A KAMADOOR

Most restaurants in India and many Indian restaurants in the US use an oven called a tandoor to roast and grill. A tandoor is a very well insulated barrel-like oven often made from ceramic. Charcoal sits in the bottom, as much as 30 inches from the top. Put the lid on and it gets extremely hot in there.

Tikka is usually long skewers crammed with hunks of meat and/or vegetables marinated in spiced yogurt. The skewers are hung in the tandoor. Naan is a marvelous yeasted flatbread cooked by slapping the wet dough onto the hot ceramic walls of the tandoor where it sticks, bakes, bubbles, and chars. It can be plain, buttered, or studded with garlic, onion, and other flavorings. Naan can be used for holding tikka like a taco, or can be slathered in smoked eggplant spread, or just used for mopping up.

A very close cousin of the tandoor is the kamado, originally designed in China.

Commercial kamados are very popular in the US, especially one called the Big Green Egg. They are excellent ovens for roasting and baking, are good for pizza, are fine smokers, and can be used like tandoors. It isn't hard to make your kamado work like a tandoor, or, as I call it, a kamadoor. See my Tandoori Chicken (page 277).

▶ I had a metal worker make a ring to hang skewers for when I want a lot of tikka. I bought long skewers with hooks from Amazon.

DON'T COOK WHOLE ANIMALS

I don't like to cook whole animals. A turkey breast is much thicker than a thigh and while breasts are best when cooked to about 160°F a thigh is best at 170°F+. If you cook a whole turkey, it is hard to get all parts to optimum temperature at once. And let's not talk about those overcooked wings. Same goes for chickens.

A whole hog is one of the world's great festive rituals, but the shoulders are best when cooked up to about 203°F so the tough connective tissues can gelatinize and the fats can render and the meat falls apart. The lean loins and tenderloins are much thinner and more tender and they are best at 135° to 140°F. No way you can get all parts cooked properly.

For the best results, break down animals and cook different muscle groups and thicknesses separately, staggering their start times or proximity to heat and monitoring them with a thermometer.

Leftovers

Perhaps you've noticed that some leftovers seem to taste better than when the dish was fresh. What gives?

Cooking may have stopped, but not all the enzymes have stopped. The most notable change is texture. Kiss any crunch good-bye. Meats get softer, even mushy, as proteins break down and fats oxidize. Occasionally you might notice more umami due to the release of amino acids. Oxidation kicks in. In general, meats don't improve much and can even develop what food scientists call WOF, warmed over flavor. Also, smoke flavor dissipates.

On the other hand, stews, soups, and dishes with a lot of sauce often do seem to improve. That's because flavors evolve as spices mellow, marry with other compounds, and are absorbed by fats. Starches break down and sometimes generate sweetness. Dishes like chili are among the ones that seem to benefit from aging.

The big problem BBQ lovers have is what to do with leftover brisket. You start with an 18-pound brisket, trim off 4 pounds of fat, and during cooking it shrinks another 3 pounds. You still have 12 pounds of meat. There are almost always leftovers. Here's what I do the next day: Slice it, spread it on a plate in a single layer, dip my hand in a bowl of water, and sprinkle water lightly on the meat, both sides. Then I pop it in the microwave. Sous vide is also a great way to reheat meats (see Sous-Vide-Que, page 107).

THE TRUTH ABOUT MARINADES

I don't use marinades very often. Nothing on your spice rack other than salt can penetrate deep. Most molecules are way too large to penetrate beyond a tiny fraction of an inch. If there is oil in there, fogeddaboutit. Oil and water don't mix. Here's a salmon fillet soaked a few hours in a typical marinade with green dye added. You can see it got into cracks and cuts, but not very far. You might notice a slight discoloration just below the surface. That's salt. It penetrates and the rest of the stuff just flavors the surface.

Marinades are most effective on thin pieces of food where penetrating just a little is penetrating a significant percentage of the food.

Marinades also make the surface wet and inhibit browning. You want a dry surface to get good browning. That's why it's a good idea to pat meats dry before cooking. When I want to flavor something, rather than marinate I usually use a rub, a blend of herbs and spices. When you chew, you mix the flavorful surface with the interior in your mouth. In the recipe section I discuss herbs, spices, and rubs in more depth starting on page 157.

Rub/Marinade Penetration

¾-inch Skirt Steak

⅛-INCH PENETRATION EQUALS
¼-INCH TOTAL = 33%

2-inch Ribeye Steak

⅛-INCH PENETRATION EQUALS
¼-INCH TOTAL = 12.5%

KEEPING FOOD JUICY

Juiciness is primarily four things: Myowater, melted collagen, melted fat, and saliva. Yes, saliva. When we see a sizzling steak our six salivary glands excrete a lot of water.

A water pan will raise the humidity inside the cooker and help keep meat juicy

BUSTED. Many cooks think that if they put a water pan in their cooker it will raise the humidity and keep the meat juicy. Not so much. Water pans are a good idea, but not for humidity.

To combust charcoal, logs, pellets, or gas, you need a lot of oxygen. Air is only 21% oxygen. While you are cooking, an enormous amount of air is moving through the cooker. A typical backyard charcoal smoker can move through more than 150 cubic feet per minute. That's more air than in an SUV. Every minute. And because of the heat, the humidity is very low in there.

A water pan gets hot and even steams a bit, but it takes a long time for it to evaporate. The small amount of moisture it contributes to the air in the cooker barely raises the humidity. The airflow direction and the cool air boundary layer around cold meat might mean the water vapor never contacts the food. Meanwhile, water also evaporates from the food. To prevent the water from leaving your food you would need to raise

the humidity to near 100%, an impossibility without a special combi-oven that is designed to raise the humidity in the cooker.

A water pan does have an interesting impact on the food. Let's say the air temperature in a smoker is 225°F and the meat is fresh from the fridge at 38°F. In practice, the water in a water pan rarely goes above 190°F because it cools as it evaporates.

Some of the moisture from the pan condenses on the meat the same way moisture condenses on a cold glass of beer because at 38°F it is 187°F cooler than the air. The condensation cools the meat, prolonging the cook, and that's good, allowing more time for the fat and connective tissues to melt.

Meanwhile, the condensation makes the food surface sticky. Small particles in smoke stick to it improving the flavor.

Since water cannot go higher than 212°F, a water pan is also good for moderating the average temperature on a grill or smoker. If the water gets up to about 190°F and holds there, it becomes a heat sink and stabilizes the cooking temperature, because water temperature takes longer to rise and fall than air temp. And that's good.

A water pan can also block direct flame when you need to cook with less energy. Just put it over the heat source and you go from powerful direct infrared energy to milder indirect convection heat. It also evens out hot spots.

LOW TEMP HELPS. When we cook, hot air causes a lot of evaporation and drip loss. How much depends on the meat, cooking temperature, and how long it cooks. For a filet mignon, which has very little fat and cooks quickly to an optimum temperature of 130°F, there might be only 10% loss. For pulled pork made from a fatty pork butt with lots of connective tissues cooked for 8 hours all the way up to 203°F, there can be as much as 30% loss.

This much we know: Muscle fibers and proteins expel water when they get hot, so rule #1 when cooking most foods is keep the temperature down. That's why many of my recipes tell you to reverse sear.

BRINING HELPS. Another proven technique is brining. When protein hooks up with salt it changes its shape and helps it hold on to moisture.

COMBI-OVENS. The ultimate solution is the new breed of combi-ovens that inject steam

MYTH

Sand in the water pan helps stabilize the temperature

BUSTED. Some folks like to put sand, dirt, or gravel in the water pan instead of water. But the advantage of water is that it cannot go higher than 212°F so it keeps the temperature down. Sand can get much hotter than that.

MYTH

Put beer, wine, apple juice, an onion, or spices in the water pan to add flavor

BUSTED. Flavor compounds in wine, apple juice, an onion, or spices are so few and far between that they cannot impart any detectible flavor on meat. For example, by weight, regular beer has about 3.5% of various sugars, proteins, minerals, acids, esters, etc. Not nearly enough to add flavor. Drink the beer. Don't waste your money. Just use hot water in the pan. If you want flavor, use a rub directly on the meat.

into the atmosphere. There are even a few with smoking capability manufactured for commercial food service. Another technique is to smoke on a normal smoker for an hour or two and then finish the cook in a combi-oven. Restaurants love them because there is less weight loss, which means more money.

WETTING THE MEAT HELPS. Putting water on the surface of food by basting, mopping, or spritzing can keep the surface wet and reduce dehydration slightly. On the plus side, there is less weight loss, and there is more juiciness. On the minus side, you don't get the crispy, chewy, concentrated flavors of a good crust. Surface water is the enemy of crust. One solution is to moisten for only the first part of the cook and then stop.

THE TEXAS CRUTCH. Wrapping the food in foil, called "The Texas Crutch," can help meat retain moisture (see Beware of the Stall, page 105).

SOUS-VIDE-QUE. This clever method is great for retaining moisture, although there is still a lot of water loss, or purge. Read more on page 107.

ROTISSERIE. Cooking on a rotisserie spit helps retain juiciness because the motion breaks up the cold boundary layer around the food, allowing more heat to get to the food. It also bastes the meat with its own juices as the food rotates.

ELECTRIC SMOKERS. In electric smokers there is little air movement because there is no combustion, so meat doesn't dry out as much.

BURGERS. Ground meats are a special problem. Muscle fibers are torn open and fat is ground up into bits. As soon as heat hits burgers they begin to cry. The good news is that there is a lot of water and fat left in the meat to tickle your tastebuds. A raw 8-ounce steakhouse burger is usually about 20% fat. But if your burgers are dry, ask your butcher to custom-grind some burger meat for you and make the blend 25 to 30% fat. That's more effective than adding ice chips to the raw patty as some people recommend.

DRIPPINGS AND FLARE-UPS

When foods cook, they drip. Water and fat drip, and if there are spices and herbs on the surface, they drip off, too. The good news is that drip loss helps concentrate flavor, and when the drips hit hot coals or hot metal they vaporize and land on the food, creating that marvelous grilled flavor we love. Also, flames licking meats add flavor. Yep, Burger King was right, flame-broiled is better.

In general, small infrequent flame flare-ups are good, long persistent fires are bad, and smoky or soot flare-ups are bad. If you have a fire and attempt to tame it with a squirt gun,

you can make soot. That's another reason why you should always set up in two zones. If you have an out-of-control fire, just move the meat to the indirect zone and close the lid.

Here's a neat trick: Make a burnt offering. Start a steak on the indirect side as you would when doing a reverse sear. Throw some fat trimmings onto the coals or deflector bars so it flares up. Then move the meat over the flames for a short while.

For some recipes you want to put a drip pan underneath the food with water in it to catch the drippings, which you can use later to make a stock for gravy or sauce. Amp it up to 11 by adding aromatics. You can use the same method on roasts, such as prime rib or crown roast of pork.

Here is a drip pan underneath a turkey. It is filled with carrots, celery, onions, apples, herbs, wine, water, and turkey trimmings, and makes an amazing stock to be used as a gravy when it is strained.

Here the drip pan under these duck halves has trimmings, a little oil, potatoes, and grapes. I start with a few tablespoons of oil and ¼ inch of water or chicken stock, and that cooks the potatoes et al. When the water evaporates, they fry in the fat rendered from the duck skin during cooking. Duck fat is not quite as tasty as bacon fat, but it's up there.

BEWARE OF THE STALL

My bride bought my first smoker for my birthday and it came with a small booklet with recipes. Next day at about 10 a.m. I started with a 5-pound pork butt, trimmed most of the fat, gave it some salt and spices, inserted a probe thermometer, and popped it on the smoker at 225°F aiming for fall-apart unctuousness at dinnertime, about 6 p.m. It motored along getting darker on the outside and warming in the center until it hit 157°F. Then. The. GD. Temperature. Stopped. Rising.

And it stuck there for almost 2 hours! And the guests were arriving. Panic. I finally had to take it off at about 170°F. No way would it shred. So I sliced it sorta like a ham, and the fat hadn't fully rendered, so I had to carve out hunks of it. I plopped the chunky slices of meat on squishy buns, dumped barbecue sauce all over, and I think I got away with it.

Here's what happened. The meat went into the heat from the fridge at about 38°F. The temperature on the surface rose and the temperature in the center rose very slowly because water is a good insulator. As it rose, the hot air caused the water on the surface of the meat to evaporate, and as it did, it sucked up energy and cooled the surface so there was less energy available to move to the center. The meat was sweating like you do cutting the lawn on a hot day. It cooled at the same rate as it heated. Stasis. It stalled.

The internal temperature couldn't go up until the outer layer dried out. But with the bad comes some good. This surface dehydration also built a crust studded with smoke, melted sugars, and flavors from the rub. And the cooling gave the center more time to render fat and melt connective tissue. The good news is that the stall is only a problem when cooking meats to high internal temperatures, meats like pork butts, beef briskets, beef ribs, and pork ribs.

For years pitmasters didn't understand the stall. They thought it must have something to do with fat or collagen and there were a dozen theories floating around. Way back in about 2007 I brought the question to Professor Blonder and he set up some simple but clever experiments. He put a giant glob of beef fat with collagen alongside a soaking wet sponge in an oven, stuck thermometers in both, and there was the answer: The fat/collagen blob never stalled, but the sponge stalled the same way a pork butt does. Evaporation cooled it. When I shared this in some barbecue message boards, they chewed me up like a pulled pork sandwich. Today, evaporative cooling is generally accepted knowledge.

The stall shows up when it feels like it, usually between 150° and 165°F, and it usually lasts 2 to 3 hours. There are three solutions:

WAIT IT OUT. Plan on a 3-hour stall, and if it gets done earlier, lower the heat on your cooker and hold it. Do not slice or pull it or it will dry up. The meat should be fine and you'll have a great bark.

THE "TEXAS CRUTCH." Here is an idea derived from the aboriginal idea of wrapping foods in banana leaves. When the meat has something close to the color you will want, and the stall hits, wrap it *tightly* in foil or uncoated butcher paper. There is a lively debate about which is better, foil or butcher paper. I prefer foil. Some folks add butter, margarine, apple juice, soft drinks, or barbecue sauce. Inside the wrap, moisture cannot escape, evaporative cooling cannot slow the cooking, and the meat temperature will power through and be done much faster. It will also be more tender because juices do come out and the meat braises and steams. Alas, you won't have a crunchy bark. If you want a better bark, when the meat comes out of the foil, toss it on a hot grill to dry out the surface.

CRANK THE HEAT. Increase the cooking temperature to 300°F or more so the meat warms faster than it cools and it powers right through the stall.

SOUS-VIDE-QUE

By marrying water and fire, by marrying sous vide (pronounced *soo veed*) with the grill and smoker, we can achieve extraordinary results, in many cases better than with either cooking method on its own. You can get extraordinarily tender, juicy, safe, and flavorful foods. It is especially useful for tough cuts that will benefit from low and slow cooking followed by a sear on the grill, like beef flank steak, beef short ribs, corned beef, pastrami, poultry breasts (including duck and goose), lean pork loin chops, and frozen meats.

Most conventional cooking is done at a temperature much hotter than the target temperature of the food. If you are grilling a steak, typically the air temperature is in the 400°F or higher range and the meat is right over intense infrared radiant heat.

Since I have already written a whole e-book on the subject, *Sous Vide Que Made Easy*, there's no need to go deep on this deep subject here. But I will cover the basics. If you want to know more, it is available for $3.99 wherever you buy e-books or free to our Pitmaster Club members.

Sous vide means "under vacuum," because when it was invented the food was sealed in plastic bags with a vacuum machine. A better name would be Precision Heat Cooking. Many cooks still use vacuum sealers because air is an insulator. We want the air out so the meat is in contact with the bag and the bag is in contact

Conventional Cooking

Sous Vide

Grilling a steak properly is like trying to catch a moving bus. You've got to jump on as it goes by. You have a very short window between undercooked and overcooked. But with sous vide, the bus stops at the perfect finishing temperature and waits for you.

You can buy special tubs designed for sous vide, but I use a stockpot with a silicone lid. I cut a slot in it for the immersion circulator.

with the warm water. But you don't need a vacuum sealer. Any food-grade zipper bag will do. You can push air out by slipping the lower part of the bag under water and zipping it as the water approaches the top. Sometimes the food tries to float. You can sink it by putting a table knife in the bag.

A special heater called an immersion circulator is placed in a pot where it warms and circulates the water. This device heats the water *precisely* to a temperature you select, say 131°F for a steak, perfect medium-rare. The food in the bag slowly heats to that temperature and *it can never get to 132°F or higher, so the meat can never be overcooked.* You get predictable results with precision temperature control. Home units sell for as little as $50. Two of my favorite brands are Joule and Anova. Both have models that can use either an app and Bluetooth or Wi-Fi.

At that temperature you can leave the food in the bath for hours. It heats the food to a uniform temperature from bumper to bumper so nothing is overcooked. The result is very tender and juicy. Sous vide low and slow even tenderizes woody vegetables like carrots by softening tough structural components. If you leave it in the water bath for several hours at 131°F or more, the food is pasteurized. And yes, you can use sous vide to make medium-rare burgers and "raw" eggs safe.

Depending on time and temp, there is still significant water loss. It builds up in the bag. I have tried making a sauce from those liquids

Pasteurization vs. Sterilization

The goal in cooking is pasteurization not sterilization.

Sterilization kills or removes all microbes and their spores by using one or more of the following: Heat, irradiation, chemicals, pressure, or filtration. Spores are dormant fortress-like forms that some microbes assume to withstand adversity. The cost and effort to sterilize is high and the impact on some food is deleterious. Sterilization is used for pharmaceuticals and medical supplies. In the food industry, the term "commercially sterile" can be used, but it means that the kill process resulted in a product with acceptable quality and that will not support growth of any microorganisms that may slip in during packaging and storage.

Pasteurization reduces the population of pathogens to a level deemed safe. That means that the number of survivors is so small, chances are you won't encounter any, and if you do, there will be so few as to be harmless. But pasteurization cannot kill bacterial spores. Pasteurization can be done quickly at high heat, or slowly at lower heat, usually above 131°F.

with poor results. Too much protein that coagulates.

Many sous vide chefs sear or smoke meats before sous vide and then put oil, spices, herbs, even slices of citrus in the bag. I rarely do. Most of the smoke gets washed off with the purge in the bag, and remember, meat is mostly water, so oil can't penetrate. I do all my flavorizing after it comes out of the bag. That's also when I add the rubs and do the searing or smoking. All it takes is 30 minutes in smoky air and you will taste it. And the internal temperature of the meat barely rises. Another trick is to chill the food when it comes out of the water bath,

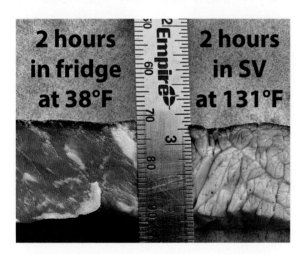

In this experiment, Professor Blonder coated the surface of two steaks with a dye as a surrogate for a marinade. He placed one in the fridge for 2 hours and placed the other in a sous vide bath for 2 hours at 131°F. As you can see, the dye did not penetrate far in the fridge, 1 to 2 millimeters. The dye penetrated just a bit farther in the sous-vide bath. But it also dissipated as some of it came out in the purge in the bag. Lesson? As we have shown in other experiments, marinades do not penetrate much. That's why I recommend that you smoke and add spices after sous vide and not before.

Sous-Vide-Que . . .

Sous vide allows you to cook at just about any temperature above 131°F. Here are some suggested starting points, but feel free to try other temp and time combinations.

. . . for Tender Cuts (such as Steaks and Chops)

1. **SOUS VIDE.** Salt, then sous vide for 2 to 4 hours at no less than 131°F, longer for thick cuts.
2. **RUB.** Remove from bag, pat dry, sprinkle generously with salt-free rub.
3. **FINISH.** Sear in a hot pan, griddle, or grill; or smoke and then sear. Bring to a finishing temp that you like. Glaze or sauce if you wish.

. . . for Tough Cuts (such as Ribs, Beef Brisket, Pork Butt)

1. **SOUS VIDE.** Salt, then sous vide at 145°F for 24 to 48 hours.
2. **RUB.** Remove from bag, do not pat dry, sprinkle generously with salt-free rub.
3. **ROAST OR SMOKE.** Roast or smoke at 225°F until 155°F.
4. **OPTIONAL.** Sear in a hot pan, griddle, or grill. Glaze or sauce if you wish.

. . . for Chicken, Turkey, and Ground Meats

Same as tough cuts.

put it in the fridge, and sear or smoke a day or three later. This chill step even seems to improve the texture.

There are a couple of problems with sous vide. First of all, you don't know what is the optimum time to leave the food in the sous vide bath and what the optimum temperature is. Fortunately, there are books and guides with tables. Sadly, no two agree. So you need to experiment to find what you like best. Start with the rules of thumb in the sidebar on page 109. Experiment!

After sous vide I like to sear meats on top of an incredibly hot charcoal chimney.

The other problem is that when it comes out of the bag it is butt-ugly and it lacks the rich flavors created by the Maillard reaction. Many people throw the meat in a hot pan to sear it. Some use a torch. But I love what the charcoal does to meat, so I sear the meat on a scorching-hot grill, lid up so that the energy is concentrated on only one side at a time. I am also fond of searing the meat on a charcoal chimney where a deep pile of charcoal produces massive amounts of IR. I call it the "afterburner." This is the ultimate reverse sear method. I use the afterburner for my Smoked Sous-Vide Steak (page 221), Japanese Beef (page 223), and Afterburner Fajitas (page 227).

Many websites recommend sous vide in the 120° to 130°F range and many people believe that over 120°F bacteria cannot grow. Not so, says the International Commission on Microbiological Safety for Foods. It is true that most foodborne pathogens such as *Salmonella*, *Listeria*, and pathogenic strains of *E. coli*, cannot grow at 120°F, but others such as *Clostridium perfringens* and *Bacillus cereus* can grow at higher temps. So, to be safe, after consulting with a prominent microbiologist, I recommend you cook red meat, pork, and poultry at 131°F and above. By great fortune, 131°F is medium-rare, the temperature at which most beef steaks are optimum tenderness and juiciness, and below the optimum temps for pork and chicken and turkey. Even after the sear step, the temps are usually under 135°F, still medium-rare and very tender and juicy. So that's the minimum temperature I shoot for.

SAFETY TIP. Botulinum spores are impossible to kill even at boiling temperature. Raw garlic and onion, which grow underground where bot thrives, should never be in a sous vide bag where the spores can flourish.

SAFETY TIP. Heat the water to the desired temperature before putting meat in.

ARE THE BAGS SAFE? People are justifiably concerned about chemicals in some plastics, especially Bisphenol-A (BPA), a compound that is believed to be a carcinogen. There are two types of bags we commonly use for sous vide, plain old zipper bags and special bags designed for vacuum sealing. Most are made from polyethylene (PE). PE does not contain BPA. Professor Blonder says, "A good quality PE bag from a reputable supplier does not contain plasticizers. Some of the polyethylene monomers might not be fully reacted and could leach out. There is a faint smell from a warm polyethylene bag, so something is volatile, but we don't know if it is riskier than smoke." I use zipper bags most of the time except when cooking at temperatures above 160°F. At these higher temps, the seams sometimes fail and water can get in. I use the heavier duty vacuum sealing bags then.

KŌJI

Kōji is the Japanese name for a fungus, *Aspergillus oryzae,* and it is an enzymatic Swiss Army knife packing loads of umami.

You ask, "Fungus on my food?" Yeast is a fungus, so without fungi, no bread, wine, or beer. Kōji is used in making sake, miso, tamari, soy sauce, bean pastes, and even some yogurts. It can transform a wide range of meats, vegetables, and many other foods with exquisite results. A word of caution: Kōji has sugars in it and if you use it on chicken skins or turkeys for instance, they will burn if you don't keep an eye on them.

There are five forms of kōji for home cooks:

KŌJI SPORES. The spores can be inoculated into cooked rice to get started. A lot can go wrong with this process, so I don't recommend it.

KŌJI RICE. Kōji rice (*miyako kōji*) is cooked rice that has been inoculated with spores of kōji, incubated, and then dried. Some cooks simply coat meat with the rice and lay it on a wire rack in the fridge for 3 to 7 days, and the wet meat induces the fungus to grow.

SHIO KŌJI PASTE. This is a better strategy for home cooks than kōji rice. It is a goopy white porridge-like umami-bomb marinade made by pureeing kōji rice. It will keep in the fridge for months. Because it is wet, it takes less time to work its magic on meat than does dry kōji rice. You can buy it premade or make it yourself from kōji rice. It takes only a couple of weeks.

AMAZAKE. This milky liquid is the base for sake, a rice wine. Sophisticated chefs like it because it is not salty and you can make it sweet or sour depending on how you produce it. Jeremy Umansky, the genius behind the James Beard Award–winning Larder Delicatessen & Bakery in Cleveland and co-author of the book *Kōji Alchemy,* uses it almost exclusively in everything from meats to breads and pastries.

LIQUID SHIO KŌJI MARINADE. This is my preference because it is so simple. It is a golden salty liquid. It is made by pressing and filtering shio kōji paste. Hanamaruki is the brand a lot of chefs use. A few tablespoons, about 10%

Shio Kōji Paste   **Shio Kōji Marinade**

of the food's weight, added to a zipper bag of chicken, pork, or beef and refrigerated for a few hours or overnight is surprisingly good. For seafood, just a few hours are all you need. It can even be used as a finisher, added just before serving.

PIZZA

You can cook pizza magnifico in any oven or on just about any grill with a lid. Pizza is a huge topic that can fill a book, so I will touch on just the basics. Making the dough is easy, but there are many different kinds of dough to select from. If you want more on this topic, please visit AmazingRibs.com.

BAKED PIZZA. This is pizza like they make downtown. It is baked on a pizza stone. Modern pizza stones are made from a special heat-tolerant refractory material called cordierite. They are good at distributing the heat evenly. On the downside, they can break easily. Baking steels are gaining popularity. They are unbreakable heavy slabs of specially

formulated steel. If you are baking pizza, I think it is pretty important to have an infrared gun thermometer. I won't start a pie until the stone is at least 600°F, and for some dough recipes, I want it hotter.

GRILLED PIZZA. You make the dough, stretch it, oil it generously, throw it on really clean grates or on a griddle, toast it on one side and get some grill marks, flip it over, top the hot cooked side with a few light and thin toppings—no raw meat—close the lid, and when the bottom is done and the toppings warm, it is time to eat.

Photo by Nate Maliwacki

ON A WEBER KETTLE. I spread hot briquets in a ring around the outside of the charcoal grate. This prevents the bottom of the pizza from burning. I put bricks on the grates and put the pizza on a pan on top of the bricks. This moves the pizza closer to the dome to help the top cook. I close the vents at the top so hot air can collect under the lid.

ON A GAS GRILL. I MacGyver a pizza oven. I place a pizza stone on the grates, surround it on the sides with bricks, and place another pizza stone on top of the bricks. I leave the back open and place a metal pan hanging over the back so hot air can be deflected into the oven.

Photo courtesy of Ooni

COMMERCIAL PIZZA OVENS. In the past ten years, there has been a tsunami of lightweight pizza ovens capable of making a pie as good as any pizzaiolo in Napoli, such as the Ooni Koda in the image above. Several sell for less than $500. We have reviews of many on the website. They are also good at baking breads, naan, tortillas, and cookies, and they can even broil meats. Some burn chunks of wood, some burn pellets, some burn charcoal, some burn gas. For me the choice is simple: Gas. Why not wood? There just isn't time for smoke to flavor the pizza. And who wants a smoky pizza anyway? Besides, managing wood ad pellet fires is a pain. Gas can be set to a temp and hold there.

THIN PAN. My wife preps and cooks most of her pizzas on a 14-inch round thin metal pan. You want a thin pan because it heats up quickly. Just oil the pan generously, stretch the dough onto the pan, add toppings, and pop it into a hot grill, lid down, so it bakes in the pan. Works great indoors, too. Another strategy is to build the pie in a well-oiled pan or on a metal pizza peel and put it in the oven. After a minute or two the bottom starts to toast and it is easy to remove the pan or peel and slide the pie onto a stone.

CAST-IRON PANS. These work, too, especially for deep-dish pizza. Give the pan a good coating of oil. Stretch the dough to fit the pan right up the sides, lay the toppings on, and bake at a really high temperature. When the top is done, peek at the bottom. If you want it darker, put it on a side burner or burner on your indoor stovetop.

COLD-SMOKING MEATS— DON'T DO IT

I know your Uncle Piotr from the Old Country has been cold-smoking sausages for fifty years and he's still alive. Well, I've been driving a car for more than fifty years and I'm still alive. Alas, over those 50 years 1.5 million Americans have died in cars. Both Uncle Piotr and I are lucky.

Cold-smoking is a technique that professional food processors use to flavor food with smoke without cooking it. It must be done in concert with proper humidity, salt, sugar, preservatives, and precision temperature control. Without cold-smoking, no Nova lox, Lebanon baloney, kielbasa, or smoked chubs. But cold-smoking at home is risky and my unswerving advice is don't do it. Done improperly it can kill. Commercial jerky and smoked fish makers know what they are doing and still recalls are common. I have excellent equipment and extensive knowledge, yet I never cold-smoke. I love my family too much. Buy your Nova lox. And if Uncle Piotr gifts you a chub of cold-smoked sausage, thank him profusely and when you get home throw it out and wash your hands.

COLD-SMOKING CHEESE AND CHOCOLATE—DO IT

Even though I don't want you cold-smoking meat, I do want you to try cold-smoking cheese or chocolate. Much less risky. When smoking cheese, you don't want to melt the cheese because it can separate. Many cheeses begin to melt when the temperature gets to about 90°F and chocolate melts around 85°F, so you need to keep the temperature below this threshold. So a self-sustained white smoke generator that doesn't produce much heat—like those discussed in Smoker Generators (pages 64–65)—is the way to go.

On a cool day or night, I open my gas grill, place a pan on the grate to hold the cheese, place a thermometer probe nearby, dial in a low temperature, and place one of the smoke generators near the pan. Sometimes I'll put a cardboard box over them just to capture more smoke. On hot days, I sit the cheese or chocolate on a plate above a pan of ice. For a quick, cheap thrill, get a block of cream cheese, sprinkle it with my Memphis Dust (page 165), cold-smoke it, and serve it on crackers or Smoke-Roasted Garlic Bread (page 367).

You can use these methods to smoke oil for sautéing, vinaigrettes, and mayo. You can also smoke chocolate and cream for whipping. The Smoking Gun is especially good for this.

SOME MORE IMPORTANT TOOLS

"I discovered that summer is a completely different experience when you know how to grill."

—TAYLOR SWIFT

RECOMMENDED CHARCOAL GRILLS

If you can only afford one cooker, a good charcoal grill is a good choice. It sears well and is usually easily adapted to smoking. You want one large enough to be divided into two zones so you can roast and smoke via indirect convection heat, and sear via intense direct infrared radiant heat. It should have excellent airflow control, because by controlling oxygen you can control the temperature and smoke. That means a snug lid and minimal leaks. It must be large enough to handle the biggest party you give. When it comes to grills, surface area is like RAM on a computer. You can never have too much.

Some of our favorite charcoal grills are, listed alphabetically: Burch Barrel,

HOT TIP

Figure a minimum of 80 square inches of cooking surface per person when buying a grill.

Engelbrecht, The Good-One, Grillworks, Hasty Bake, Kalamazoo, M Grills, Napoleon, Nuke, Portable Kitchen, Primo, Slow 'N Sear Kettle, and the good old Weber Kettle. We'll get to smokers in a little bit. For our other favorites and a searchable database of hundreds of ratings and reviews by Max Good and my team, go to AmazingRibs.com/mm.

RECOMMENDED GAS GRILLS

When it comes to gas grills, there is a dizzying array of options, and they change with each model year. Here are some of our favorite manufacturers, listed alphabetically: Broil King Signet, Char-Broil Commercial, Fire Magic, Hestan, Kalamazoo, Napoleon, and Weber.

RECOMMENDED SMOKERS

Many grills can be set up for smoking, but dedicated smokers generally make better-tasting smoked food. Below are some tips for shopping and some recommendations.

New models come out every year, so check AmazingRibs.com before you write a check.

LOG SMOKERS (AKA STICK BURNERS)

The smokers that make the best-tasting meat burn logs, but they are expensive and require a lot of practice and patience. Most log burners are called offset smokers because the firebox is set off from the food chamber. Offsets come in two basic flavors, direct flow and reverse flow.

DIRECT FLOW. If the chimney is on the opposite side away from the firebox, then warm air and smoke have a direct pass through the cooking chamber and across the food. Direct flow smokers tend to be hotter on the side nearest the firebox.

REVERSE FLOW. Some have the chimney on the same side as the firebox. They have a solid metal plate under the food that channels the air from the firebox the entire length of the cooking chamber to the far end where an opening allows it into the cooking chamber and

it doubles back over the food to the chimney. This reverse flow cools the smoky air and heats the metal plate under the food which evens out the temperature within the cooking chamber.

Both designs produce excellent smoked meats and even though they are designed to be log burners, they can burn charcoal, and even though they are designed to be smokers, you can put charcoal in the cooking chamber and you have a large capacity grill.

There is a big difference between the cheap offsets at the hardware stores for $300 or less and the serious pits made for competition teams and caterers in the $1,000 and up range. The cheapos are headaches. The doors don't fit tight, so you can't control oxygen; the walls are thin, so they don't retain heat; and they rust. If you own one, you should stick with charcoal for heat and wood chunks for flavor.

Here are some top-quality offset pits, listed alphabetically: Franklin, Horizon, Jambo, Klose, Lang, Meadow Creek, Oklahoma Joe's Bandera, Peoria, Pitmaker, and Yoder. There are reviews of these and others on the website.

KARUBECUE. The Karubecue produces the best-tasting meat of any smoker on which I have cooked. Logs are loaded above the cooking chamber and smoke is pulled down through the glowing embers before it enters the cooking chamber. In the process, the hot embers burn off impurities and the taste of the blue smoke is unparalleled.

Reverse flow offset smoker designed for logs.
Courtesy of Lang BBQ Smokers

This is the most unusual and clever design. The firebox on the Karubecue is above the food and smoke is drawn down through the hot coals. Photo by Bill Karau

I know you want a stick burner, but a word of caution: They require skill and practice. Managing the fire, airflow, and smoke is tricky. I advise you to wait until you have mastered something simpler before you get any kind of "stick burner." You can get excellent quality with smokers that burn charcoal or gas for heat, and you add wood for flavor, and they start at about $400.

CHARCOAL SMOKERS

Hasty Bake Grills and The Good-One are two models I love. They are grills that are easily set up for smoking. Slow 'N Sear makes a kettle that smokes small quantities well. Masterbuilt has a Gravity Series that you load up with charcoal and it feeds the fire with the right amount needed automatically. Weber Kettles can be adapted for smoking small quantities, especially if you add the Slow 'N Sear insert. PK Grills are easy to set up in

two zones for smoking. Most kamados are fine for smoking small quantities. There are several drum smokers to consider, among them Pit Barrel's lineup and Oklahoma Joe's Bronco. The Weber Smoky Mountain is extremely popular.

GAS SMOKERS

Gas smokers are among the best bargains on the market. There are several cabinet-style gas smokers that look a bit like small dorm refrigerators with a compact footprint and large capacity. You turn on the gas, adjust the temp, toss some wood into a pan above the flame, load in the food, and walk away. Temperature control is pretty good. They are inexpensive and make fine food. Look at the Camp Chef, Char-Broil, and Masterbuilt. If you are a purist and want to sneer at the idea of a gas smoker, keep in mind that most the nation's best barbecue restaurants use smokers that get heat from gas and flavor from wood. In addition, most gas grills can be set up in two zones for smoking.

WOOD PELLET SMOKERS

Although many manufacturers call pellet cookers "grills," most are not very good at direct infrared radiant heat searing, so I think of them as smokers, not grills, and they are very good at smoking, although the smoke flavor tends to be mild. These are some of our faves, but you should consider them smokers, not grills, listed alphabetically: Camp Chef, Green Mountain Grills, Louisiana Grills, MAK, Pit Boss, and Traeger. Weber's Pellet Smokers are among the few that can both smoke and sear well.

GRILL GRATES

Grill Marks

Proper Sear

Yes, I know you've been indoctrinated to start drooling when you see grill marks, but the truth is that grill marks mean much of the meat's potential has been lost because the stripes are the only part that has had Maillard reaction. The rest is tan, and tan meat is meat that has less flavor than brown meat. So the goal should be an even dark brown all across the surface. I want the radiant heat to do the job. If I want the metal itself to sear my steaks, I could do it better in a frying pan. Skip the grill marks.

There is a mystique around cast-iron grates. I am not impressed. They absorb and hold a lot of heat, and they brand the meat surface with that heat, but I don't want my grill grates to cook just parts of my meat surface. Besides, cast iron rusts easily and is a pain to maintain.

OK, sometimes grill marks do come in handy. When you are cooking thin or small foods like skirt steak, asparagus, kebabs, or shrimp, it is almost impossible to get uniform dark-brown surfaces with just radiation or convection before the center is overcooked. The only thing that works is conduction. And that means grill grates that transmit heat well.

My two favorite grill grates are polar opposites. For charcoal grills, I prefer a wire grid. They allow maximum exposure to infrared and they are easy to clean and

maintain. Most of the thin wire grates are chrome- or nickel-coated, and eventually the coating wears off and the steel rods underneath rust. Enamel coatings last longer and stainless is the most durable.

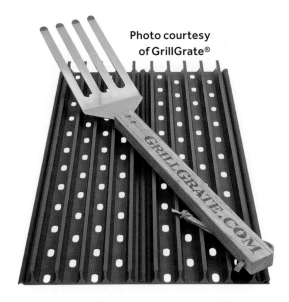

Photo courtesy of GrillGrate®

For gas grills, I recommend a product called GrillGrate®, the best thing to happen to gas grills since the propane tank. GrillGrate®

is an amazing invention featuring reversible interlocking aircraft aluminum panels. Each panel is a flat plate with raised rails and the panels lie on top of your current grill's grates or they can be used to replace them. They increase the grill surface temperature, reduce flare-ups, and minimize hot spots. GrillGrate® comes in a variety of sizes so you can customize your order to fit the size of your grill. They can be easily moved from one grill to another.

Each plate has holes in it so hot air, smoke, and combustion gases can rise through them while most of the heat is

Photo from Adobe Stock
by Kai Beercrafter

trapped below and builds. The bottom plate, ¾ inch below the food, becomes the main heat source. Juices drip into the valleys between the rails where they are vaporized and the vapors land on the meat, enhancing flavor. The holes in the base allow some juices through where they burn, and smoke and combustion gases travel up through the holes to reach the food, adding more flavor. The solid bottom plate spells the end of asparagus falling into the fire. The flat underside can also be used as a griddle.

You can throw wood chips, pellets, and sawdust into the valleys between the rails and then put food on top. The wood begins smoldering almost instantly and imparts a delicate wood smoke flavor, even on fish and other quick-cooking foods. There's a picture of a GrillGrate® in my recipe for Close Proximity Smoked Fish with Poblano-Basil Cream Sauce (page 308).

The sturdy stainless steel custom-designed spatula that comes with GrillGrate® has fingers that slip between the rails and lift even the most delicate pieces of fish with ease. There are even special stainless steel tongs with fingers available as an option.

GrillGrate® works fine on charcoal grills and pellet smokers, but I don't find them necessary there.

GRILL TOPPERS

You need grill toppers for small foods that would otherwise sacrifice themselves by diving through the grates into the inferno like Aztec virgins. Grill toppers prevent this loss. They come in really handy when grilling things like mushrooms, carrot coins, onion slices, string beans, nuts, shrimp, scallops, and small chunks of meat.

I prefer grill toppers to skewers because I can cut big chunks of meat about 2 inches in diameter, large enough so the exteriors can sear and the interiors have half a chance of remaining pink. Then I put them on a metal grill topper where the hot metal can help with the searing. I can then roll them around and get the color I want on all sides, even if they are funny shaped. But most important, if they are irregular sizes, I can get them off the heat when they are ready, one hunk at a time.

Photo courtesy of Weber Grills

METAL PLATES AND PANS. Flat stainless steel or enamel-coated plates with holes are great toppers. So is GrillGrate® (see page 121). Perfect for shrimp and onions, things that benefit from searing by contact with hot metal. There are disposable aluminum versions, but they don't sear as well as heavier metal. The shiny unit on page 122 is the Weber Deluxe Grilling Pan for about $35.

MESH BASKETS. Several companies make bowl-shaped "grill baskets" made of stainless steel mesh, perfect for stir-grilling veggies. They cost between $25 and $40.

FLEXIBLE MESH. Big Poppa BBQ Mats are a screen-like woven mesh with holes about ⅛ inch wide. The mesh is nonstick Teflon-like coated fiberglass. I love these for smoking small things like oysters, cherry tomatoes, and nuts. I use them only on my smokers, because you just can't get them very hot or the PTFE coating can produce hazardous gases.

GRILL MATS. There are a number of solid, flat, flexible nonstick mats that lie on top of the grates. But there are no holes so no smoke reaches the food. You might as well be cooking indoors in a frying pan. And beware, these nonstick surfaces can off-gas noxious fumes at high temps.

SALT BLOCKS. You can buy blocks of Himalayan pink salt. Mine is 9 × 9 × 2 inches thick. The pink salt comes from the Salt Range in the Punjab region of Pakistan and it is absolutely not the same as pink salt used in curing bacon or other meats. It is just plain old salt with some pink color that comes from some small amounts of iron. When heated in an oven or grill, the blocks retain heat exceedingly well. Get one hot, place it on a trivet on top of a thick stack of cloth place mats or kitchen towels in the center of the dining table, and plop strips of meat and vegetables on them. They sizzle and suck up just the right amount of salt. Cleanup is easy, just rinse and scrub any chunks off and that's it. A little elbow grease. No soap needed, since no common pathogens can grow on solid salt. Eventually meat juices will color the pearlescent block, giving it a beautiful glowing amber patina. Be careful in handling them, as they can crack and break. I have gotten dozens of uses from my block. But if they do break, you can grind them up and use them for seasoning food or melting snow on your sidewalk.

PLANKS. I'm not a fan. The wood, usually cedar, does not get very hot so you can't sear the underside, and the skin on salmon, when crisped, can be very tasty. They really don't produce much smoke. Wood chips and chunks do better.

GRILL BASKET. There are several large perforated tubes like the one pictured here from BBQ Dragon that work well. Some attach to your rotisserie spit and others you just roll around on the grill grates.

FLIPPING GRILL BASKETS. The number-one problem with cooking fish is that they stick to metal. Coating with mayo or oil helps a lot, but they still stick. You can leave fish skin on and be prepared to sacrifice it, put slices of citrus under the fish, put fresh herb branches under it (see Peter Mayle's Flaming Fish: A Wonder, page 316). I reach for a two-sided fish basket. Most are thin metal grates hinged on one side. You oil them, put the fish in, and clamp down. Then you can flip the basket with the fish inside. The one shown here is the Weber Grilling Basket. There are two sizes from about $40 to $50.

SKEWERS

I don't use skewers very often and I never use round bamboo skewers. I prefer cooking chunks on grill toppers. Things on skewers need to be similar in size for them to cook at the same rate.

If you must use skewers, let me recommend flat sword-shaped skewers. When you want to turn meat on round skewers the skewers often turn but the food doesn't. To prevent this spinning, you need to use two round skewers side by side. But flat metal skewers turn things better. And don't worry, the metal never gets hot enough to cook the interior of the meat. The cold wet meat keeps the metal cool.

Wooden skewers also tend to burn. I've even seen them catch on fire. Most books and recipes say to soak bamboo skewers for 30 minutes or more to keep them from burning during cooking. So I got two skewers. I soaked one for an hour, and one not at all. Both burned after 30 minutes at a surface temperature of 450°F on a gas grill. Soaking made little difference because wood does not absorb water and the water on the surface dried almost instantly.

CAST IRON

I will be excoriated for saying this, but I have fallen out of love with my bare cast-iron pots and pans, and I have a lot of them. I know they are beloved by many. Not me.

First, bare cast-iron pans are more difficult to clean and maintain than my other pans. And no matter how well I clean them, they are never really clean. You're not supposed soak them in the sink or wash them in the dishwasher. If you have a cast-iron pan, right now go wipe the surface with a paper towel. It came away brown, didn't it? That's oxidized/rancid oil, rust, and flavors from the last thing you cooked. If you cooked fish in it, it still smells like fish, doesn't it? Because the pans are black you can't see the dirt. Forget to give it a coat of oil after cleaning and it rusts. Getting the rust out is a major pain. They are heavy, too heavy for a lot of old-timers. They scratch glass cooktops. Don't drop one on that glass cooktop or on your toe. And, although a well-seasoned cast-iron pan is slippery, it is not really nonstick. A new ceramic pan kicks its butt.

For most of our cooking we use stainless steel pans with heavy bottoms made of two layers of stainless with a layer if copper in between. These heavy bottoms distribute and retain heat well. Stainless is easy to clean and you can see when it is clean. You can also see colors in foods as they cook. You can even use them on the grill. Our set was a wedding gift fifty years ago and they work like new. We also use new ceramic-coated pans that are as slippery as a ski slope. All this said, I use a large cast-iron Dutch oven for deep-frying because it holds heat so well, and occasionally I grab a cast-iron skillet or griddle for shallow-frying on the grill because they hold a lot of heat. Just make sure the handles are oven safe.

Cooking with cast iron is healthy

Cooking with cast iron is unhealthy

Health nuts say that some of the iron gets into the food and that is good for our hemoglobin. Others say that if you have hemochromatosis, a metabolic disease where your body can absorb too much iron, tiny amounts of iron can leach into food and make you sick.

BUSTED. Wellllll, in order for iron to leach out, it has to be in contact with the food. But if you have seasoned the pan properly, how, pray tell, does the metal touch the food? And how much benefit out of the 80,000 meals you will eat in an 80-year lifespan are prepared in cast iron? Will these phantom benefits add one minute to your life?

A bowl of Cheerios contains more iron than a serving of tomato sauce heated in a cast-iron pan. Plus, the chemical form of the iron in Cheerios is more easily absorbed by the body. You want iron? Just eat some meat and vegetables.

GLOVES

You need insulated gloves to handle hot objects. Get gloves with fingers, not mitts. You need articulation. I'm not a fan of the all-silicone gloves. Too slippery. There are several gloves made from synthetic fabric called aramid fiber, such as Kevlar. Many have silicone pads on them and some are lined. They offer good heat resistance, flexibility, and grip, but they cannot be used when wet. My favorites are suede/leather welding gloves. They are long, almost up to my elbows, flexible, and can be washed. I have picked up glowing coals with them. If you are doing a lot of oyster shucking, get cut-resistant gloves.

DIGITAL SCALE

Many recipes call for ingredients by weight, especially baking recipes. When you are baking, you need to weigh the flour because a loosely packed cup can have a lot less flour than a tightly packed cup. Ditto for shredded cheese, brown sugar, and salt. If you want the perfect 6-ounce burger, you need a digital kitchen scale. I like the one from OXO. It is accurate up to 11 pounds, and the screen can be pulled out in case the thing you are weighing hangs over.

TONGS

You need tongs at least 18 inches long with the business end designed to grip. Many have a locking mechanism that allows you to store them in the closed position. A few are like two spatulas hinged together, great for burgers and fish.

OTHER TOOLS

There is an amazing array of other gadgets and gizmos, many of which are practically necessities. In addition to those mentioned above, consider a rotisserie, large wide heavy spatula with a thin sharp edge, knife set, knife sharpener, several sizes of disposable aluminum pans, heavy-duty foil, a hypodermic injector, meat shredders, silicone basting brushes, headband flashlight, rib holder that has slots for 6 or more slabs of ribs standing on end, propane tank weight gauge, squirt bottles, apron, spot remover (because you *will* get grease on your shirt), adjustable pepper grinder, garlic press, premium beer cooler, stand mixer with meat grinder and sausage stuffer, steamer basket, blender, food processor, spice grinder, hinged grate for your Weber kettle, and of course, corkscrew and can opener. When it comes to chopping onions, I keep a pair of swimmer's goggles in the kitchen.

Because prices differ from store to store and month to month, and models come and go, we have produced a more comprehensive list on AmazingRibs.com/mm. Please check there before making a buying decision. My team and I rigorously test, review, and rate products on an ongoing basis and there are searchable databases of more than 1,000 grills, smokers, thermometers, accessories, books, and more. It's a dirty job, but somebody's got to do it.

SAFETY

"Food is expensive. It is costly and embarrassing to overcook it. Friends and family are priceless. It is not nice to kill them."

5

There is no such thing as risk-free living. Or eating. Eating pathogens is pretty high risk. Eating undercooked chicken is less risky, but still risky. Eating a medium-rare hamburger is lower risk still, but still risky. Eating a slightly pink pork chop is not very risky at all. The greatest risk you take is getting in a car.

Do you buy only organic food and grass-fed beef, dote on whole-grain bread, eat yogurt daily, and then use your cell phone when driving home? That cell phone negates all the other risk reduction. If you use it when driving, you are hereby authorized to eat bacon with every meal for the rest of your short life.

According to a 2013 study in *Emerging Infectious Diseases*, more than half of all foodborne illnesses were caused by plant foods. Leafy greens like lettuce led the list. That's because fields of vegetables are exposed to contamination from Tweety Pie, Thumper, Bambi, Pumbaa, and Mickey as well as irrigation water contaminated by Porky and Elsie, not to mention the hands of pickers after they use the toilet. Then we eat them raw.

The CDC estimates that every year roughly 1 in 6 Americans gets sick from foodborne illnesses, 130,000+ are hospitalized, and 3,000+ die. That's more than who died on 9/11/2001, more than US military who died in Afghanistan.

Contamination can come in the form of bacteria, viruses, or parasites. The air, water, and soil around us are teeming with bacteria. Almost all food has contamination, either from the air, soil, fertilizer, water, slaughterhouse, processing plant, butcher shop, store, or your kitchen.

If you ingest enough of them, they can leave you sitting on the toilet for hours, plant you on your knees in front of the porcelain god, or propel you to the emergency room or even the cemetery. These bad guys can often take a day or more to grow in your gut before they knock you down, so figuring out what got you sick is hard.

GRILL AND SMOKER SAFETY

- Save the grill manual and remember where you put it.

- Handle hot grills, coals, and liquids with respect. Be alert. Keep children and pets away from grills and smokers, uncooked meat, hot liquids, and sharp objects. No horseplay near cookers.

- Before you use a new grill or smoker, fire it up on high and let it run for about 30 minutes to burn off any oil, grease, or packing materials from the manufacturing process or from shipping.

- Before cooking, heat the grates to Warp 10 to carbonize grease and scraps from your last cook. Then clean them thoroughly. Inspect for bristles left behind.

- Fire is always a risk. Don't keep a hot grill close to your house or deck railings. Beware of overhanging rooflines, wires, or trees. Keep an ABC-rated fire extinguisher nearby. Water will only spread grease fires.

- Never use grills or smokers indoors or in garages. They produce invisible carbon monoxide that can kill. Don't place your grill next to a furnace air inlet or even a window. Your house is often under a negative pressure and can suck in carbon monoxide.

- Don't cook near gasoline or other flammables. Keep propane tanks outdoors in an upright position at least 2 feet from burners unless there is shielding.

- Never use gas, paint thinner, solvents, or kerosene to start your charcoal.

- On gas grills, always lift the lid when you ignite the burners. If you have one burner lit and want to add others, open the lid. A gas buildup under the hood could blow it open and flash in your face.

- On kamados the lid seal is very tight, so when you open it, air rushes in and it can flash flame in your face. Stand back and open the lid slowly.

- If you smell gas, turn off the grill immediately.

- Use potholders and/or insulated gloves. I recommend leather welder's gloves. Silicone gloves are too slippery.

- Hot liquids and steam are as dangerous as hot metal and flame. If you or someone else is burned, don't take it lightly. Doctors say to rinse the burn in cold water but never use an ice pack. It can make a burn deeper and more serious. If the burn is over a large area or if the pain is so bad you can't control it with over-the-counter painkillers like Tylenol, Advil, or aspirin, you need to go to the emergency room. If you lose sensation and you can't feel any part of the wound, you need to go to the emergency room.

- Check the bottom of your grill for rust. You don't want coals or hot grease falling through.

- Do not discard ash or coals until the coals are thoroughly dead. Let them sit overnight or dump water on them before you put them in your trash can. If you pour water over hot coals, it will produce enough steam to melt your nose, and enough hot water might come out of the bottom to melt your toes.

- Bare feet, sandals, flip-flops, and loose clothes are a bad idea around grills. If you have long hair, tie it in a ponytail. Grilling is yet another great excuse to not wear a tie.

- Don't put small grills on flammable surfaces or glass tables.

- Drink with moderation. Drunk grillers have accidents.

- When you are chopping, slicing, or dicing, hold the knife firmly and hold the food with your fingertips curled in toward your palm so they can't get under the blade. Called "the claw" grip, it seems awkward but you must master this skill.

- Dull knives can slip and hurt you. Get a good sharpener (we tested several and recommend some on the website).

- Never open cans with a knife. I don't care what you saw on *Iron Chef*.

CLEAN

- You may handle uncooked food with your bare hands, but you must first wash your hands past your wrists thoroughly with hot water and soap for 20 seconds by rubbing them vigorously. That's how long it takes to sing the "Happy Birthday" song twice. Keep a brush to scrub under your fingernails near the sink.

- There is a conservation movement that views paper towels as wasteful. Disregard them. Paper towels are biodegradable and they are an essential part of a healthy kitchen. Use paper towels a lot, especially if you drip meat juices on the counter. Cloth towels are easily contaminated on first use and then they hang around for days. Toss them in the wash often.

- Buy an empty spray bottle at the drugstore and fill it with 1 ounce (2 tablespoons) of unscented liquid chlorine bleach and 1 quart of water. Use it to sanitize surfaces after cleaning them with soap and water, especially surfaces that have touched raw meat. Leave it on the surface for about 3 minutes before wiping it up.

- Clean your sink daily with Comet or another cleanser that contains chlorine or use your chlorine spray. The underside of the rubber flange in your disposal is one of the nastiest things in the kitchen and every time you turn it on it disperses contaminated aerosols into the sink and air. Glove up, lift the flange, and clean it thoroughly with dish soap. Bleach could harm the rubber.

- Wet sponges and dishcloths can give your disposal a run for its money. They grow bacteria, yeast, and mold in a hurry. Putting a wet sponge in the microwave for 2 minutes will kill bacteria. Do it daily. Do not microwave a dry sponge; it can catch fire. If you don't have a microwave, boil them for 5 minutes or run them through the dishwasher. Replace sponges often.

- Wash all dishes, knives, tongs, and brushes that have touched raw meat in hot soapy water. This means that if you use tongs to put raw meat on the grill and turn it, you must clean them before you use them to take food off the grill.

- A hot dishwasher and its detergent will make things safe. Throw the sink strainer in there every day or two.

- Be sure to clean the probe on your thermometer after you are done using it.

- Always use a cutting board. Never cut anything in your hand. There is a lively debate over which is safer, wood or plastic cutting boards. Science says both can be safe if they are cleaned thoroughly. Scrub them well with warm soapy water, rinse, and then scrub again with a chlorine-based cleanser like Comet. Hit them with your bleach spray. I prefer plastic cutting boards because they can go into the dishwasher, where they get exposed to high heat and detergent. Thin plastic

Washing Meat

A few years ago, food safety experts offered a wise warning: Don't rinse food under the faucet. That's because the surface of the meat is often contaminated with bacteria and the water pressure aerosolizes the contamination and it gets all over the sink, your hands, clothes, countertops, dish drain, and the dishes therein.

The problem is that sometimes meat is just slimy and, although cooking will kill bacteria, I want to get rid of it. It's just gross and I don't want any flavors from the slime in my meal. Also, if you drop food on the floor, you should wash it. The 5-second rule is a myth.

So I put meat in a tub in my sink and wash it by submerging it in water. If I must use the faucet, I keep the pressure down to a gentle stream. After washing, I dry the meat with a paper towel so it will brown better.

can warp, so get them at least ¼ inch thick. When cutting, put a wet paper towel under it to keep it from skidding. OXO Good Grips makes nice ones that are nonskid with silicone feet. When boards get deep cuts, sand them smooth or throw them out.

SEPARATE

- Be conscious of cross-contamination. Store raw meat in sealed containers so it cannot drip on other foods.

- Do not carry raw meat over the floor without having a plate under it, especially if you have young children or pets.

- You must not put cooked meat on a platter that carried raw meat.

- Cut meat and veggies on separate boards.

- If you want to use a marinade as a sauce, make extra, set some aside for use as a sauce, and don't let it touch raw meat. Be sure to discard marinades, bastes, or mopping solutions after you're done cooking. You cannot save them for future use.

- Even if the meat is browning, the juices bubbling to the surface may be contaminated. Painting meat with a brush and dipping it into a marinade or sauce contaminates the meat, brush, and marinade. You cannot use a used marinade as a baste during the last 30 minutes of cooking or as a dipping sauce at the table.

"When in doubt, throw it out."

- Don't leave the sauce bottle sitting on the grill for hours. To prevent waste, pour the sauce you need into a cup or bowl. The best way to baste or apply a barbecue sauce is to spoon, pour, or spritz the liquid onto the meat. If you must use a brush, use a silicone brush because it holds a lot of sauce and because it is easy to sanitize in the dishwasher. Discard bristle brushes.

- If you are a guest in someone's home and you see them using an unsafe method, such as putting cooked chicken on a platter that has had raw chicken, politely, quietly, but firmly, speak up.

COOK

- All raw food is riskier than cooked food. Cooking kills pathogens. Yep, romaine is riskier than hamburger. I do not eat steak tartare (which is raw beef) or any dish made of raw fish, such as seafood sashimi, poke, or seviche.

- Use a thermometer and my Basic Food Safety Temperature Guide (page 41), not a clock. Meat color, juices, and texture are not reliable indicators of doneness and safety. Poultry, ground meats, and leftovers are especially risky.

- Do not stab meat with a fork or a Jaccard to tenderize it unless you will be cooking it past 165°F. When you puncture the surface, you push any surface contamination down into the center of the meat.

- Your grill, smoker, oven, frying oil, or water must be above 175°F unless you are cooking sous vide.

- When the meat is done, if you aren't serving it within 60 minutes, you must keep it warmer than 131°F.

CHILL

- Meat kept in the fridge can still host and grow dangerous microbes, albeit slowly, so just because it is chilled doesn't mean it is safe. Put a thermometer in your fridge and keep the temperature between 33° and 39°F. Shoot for about 38°F.

- Keep in mind, many meats you buy may have already been stored in the grocery for several days. So it is best to cook meats soon after you get them home or freeze them. Five days is the longest you should keep raw meat, although vacuum-sealed meat can keep a few days longer.

- FIFO means first in, first out. That means if you buy a slab of ribs on Monday, and then they go on sale on Wednesday so you buy another slab, cook the slab you bought on Monday first. FIFO also applies to canned foods and dry goods. Write the date of freezing on frozen food packages with an indelible marker. Put a date on leftovers, too. Date everything.

- Look carefully at anything that has been aging in the fridge, and if it has any sign of mold or slime, throw it out. That goes for cheese, too. Smell anything that has been kept in the fridge for more than 3 days.

- Be thoughtful about putting hot foods in the fridge. A burger or two, no biggie. But a quart of freshly made hot BBQ sauce can raise the temperature of the entire fridge. Cool it first in a cold-water bath.

- Do not leave leftovers on the table for more than 1 hour. Divide them into small portions so they cool quickly. Refrigerate leftovers promptly on the cooler shelves of the fridge. FDA requires commercial food processors to get the temperature down to 41°F in 6 hours.

- Cooked foods should be used within a week if they are stored in the refrigerator, regardless of how they have been cooked, even if they have been smoked.

- There is no culinary benefit to leaving food at room temperature before cooking and there is a risk. It can take hours for a steak to come to room temperature. Bacteria replicate rapidly at room temp. Since smoke sticks to cold food better than warm, go right from the fridge to the grill or smoker.

- Oxygen in the packaging can change the flavor and texture of the meat. Ground meats have more oxygen mixed in, so they start oxidizing and tasting funny sooner than solid muscle meats. Pork fat gets funky faster than lamb fat, which gets funky faster than chicken or turkey fat, and beef fat is the last to rancidify. Consider getting a vacuum sealer. They really work. There are some nice inexpensive handheld units.

- Freezing stifles bacterial activity. Put a thermometer in your freezer and keep it 0°F or below. At that temperature most dangerous microbes cannot grow, but freezing does not kill microbes, it just lulls them. Try to use frozen meats within 2 months. Any longer and quality can suffer. In general, the bigger the hunk of meat, the longer it will keep in the freezer. Label all frozen foods with a date.

- Freezing is also good at preserving flavors and nutrients, but as water freezes it expands and ice crystals are sharp, so they can pierce cell walls, causing purge. The faster and colder you freeze, the less purge. Beware of freezer burn, discolored pale patches of dried foods. Freezer burn occurs because the food was not wrapped tightly and air saps the food of moisture. It's like frostbite. Patches of freezer burn often are accompanied by ice crystals. Freezer burn is safe, but the burned parts will probably be dry and bland. Trim it off and cook it, but don't serve it to Mom or the boss.

THAWING

- The secret to thawing is to warm frozen food but not leave it in the "real danger zone" of 40° to 130°F in which bacteria multiply rapidly. Never thaw meat at room temp. Bacteria multiply too fast. That is

a recipe for fluid loss via all your apertures.

- **REFRIGERATOR THAWING.** This is the easiest method. Leave the meat in the fridge in its packaging in a pan deep enough to catch purge. Allow at least 1 day for every 3 pounds, so if you have a 20-pound turkey, you will need at least a week.

- **COLD WATER THAWING.** Water conducts heat faster than the air in the fridge even at the same temperature. Fill the sink or a pot big enough to hold the meat with *cold water*. Put the meat in a watertight plastic zipper bag and get as much air out as possible. Stir the water occasionally to break up the boundary layer of cold water surrounding the meat. Use a digital thermometer to monitor the water temp and keep it under 40°F. Add cool water or ice cubes if needed. Allow 30 minutes per pound, so if you have a 20-pound turkey, you will need at least 10 hours.

- **HOT WATER THAWING (FOR THIN CUTS ONLY).** A USDA-sponsored research project published in mid-2011 showed that you can thaw a 1-inch-thick steak in 102°F water in 11 minutes because the meat moves rapidly through the danger zone. Their tests also showed less purge than traditional thawing methods. If you want to try this at home, use hot water, lots of it to absorb the cold (if you have a sous vide machine, use it), weight the meat under with a plate, stir it occasionally, and set a timer so you don't leave it in hot water too long. Thawing times will vary depending on the thickness of the meat and the actual heat of the water. This method is best for steaks, chops, and chicken parts, but it will not work on thick cuts like roasts, pork butt, brisket, or turkey, because the exterior will stay in the danger zone too long.

- **DEFROSTING TRAYS.** When two objects of different temperatures are in contact with each other they try to get to the same temperature. So if you place frozen food on something that conducts heat easily, like an aluminum plate, it will mediate the difference between the frozen food and the room-temperature air. They work, but not as miraculously as the commercials on late night TV would have you believe.

PROCEED WITH CAUTION

- **RAW OYSTERS AND OTHER BIVALVES.** Anything raw is risky, and because the oceans are warming and they are more contaminated, and because bivalves are filter feeders, they suck in all sorts of contaminants. I love raw oysters but I have had to break the habit.

- **UNPASTEURIZED RAW MILK.** Unpasteurized raw milk is easily contaminated by dirt on the animal's teats from the field and barn. It is *not* healthier—I don't care what you have read in social media. About two hundred people are sickened by raw milk every year and half are under age nineteen. The results are very unpleasant. When you encounter someone who says it is healthier, ask them to name someone made sick by pasteurized milk.

- **UNPASTEURIZED PACKAGED JUICES.** There are pathogens in the air. Fruit is easily contaminated. Washing does not get it all off. After squeezing, their juices must be pasteurized.

- **GROUND MEAT THAT ISN'T COOKED WELL-DONE.** Meat is easily contaminated in the slaughterhouse, but it is all on the surface of a piece of meat, so a hot grill or pan makes it safe. However, grinding the meat distributes the contaminants throughout. This applies to sausages as well as burgers. Cook ground meat or sausages to 160°F throughout. On page 289 I share a method for making medium-rare burgers safe.

- **RAW SPROUTS.** Sprouts are grown from seeds. Seeds are easily contaminated in the field, shipping, and storage. The conditions for growing sprouts are exactly the same conditions for growing pathogens: Warm and moist. Raw sprouts are the riskiest food in the store. Far riskier than undercooked meat. Cooking makes them safe.

- **NEEDLE-TENDERIZED MEATS.** Some processors tenderize tough cuts by stabbing the meat with needles or small blades to cut tough connective tissues. This also pushes contaminants from the surface to the inside.

- **RAW EGGS.** Experts say the risk of egg contamination is much lower today than it was twenty years ago, but it is still there. Make eggnog, hollandaise, Béarnaise, and mayo only from pasteurized eggs. I share a method for pasteurizing them at home on pages 42–43.

- **WHY IS MEAT IN MY FRIDGE TURNING BROWN?** At first, oxygen reacts with myoglobin to turn meat red. After a while water rusts the iron in myoglobin, which turns it brown.

Cook chicken until the juices run clear

BUSTED. Chicken is higher risk than any other whole muscle meat and only a thermometer reading of at least 160°F can tell you if chicken is safe.

In February 2014 the cover story of *Consumer Reports* said that they had tested chicken breasts from supermarkets across the nation and 97% of them contained bacteria that can make you sick, and almost half contained antibiotic-resistant strains that make infections harder to treat. In 2022 they told us that one-third of all ground chicken samples they tested had salmonella and every strain they found was resistant to at least one antibiotic. The chicken-processing system clearly needs improvement. Handle chicken as if it is radioactive.

I began to wonder about the clear juices rule of thumb when I accidentally overcooked a chicken to 175°F and there were still pink juices. Nothing is more embarrassing than having to take meat off my guests' plates and run it back out to the grill (or microwave) while they discard the "contaminated" veggies and get clean plates. Been there done that?

Pink meat and pink juice in cooked chicken, turkey, and even pork is due to two proteins, myoglobin and cytochrome c.

Normally, above 140°F myoglobin begins to lose its pink color. When muscle is high in pH (low in acid), the meat may need to be 180°F or more before the myoglobin is denatured sufficiently to see clear juices.

What causes the pH to be high? According to my source at a big chicken processor, "pH fluctuations are typically a function of both genes and preslaughter stress conditions. Stress may occur during catching, transportation, holding at the plant, and unloading the birds. Climatic conditions can also have an impact. These are all things we try to control since meat will not retain moisture with further processing. This leads to a less juicy product for the consumer and yield loss, which is money to us."

Cytochrome c changes color somewhere around 220°F. Some nitrogen compounds can also prevent the pigments from changing color. Sodium nitrite and sodium nitrate from celery and other vegetables can keep your meat pink.

Red bones are common in chickens because red is the color of bone marrow, where blood is made. Chickens are only about 6½ weeks at slaughter and the bones haven't fully calcified, so they are translucent and porous. Color leaks out. If you

grew as fast as a chicken, you'd weigh 350 pounds by age 2!

Also, sometimes chicken and turkey can be pink because of chemical reactions that happen during the formation of the smoke ring (see page 69). Every BBQ joint can tell you stories of irate diners sending back chicken because of the smoke ring.

USDA says, "All the meat—including any that remains pink—is safe to eat as soon as all parts reach [a safe temperature] as measured with a food thermometer."

THE BOTTOM LINE. The old rules that said cook until the juices run clear are (ahem) clearly a myth.

- **WHY DOES MY MEAT SHINE LIKE A RAINBOW?** It is simply a fluke of lighting that strikes the surface just the right way when the surface has been cut on a certain angle. Not bacteria or an oil slick.

- **WHY IS MY MEAT GREEN?** Bad bacteria. Throw it out.

- **MY MEAT SMELLS FUNNY, WHAT SHOULD I DO?** Often meat will smell a bit odd when you take it out of a vacuum-sealed plastic bag, but the smell should dissipate within a few minutes. If it still smells funny, then chances are it *is* funny.

USING MY RECIPES

6

"If you always do what you've always done, you'll always be what you've always been."

—MEATHEAD

Here is my recipe for cooking success:

PRACTICE BY FIRING UP YOUR COOKER WITHOUT FOOD. And learn how to dial in 225°F and 325°F in the indirect convection zone. Practice in hot weather and cold weather and rainy weather. Master your tools. Get to know the Warp Scale (see page 47).

READ THE RECIPE THOROUGHLY. Print or photocopy the recipe so you can make notes on it. Take it shopping. "2 teaspoons minced fresh rosemary" means that you mince the rosemary before measuring it and "2 teaspoons fresh rosemary, minced" means that you measure 2 teaspoons of whole rosemary leaves and then mince them. In the same way athletes visualize the game, sit down, close your eyes, and do a mental walkthrough before you start. Say to yourself: "I am smarter than a slab of dead animal."

LISTS OF INGREDIENTS. Ingredients in my recipes are listed in the order in which the recipe calls for them. If order doesn't matter, I list them in descending order of quantity.

MAKES AND TAKES. Most of my recipes have "Makes" (how many servings a recipe makes) and "Takes" (how long a recipe takes to cook). I consider a serving of meat to be 6 ounces. When I guesstimate how long a recipe takes, I try to include all the washing, measuring, chopping, and peeling. But times can vary drastically depending on how fast you work and temperature variations.

BEWARE OF OUNCES. Ounces can be used for measuring both volume and weight, but they are very different. An ounce of water by volume will weigh much more than an ounce of flour by volume. When ounces are called for in a recipe for liquids it means fluid ounces, which are measured by volume as in a measuring cup, and as it happens, 1 fluid ounce of water, wine, stock, and milk also weighs 1 ounce by weight. But when ounces are given for solid ingredients in a recipe, they should be measured by weight. So when a recipe calls for 4 ounces of grated cheese it means 4 ounces by weight, not the 4 (fluid) ounces marked on a measuring up. If you grate cheese and pour it into a measuring cup to the 4-ounce mark, you will have only about 2 ounces—half the amount of cheese you need.

MEAT. I define meat as edible muscle and that means poultry and seafood are meat. Many people do not consider fish to be meat. I disagree. Chefs sometimes refer to the edible interiors of fruits and vegetables as meats. Although technically incorrect, it is a useful metaphor.

DO THE RECIPE AS WRITTEN THE FIRST TIME, NO CHANGES. You will then have a memory of what it is *supposed* to taste like. Then you can riff on it. Funny things can happen if you swap. It may seem safe to swap granulated white sugar for molasses when making beans, but molasses is acidic, and beans can get mushy if cooked in an alkaline environment.

START WELL IN ADVANCE. If you are in a rush, you will make mistakes. If you have multiple dishes to prepare, try to do as much prep in advance as possible or assign them to someone else. Multitasking is risky. Time estimates in recipes don't consider interruptions from kids or the phone, or because the charcoal didn't start properly. It's no fun panicking and zooming around when company arrives.

MISE EN PLACE. This translates roughly to "put everything in its place" and it's the best thing from France since Champagne. Anthony Bourdain called mise his religion. Before you get started, clear and clean your workspace, gather all your tools and ingredients, and arrange them in the order you will need them. Make sure you have plenty of fuel for your cooker.

This happened to me before I learned mise: I browned some meat in a pan. The next step was to add some onions and mushrooms and they would shed water and start to dissolve (deglaze) the delicious browned bits (fond) stuck to the pan. But I hadn't chopped the onions and mushrooms yet, so I hurried and by the time I was done the fond had burned and

was wasted. Far better to get these ingredients all together in advance and do the chopping, slicing, and dicing before you fire up the stove or grill. I have seen several articles lately by cooks I respect pooh-poohing mise. Ignore them, please.

ALWAYS PREHEAT THE COOKER. We want the metal in the cooker to absorb heat so when you lift the lid and then close it, the inside temperature bounces right back. Then clean the grates.

THERMOMETERS ARE REQUIRED. You need them for almost all recipes. Cook with a thermometer not a clock. More on thermometers in Thermometers Are Vital (page 37).

2-ZONE ALMOST EVERYTHING. For most recipes on a grill, I recommend 2-zone cooking for temperature control (see 2-Zone Setup for Almost Everything, page 49).

LID POSITION. Almost all the recipes in this book require you to cook, roast, bake, and smoke with indirect convection heat with the lid down. When I ask you to sear, the food is over direct infrared radiant heat, and most of the time the lid is up.

TASTE TASTE TASTE AS YOU GO. Even if you follow a recipe precisely, never forget that taste is a matter of taste. Especially when it comes to salt, sugar, and hot stuff. If you cook a sauce or other liquid, it can reduce in volume as the water evaporates, but salt and sugar remain the same. So what tasted perfect at one point can be too salty or too sweet by the time the sauce has reduced.

CLEAN AS YOU GO. Before you start cooking, start the cleanup. Wash the cutting board and knives now.

DON'T BE DISCOURAGED BY MISTAKES. Don't be afraid to make mistakes. Don't be afraid to challenge yourself. Everything is difficult and then it becomes easy. Try new things. Do not accept defeat. Just keep a nice collection of delivery menus on hand.

TEST NEW RECIPES. Try them out for your family before you cook them for guests.

HOW MUCH WOOD? I have not specified precisely how much wood you need for smoking because the strength and flavor of wood depends on many variables, not the least of which is your preferences. Every cooker is different. Woods are different. Go easy at first. A meal is never ruined by too little smoke. I strongly recommend that you use the same wood and fuel for a year until you have all the other variables under control. Keep chunks on hand for long cooks and pellets for short ones.

SOME RECIPES CALL FOR COOKING IN A PAN OR POT. You can do that on a side burner, on the direct heat side of the grill, or, horrors, indoors. I recommend that you have a skillet, a saucepan, and a baking pan set aside just for outdoor cooking. Make sure the handles won't burn or melt.

ROOM TEMPERATURE. When an ingredient should be at room temperature, that means 70° to 72°F.

KEEP RECORDS WITH A COOKING DIARY. Be specific. Record times and temperatures. Weigh the amount of wood. You can download this cooking diary/log from AmazingRibs.com/diary.

MEATHEAD'S COOKING LOG		Date			Cooker	

Time	Cooker Target Temp	Cooker Actual Temp	Meat Temp Goal	Ambient Temp, Weather	Action taken: Vents, wood (type and amount), fuel, water, mop, turn, etc.

Finishing and serving (seasoning, sauce, accompaniments)

Finished product (aroma, bark, color, smoke ring, texture, tenderness, smokiness, flavor)

Next time

3/31/23   or more great barbecue and grilling info visit https://amazingribs.com

NOTES ON SOME OF MY INGREDIENTS

CHILES (with an "e") are the colorful fruits of the *Capsicum* plant, also called peppers. Chiles get heat from capsaicin, a chemical irritant. Most of the heat (85%) comes from capsaicin glands in the ribs, not the seeds. This means that you can get the flavor of a chile without intense heat by removing the ribs.

Chili (with an "i" at the end) has multiple definitions. In the US, chili powder is a powdered seasoning mix made from dried chile peppers, cumin, garlic, and other spices. In Europe, chili is usually just ground dried hot chiles. Chili is also savory meat stew, usually beef, seasoned with American chili powder.

Chilli (with "ll") is a misspelling.

The amount of heat in a chile pepper is measured on a culinary Richter scale called the Scoville Heat Units (SHU) scale. On page 145 is a listing of some common chiles and sauces and their average SHU, but keep in mind, this is not written in stone. I have harvested two jalapeños from the same plant in our garden that are significantly different in heat.

When chopping hot chile peppers it is wise to wear gloves, and be sure not to rub your eyes, pick your nose, use the urinal, or make love before you wash thoroughly.

Capsaicin oils are not water-soluble, so if your mouth ignites when eating hot peppers or hot sauce, beer and cold water are not going to put out the fire. They only slosh the capsaicin oils around. On the other hand, lipids bind with capsaicin, so cream, milk, yogurt, butter, chocolate, and other fats do a better job of damping the flames.

Unfortunately, most ground dried chiles, flakes, and sauces have little pepper flavor, just heat, and standard grocery store hot pepper flakes can contain a lot of hard woody seeds that have no flavor and get stuck in your teeth.

CHIPOTLES. These are also used to make a hot sauce or packed in a can with a vinegar and tomato sauce called chipotle in adobo. I use Tabasco Chipotle sauce a lot because it is tasty, not too hot, and easy to find. The canned version in adobo is great for stews and sauces. You can use the whole chiles in the adobo sauce, chop them up, or even use the adobo sauce straight. Many of my recipes use ground chipotle as a source of heat because it has no seeds, it is smoked, has a nice flavor, and it is not painfully hot (but do be careful with it). It is great on pastas and pizzas and just about any other place you want heat. Ground chipotle is made from red jalapeños that are smoked and dried, seeds removed, and ground. Easy to do at home. I keep a shaker on the dining table.

PAPRIKA. Paprika, as we know it in the US, is made from ground dried sweet chiles. But in many countries, paprika is ground from dried hot peppers. In some countries paprika means fresh sweet chile peppers, not ground. Many cookbooks call the paprika popular in the US "sweet paprika," but that can be confusing because there is no sugar in it and no paprika bottles actually say "sweet paprika." So I have my own nomenclature: When I refer to "mild paprika" in my recipes I mean the simple mild-flavored orange-red powder on most American grocery store spice racks. Then there's "hot paprika," which has some hot peppers in the blend. There's also "smoked mild paprika" and "smoked hot paprika."

SHU (ESTIMATED)	CHILE
100,000–350,000	Scotch Bonnet Chiles, Habanero Chiles
50,000–100,000	Thai Bird Chiles
15,000–30,000	McCormick Crushed Red Pepper Flakes
10,000–25,000	Serrano Chiles
2,500–8,000	Jalapeño Chiles, Chipotle Powder and Chiles
2,500–5,000	Tabasco Sauce, Cholula Sauce
2,200	Sriracha Sauce by Huy Fong (other brands are different)
1,500–2,500	Tabasco Chipotle Sauce
1,000–3,000	Anaheim Chiles, Cascabel Chiles, Pasilla Chiles
1,000–1,500	Poblano Chiles, Ancho Chiles, American Chili Powder
450	Frank's RedHot Original Cayenne Pepper Sauce
0	Bell Peppers, Banana Peppers, Mild Paprika

EGGS. I use "large."

FLOUR. This is "all-purpose flour" unless otherwise specified. Sift it and weigh it. Measuring it by volume such as tablespoons and cups is highly inaccurate.

FRUITS AND VEGETABLES. The default size is medium. They should always be fresh and scrubbed with cool water.

MAYONNAISE. Full-fat mayo only. Never substitute low-fat mayo or Miracle Whip. The chemistry is very different.

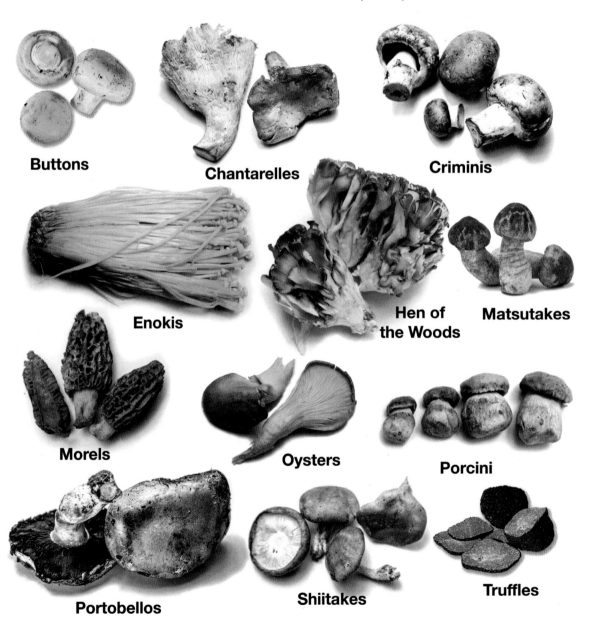

Buttons

Chantarelles

Criminis

Enokis

Hen of the Woods

Matsutakes

Morels

Oysters

Porcini

Portobellos

Shiitakes

Truffles

MILK. Milk can be whole milk or 2%, but not fat-free milk, soy milk, almond milk, or any other substitutes.

MUSHROOMS. In a few of my recipes I call for morels, a wonderful but expensive mushroom. If you can't find them, good old button mushrooms will work just fine. When tasted side by side they are not embarrassed by their uptown cousins.

Fresh mushrooms have a refrigerator life of about 2 weeks, depending on how fresh they were when you bought them and how you store them. Store them unchopped and unwashed in a paper bag in the refrigerator. If you store them in plastic bags, plastic wrap, or an airtight container, they will deteriorate more quickly. Most mushrooms do not freeze well.

When it is time to cook, if the stems are woody, remove them, freeze them, and chuck them in the pot with bones and meat trim to make stocks. Otherwise just slice off the fibrous bottom of the stem and compost it.

Dried mushrooms can be almost as good as fresh. They keep for years at room temp. I store mine in a dark cabinet in a paper bag. They can be reconstituted by soaking in hot water for 15 minutes or so, and the soaking water is laden with flavor that can be reduced for sauces or stocks. Freeze it and save it. When dry, mushrooms can be grated into a powder and added like a spice to rubs. Use

Don't wash mushrooms

BUSTED. Most chefs say you should never wash mushrooms. Just wipe mushrooms with a brush or a damp paper towel. Mushrooms are about 90% water and they absorb very little water when washed. I have tasted washed and brushed mushrooms side by side and I can't tell the difference.

Mushroom cells are mostly a compound called chitin, the compound that makes up the shells of lobsters, shrimps, and crabs. It makes it hard to overcook mushrooms. Lam Lan of *Cook's Illustrated* showed that if you put them in a bowl of water, and microwave them for 2 minutes, a lot of the water and air comes out, compressing the mushroom and concentrating their flavor and umami.

Try this: Melt some butter, toss in a sprinkle of salt, and press or mince in garlic. Then baste the mushrooms with the garlic butter and grill them on a metal grill topper. They will absorb tons of flavor.

Unwashed 2 3/8 ounces   **Washed 2 5/8 ounces**

a few pinches to really add an umami oomph to burgers and sprinkle some on meats and veggies.

Canning mushrooms concentrates their flavor and there is a rich briny liquid in the can. The flavor is different from fresh, but often the umami-loaded taste is very good. Never use canned mushrooms if the can is bulging or dented, and be wary of home-canned mushrooms.

PARMIGIANO-REGGIANO. This is the "king of cheeses." It is a salty hard cow's milk cheese made only in Parma, Italy, and aged for a minimum of two years. It is best for grating and it often has crunchy crystals that develop with age. There is no exact substitute for it, although Grana Padano and pecorino are suitable. American parmesan is a poor substitute.

SALTS. I use only two salts in my recipes: Morton Coarse Kosher Salt and Maldon Sea Salt flakes. Salt grains are different sizes and therefore are different strength when measured by volume. Morton is in all the stores, and it is easy to pinch and sprinkle. I cook with it. For sprinkling at the table I use Maldon Sea Salt flakes. I don't cook with it. Maldon grains are larger and it gives a crunchy pop, like pretzel salt. I am not paid for these endorsements. If you must use table salt, reduce the quantity called for by about half because it is a small grain.

SOY SAUCE, TAMARI, SHOYU, LIQUID AMINOS. These are all rich, dark brown, salty, glutamate/umami-loaded liquid condiments. There are many brands from many countries and a tasting can be as interesting as any wine tasting. There are some differences in flavor and in saltiness. I stock three types: a basic dark soy sauce, a reduced-sodium soy sauce, and high-quality tamari for finishing and tableside use.

Beware of Chinese restaurant take-out "Soy Sauce." No soybeans were harmed in making the one on the right.

STOCKS, BROTHS. Some cooks say that stock is made from bones and broth is made from meat. Not true. There are vegetable stocks. Some say stock will gel when cooled and broth will not. Not true. Broths can gel. For all *practical* purposes, there is no difference, and I use the terms interchangeably. Stocks and broths can be used as bases for soups, stews, gravies, and sauces. They are also great for cooking grains such as rice, bulgur, and farro; or for preparing pastas such as couscous; or for poaching eggs and meats. They are easy to make from leftovers, bones, and trim, such as chicken necks and mushroom stems. They are usually made by placing meats, bones, and skin in cold water, bringing it to a simmer but not a boil, and after a short while skimming off the foam and fat. Then vegetables, herbs, and spices are added and the stock simmers for 2 to 12 hours to extract the flavors, proteins, and gelatin. Grilling or smoking the meats, bones, and vegetables before simmering really cranks up the flavor. After simmering, the stock can be strained or filtered. I like to freeze stocks in ice cube trays and store the cubes in plastic bags.

SUGARS. When used sparingly, sugar can create a richer middle flavor, like adding a cello to the orchestra. And if sprinkled judiciously on the surface, it can help with browning, a handy trick if you are using a gas grill or pellet smoker that just doesn't get hot enough to create a great sear. When my recipes call for "sugar," I mean granulated white sugar. If I say brown sugar, you can use either light or dark. I usually use dark. Brown sugar is just granulated white sugar with molasses added, but it has a different flavor and different chemical properties. None of my recipes call for raw sugar, muscovado, Barbados, demerara, or turbinado. I find that in the context of grilling, smoking, or my sauces, that they make little difference in the taste of the outcome. And they are expensive. In baking, there can be quality differences you can taste.

VINEGARS. Vinegar brings liveliness to many dishes, because, as we know, acidity is a crucial element. There are many vinegars and I use them all.

Distilled white vinegar. This is my first choice for pickling, slaws, some dressings, and sauces. It is vinegar made from clear distilled spirits like vodka. About 6 to 8% acidity.

Apple cider vinegar. Apple cider vinegar is made from apple juice, it is straw colored, and it is more flavorful than distilled vinegar. Usually, you should not use it if the recipe calls for distilled because the flavors may clash and because sometimes it is slightly sweetened. It is typically 5% acidity.

Wine vinegar. Made from inexpensive red or white wine, it has more flavor than distilled vinegar. I keep red and white wine vinegars on hand for salad dressings. Usually around 5% acidity.

Rice vinegar. Made from rice wine, they are usually a pale straw color and not as high in acid as other vinegars with a mild hint of sweetness. They are an excellent choice when your dish needs to be livened up without becoming too puckery. I use it a lot, not just in Asian dishes. Try it in your coleslaw. Seasoned rice vinegar is the same as plain, except it just has a bit of salt added. I usually call for unseasoned. About 4%.

Balsamico tradizionale vinegar. True balsamics are made in Italy and the best are labeled tradizionale. They can cost more than $100 for a 100 ml (3.4-ounce) bottle. They never have an age on the label because they are a blend of vintages, but usually they average at least twelve years old and can go up to fifty. Some are as thick as maple syrup and almost as sweet. The flavor is truly extraordinary. I use it for drizzling like a syrup on ripe tomatoes in Caprese salads in August. Balsamic made in California or elsewhere may be delicious, but it is not really balsamic. 4.5% acid.

Condimento balsamic vinegar. Made in a similar method to tradizionale, but the average age is less and although it is thick it is not syrupy. Expect to pay $30 to $50 for 8 ounces. It is also great on tomato salads. 4.5% acid.

Salad-grade balsamic vinegar. This is a term I have coined for the inexpensive stuff in grocery stores. They are usually wine vinegars made in large tanks, sweetened with sugar, and artificially colored. They often display bogus age claims. Still, some are delightful tasting. I keep some on hand for marinades and sauces. I also buy salad-grade balsamic and reduce it to syrup for drizzling (see Balsamic Syrup, page 198). 5% acid.

Other vinegars. There is malt vinegar, sherry vinegar, and black vinegar from China; there are vinegars made from a base of cane juice, date juice, raisins, coconut, all manner of fruits, and with all manner of flavors. Vinegar is a strong solvent, so if you put something in it, in short order it will extract flavor. Tarragon and other herbs are often added to vinegar bottles as are chile peppers and even fruits and vegetables. Raspberry vinegar is common. Fruit vinegars are especially nice on spinach salad or dark bitter greens. One of my all-time faves is calamansi vinegar from France. It is 7.5% acidity.

WINE, BEER, SPIRITS. Wine, beer, and spirits are flavorful liquids that can be used to amp up many recipes. Wines have complex fruit flavors and whiskeys often have woody flavors. Liqueurs bring exotic flavors and sweetness. The bad news is that the alcohol is not your friend in cooking. It can taste awful in sauces and it can denature proteins and make meats mushy. Cooking with beer can make dishes bitter. For these reasons I rarely marinate with them.

But I do use them in sauces. First, I boil them to get rid of most of the alcohol. I hear you sobbing, but it must be done. If you refuse to use wine, beer, or spirits when my recipe calls for them, you can substitute water or juices. Just be thoughtful about the flavors.

Beware of young tannic red wines such as Cabernet Sauvignon. Tannin is found in grape skins and seeds and has a tactile powdery sensation. Makes your tongue feel like you've been licking dusty window blinds. It is also bitter. When making a red wine sauce with a tannic red, the tannins concentrate. Not pleasant. Chose fruity low-tannin reds like Gamays.

Never use something labeled "cooking wine." This is usually poor-quality wine that has been altered by adding salt. This is done to make it undrinkable so that restaurant kitchen help won't be nipping and so it can be sold in stores that do not have liquor licenses. Cook only with wine you would drink. You do not need to use expensive wines for cooking, but it should have lots of flavor and taste good. I rarely spend more than $15 on a bottle of wine for cooking.

ZEST. Several recipes call for the zest of a lemon, lime, orange, or grapefruit. A citrus peel has two parts: The colored, bumpy, thin, outer layer filled with fragrant citrus oil that we call zest. Beneath the zest is a white cardboardy, bitter pith. The zest is the good stuff. The best way to get small bits of zest for cooking is with a Microplane or the medium-size holes on a box grater. For wide strips for cocktails, use a vegetable peeler. You can leave the zest in a bowl and it will dry overnight and you can save it for later. Depending on its size, an orange will yield about 1½ to 2 tablespoons of fresh zest and less than 1 tablespoon dried. A lemon or lime will yield about 1 tablespoon of fresh zest and ½ teaspoon dried. Citrus skins can be contaminated during growing, harvesting, and handling, so whenever you use citrus, thoroughly wash it with soapy water, rinse thoroughly, and dry. Yes, you should use soap; just be sure to rinse the fruit thoroughly.

HERBS AND SPICES

Herbs and spices can take the humdrum and make it hum. Spices are usually made from seeds, bark, roots, berries, and fruits. Herbs are usually made from green leaves. Because dried herbs have no water, they are more concentrated than fresh. If you wish to substitute dried for fresh herbs, the rule of thumb is to use half to one-third.

Many of my recipes call for the dried herbs you can get at the grocery store or one of the excellent specialty spice merchants online. Occasionally I call for fresh herbs that can often be found in better produce sections. I live in a suburban bungalow with a small backyard and my wife has commandeered about one-third of it for veggies and herbs. We also grow some herbs indoors on windowsills. Nothing brings life to a recipe like fresh herbs, and there are three we use a lot:

THYME, which is close to an all-purpose herb for just about any dish.

TARRAGON, which is superb for eggs, fish, chicken, cream, cheese, butter, and most starches.

ROSEMARY, which goes with many meats and especially loves lamb, beef, venison, pork, chicken, potatoes, and winter squash.

If you buy them fresh, remove them from packaging, cut off and discard the bottoms of stems, wrap the stems in a damp paper towel, slip the stem ends into a plastic bag leaving the leaves out in the air, and put them in the fridge. When cooking with fresh herbs, add them two to three minutes before serving.

Dried herbs need a little time to hydrate and contribute to a dish so they need to be added during cooking.

To get the most out of your dried spices, consider "blooming" them by pouring them into a dry pan and heating them gently over medium heat, stirring often. This toasting breaks down some of the structures and releases the oils. Don't let them sit still for more than 10 seconds or they can burn. A minute or two is all that's necessary to open them up. When they start to smell strong and nutty, pour them out of the hot pan so they don't continue to cook. Blooming is a good way to breathe life into old spices and herbs.

Another method is to bloom spices and herbs in oil. Preheat a saucepan to something under 300°F, add a few tablespoons of oil to make a puddle, just enough to make a paste with the rub. Now pour in the spices, stir them around in the puddle of oil, and cook for no more than 2 minutes. You don't want to cook them too long, which would drive off the volatile compounds. There will be tiny bubbles around the spices as water vapor escapes and some seeds might pop like popcorn and spatter. That's why I recommend a saucepan not a skillet. If you use butter, beware that it can turn brown if the pan is too hot. Brown butter is wonderful but not for everything. You can also bloom whole vanilla beans and dried chile peppers.

It is a good idea to write the date of purchase on bottled herbs and spices because they lose their potency and oxidize, and oils get rancid with time. Store them in a dark space. As a rule of thumb, replace them after a year. I burn old spices and herbs in recipes like Tea-Smoked Whitefish (page 310).

Preground spices lose their fragrance and potency faster than whole seeds. I try to buy whole seeds and grind them as needed: Caraway, cardamom, celery, coriander, cumin, dill, fennel, mustard, and peppercorns. I used my coffee grinder for both coffee and spices until I poured a cup of coffee that set me aflame from the chipotles I ground in it. I then bought a Shardor Electric Grinder on Amazon. It can do coffee, herbs, spices, even wet and oily things like garlic. The cups with the blades can be run through the dishwasher.

If a recipe calls for seeds and you only have ground, cut the quantity in half. If a recipe calls for a spice or herb, you can often find a good substitute on the table in my last book or on my website. *The Food Substitutions Bible* by David Joachim is a big help.

DRIED ONION. Granulated onion has larger grains and more air than onion powder. My recipes use powdered onion.

FRESH GARLIC. Many of my recipes call for pressing or mincing garlic. That's because this extracts more juice and flavor than chopping or slicing. I never use precrushed garlic because it can oxidize. When using fresh garlic, it usually should be added last and cooked for only a minute or two or else it browns and gets bitter.

DRIED GARLIC. This has a very different flavor than fresh, so they shouldn't be swapped. Granulated garlic has a slightly larger grain, about the size of sand. If you use it instead of garlic powder, add about 25% more. I never use garlic salt. It has salt in it and we don't need extra salt.

PEPPER

BLACK PEPPERCORNS. Black pepper is the most popular spice in the world. If you buy whole peppercorns and grind them as needed, you will be rewarded with better flavor. Black pepper is the fruit of the *Piper nigrum* vine in tropical climates. It gets its zing from a compound called piperine. Tellicherry from India and

Lampong from Indonesia are considered among the best. Size matters, as larger peppercorns contain more aromatic compounds.

GREEN PEPPERCORNS. Some *Piper nigrum* berries are picked when green and treated to retain their green color. They are sometimes sold in brine. They have a short shelf life. They are very fragrant, have a milder, brighter, more floral character, and work well on white meats with herbs.

WHITE PEPPERCORNS. If ripe peppercorns are dried and the skin removed, what remains is a white seed. I use ground white pepper in recipes where black specks will be unsightly, such as white sauces, cream soups, mashed potatoes, and fish dishes.

RED PEPPERCORNS. As *Piper nigrum* berries ripen, they turn red. Some are treated to retain their color and sold as red peppercorns.

PINK PEPPERCORNS. This colorful spice is not a *Piper nigrum*. It is from three different bushes, mostly the Baies rose. Another variety is related to cashews. That makes it a tree nut and as such can be hazardous to people with tree nut allergies. They add color and a sweet citrus character. Their husks tend to clog peppermills, so it's best to grind them with a mortar and pestle.

OTHER PEPPERCORNS. There are a variety of other hard-to-find things named pepper, among them cubeb (also known as Java pepper and tailed pepper), Tasmanian pepper, Dorrigo

Pepper Grinds Make a Difference

HOT TIP Whole peppercorns aren't very flavorful; they are best when ground, releasing their oils. So it is best to store whole peppercorns in a peppermill and grind them when needed. Like most spices, peppercorn flavors come from delicate oils and other compounds that evaporate and oxidize rapidly when they are broken open. Their flavors can also be damaged by light. The size of the grind is a quality factor and I usually specify the size I recommend for a recipe. The best solution is to get a pepper mill that has an adjustable grind setting. Get one with the grinder on top. That means that it won't leave a little pile of dust on your table. Here are the various grinds I call for:

CRACKED. Corns that are barely crushed and pieces may be as large as halves. Used in dishes like steak au poivre where they provide exciting, crunchy, aggressive pops of flavor and spice.

COARSE GRIND. Large chunks, perhaps one-eighth the size of the peppercorn; they too provide distinctive texture and pops of flavor.

MEDIUM, SHAKER, TABLE GRIND. A finer grit that will fit through the holes of a pepper shaker and usually served at table. This is about the size of the stuff you get preground in the grocery.

FINE GRIND. Almost a powder. Useful in spice rubs and blends. Can scorch when subjected to high heat.

pepper, grains of paradise (also known as alligator pepper), melegueta pepper, long pepper, Selim kili pepper (also known as African pepper, Ethiopian pepper, grains of Selim, Guinea pepper, and kimba pepper).

VEGETABLE OILS

OLIVE OILS. There are legally defined label terms, including "extra virgin" and "virgin," but I rarely use them because there is so much fraud. In my recipes I simply call for *high-quality* (roughly equivalent to "extra virgin"), *good* (roughly equivalent to "virgin"), or *cooking grade* (equivalent to plain old "olive oil"). I recommend you taste several oils until you find one of each and inventory them. Age, light, warm temperature, and air are enemies. It is made in the autumn, so buy it young (it is made in mid-autumn in the Northern Hemisphere) and store it in a cool dark place. *Smoke point: The smoke point can range widely from 375° to 420°F.*

VEGETABLE OIL BLENDS, CORN OIL, SOYBEAN OIL, SAFFLOWER OIL,

SUNFLOWER OIL. These oils and blends are the most popular in groceries and they all tend to be inexpensive, mild in flavor so they let the flavor of the food show through, so they are good for salad dressings, and they have relatively high smoke points so they are good for frying. *Smoke point: About 450° to 500°F.*

REFINED PEANUT OIL. This is the oil of choice among Asian cooks, especially for high-heat wok cookery because of its high smoke point and neutral flavor. This also makes it a good choice for deep-frying. It is pale in color and the refining process is said to remove allergens, but people with peanut allergies should avoid it. *Smoke point: About 450°F.*

CANOLA OIL. Growing in popularity for shallow-frying and all-purpose cooking uses, but not recommended for deep-frying because it can develop off-flavors. There are false accusations floating around the internet that it is toxic. *Smoke point: About 400°F.*

SESAME OIL. If it is clear and yellowish, it is pretty much an ordinary seed oil. But if it is brown, the seeds have been toasted first, and the flavor is more intense than any other oil. One sniff and you will say, "Chinese food," because a few drops of it find their way into so many Asian dishes. It is for flavor only, not cooking. You really should have a bottle around the house. Add a few drops to mayo for making turkey sandwiches. *Smoke point: About 350°F.*

AVOCADO OIL. Mild taste and high in monounsaturated fat have made this cooking oil chic. That said, I have encountered a few that are very dark green and strong flavored, so beware. *Smoke point: About 520°F.*

VEGETABLE SHORTENING. Crisco is the most popular by far. It is a blend of soybean and palm oils that have been hydrogenated so it can be solid at room temperature. Recommended for: Pie crusts, cooking, frying. *Smoke point: About 330°F.*

MARGARINE AND IMITATION BUTTERS. Many are about 80% vegetable oil and they spread easily. Some are even blended with butter or yogurt. Recommended for spreading on bread as a condiment and low temperature cooking. *Smoke point: Varies significantly from brand to brand but it is usually 300°F or below.*

OTHER NUTS, SEEDS, AND GRAINS SUCH AS WALNUT, GRAPESEED, MUSTARD SEED, PUMPKIN SEED, COCONUT, RICE BRAN, AND PALM OIL. Each has its unique flavor, cooking characteristics, fan club, as well as internet rumors about its health advantages and disadvantages. *Smoke point: Varies significantly from brand to brand.*

FLAVORED OILS, SUCH AS HERB OILS, GARLIC OIL, AND TRUFFLE OIL. Many specialty stores sell flavored oils. They can be very tasty, but I prefer to simply add my own flavor to the oil when I need it rather than inventory yet another condiment. A word of caution: Garlic is exposed to the botulism microbe that lives in soil—and bot loves to grow in a jar of oil without air. And it is hard to kill but it can kill you. Commercial garlic oil is treated to make it safe. But it is not a good idea to keep bottles of homemade garlic oil around. If you need it, make it fresh. And if the Italian restaurant you're in sets a bottle on the table, ask for just plain oil. Truffle oil usually has never seen a truffle. It gets it flavor from a chemical whose taste is a dead ringer for the fungus.

ANIMAL FATS

Animal fats are hard and waxy. They are usually solid at room temperature and can oxidize and go rancid, so they should be refrigerated or frozen. Fat from grass-finished cattle tastes different than corn-finished cattle, and fats from animals that have grazed on wild sage or rosemary can taste like sage or rosemary. Fat from young animals tastes different than older animals. Many animal fats start to smoke at around 375°F, which is the magic number for most deep-frying, so don't try it.

BACON, DUCK, GOOSE, AND CHICKEN FAT. If you are careful not to burn it when you make bacon, you can pour it into a jar and store it in the fridge for months. It is unmatched for pan-frying potatoes and many vegetables. Duck fat and goose fat are also flavorful. Chicken fat, called *schmaltz* in Yiddish, is tasty and creamy. That's why the expression "He fell into a schmaltz bucket" means he fell into good fortune. *Smoke point: About 325°F.*

BEEF FAT, TALLOW, SUET. Ground beef fat mixed in with ground meat can add flavor to burgers and sausages, and tallow painted on steaks just before searing amps it up to 11. Plop a teaspoon on your griddle next time you make a smash burger like the Oklahoma Onion Burger (page 291). *Smoke point: About 400°F.*

BUTTER. Butter is always unsalted in my recipes. Let's control salt content without the wild card of an unknown quantity coming from the butter. That said, if all you have is salted butter, cut back a tad on salt in the recipe. Butter is made by the vigorous whisking of fresh cream skimmed from the top of milk that has not been homogenized. It is usually made from cow's milk, but it can be made from sheep, goat, buffalo, and other mammals. It is typically 5% milk proteins, 15% water, and 80% fat. The amounts vary from producer to producer. Some butter also has salt added, so butter is usually labeled *salted* or *unsalted*. Brown Butter (page 189) has lovely nutty notes and you might want it in some recipes. A typical stick of butter is 4 ounces, 8 tablespoons, ¼ pound, or ½ cup. Tattoo this on your arm. *Smoke point: About 300°F.*

CLARIFIED BUTTER AND GHEE. These are butter that has had the water and milk solids removed. I use them a lot because they have butter flavor and a high smoke point. Recipes for both are on page 188. *Smoke point: About 450°F.*

LARD. Lard is pork fat. "Leaf lard" from the fat surrounding the kidneys is prized by bakers because it makes especially flavorful and flaky pie crusts. *Smoke point: About 325°F.*

RUBS AND SPICE BLENDS

*"You can easily put together your own favorite spice blend. . . .
Just watch out for the sodium content."*

—EMERIL LAGASSE

There are many brilliant chefs who specialize in minimalist cooking, salt and pepper only, allowing the substrate to show off in all its glory. I lean that way most of the time. But what if you want to amp it up to 11? The answer is rubs, marinades, and injections.

Let's start with spice and herb blends, also known as rubs. You probably grew up with some of these popular commercial spice blends in the house: Taco Seasoning, Pumpkin Pie Spice, Poultry Seasoning, Lawry's Seasoned Salt, Chili Powder, Old Bay, Jerk, 5-Spice Powder, Lemon Pepper, Garam Masala, Curry Powder, and Herbes de Provence, to name a few.

In addition, there are hundreds, yes hundreds, of commercial spice blends designed for barbecue and grilling. Every good cook should have a few signature house rubs to brag on. Just steal my recipes. There are a few here, more in my last book, and more still on AmazingRibs.com. Then experiment with variations. But remember, you cannot judge a rub straight from the bottle. It tastes very different after cooking. The juices of the meat mix with the herbs and spices, dissolve them, and undergo chemical reactions catalyzed by heat. A rub may taste too spicy when raw, but keep in mind there will be a bite-sized piece of watery food underneath it, diluting it.

THE FIVE Ss OF A RUB

A good rub is like a good orchestra: It has a range of instruments to play all the notes in harmony.

SPICES AND HERBS. Not all of them taste good on all foods, but the spice rack is full of great flavors. Garlic and onion powder are common. Mild paprika is often included in barbecue rubs, not so much for flavor as for color. I urge you to experiment with others.

SPICY HOT. Hot pepper sensations, often called spicy flavors, are often in rubs because they add excitement, but go easy, not everyone likes it as hot as you do. Black pepper brings the zing, as do hot chiles, ginger, horseradish, and mustard powder.

SAVORY. Although there is an herb named savory, in most cases when cooks speak of savory they mean the earthy, meaty, or herbaceous flavors that come from umami-laden glutamic acid. So add something with glutamate like Ac'cent, nutritional yeast, dried tomatoes, or powdered mushrooms.

SUGAR. Is it necessary? Sugar is a common addition to rubs because it is a flavor amplifier, it helps browning, it helps with crust formation, and it offsets bitterness and acidity. Sugar also is hygroscopic, meaning it attracts water. So when a rub with sugar is applied to meat, it pulls moisture to the surface where it dissolves the sugar and some of the other spices and their compounds and forms a slurry. As the surface dries, the sugar and spice slurry meld with meat juices, fats, and protein to form bark. The more the sugar is cooked, the more its chemistry changes and the less sugar and sweetness remain.

SALT. Salt is important because it penetrates deep, amplifies flavor, hides bitterness, and helps meat retain moisture. It is in practically all commercial rubs. But as I explain below, if you make rubs at home, you should leave out the salt so you can dry brine.

For Those Wary of Sugar

I appreciate that many of you feel the need to reduce sugar in your diets, so I try to keep it down when possible. When I add sugar, it is because I believe it is important.

Many barbecue rubs and most barbecue sauces have sugar in them; some have a lot. Is it really an issue? Let's look at my all-purpose rub, Meathead's Memphis Dust (page 165).

There are about 2 tablespoons of Meathead's Memphis Dust on a slab of ribs. Less than half of that is sugar. Some of it scrapes and drips off during cooking. If you eat half a slab, you're eating about 1 teaspoon of sugar. The glycemic load (GL) is about 3. Compare that with a slice of white bread with a GL of about 11. Glycemic load represents how much a carb raises blood sugar.

Classic Kansas City–style red barbecue sauces are pretty sweet. But you don't want to hide the meat. One light coat should do the job, perhaps 5 tablespoons on a slab. After drip loss, that's 1½ teaspoons of sugar on half a slab, GL about 3, same as the rub.

So combined, rub and sauce, if you eat half a slab, you're getting a GL of about 6, or 54% of the GL of a slice of white bread.

If that's still too much, there are some fun sauces without sugar or with very little sugar. Try my Lexington Dip #2 (page 184).

NO SALT IN MY RUB RECIPES

Almost all commercial rubs have salt in them. Sometimes salt is half the blend. That's expensive salt! The more salt, the more the profit. Even more reason to make your own rubs without salt. Remember: You can always add salt, but you can't take it away.

If you put a commercial salt-laden rub on pork ribs, the salt will work its way to the center in an hour or two and the herbs and spices will remain on the surface. But if you put the same blend on an 8-pound boulder of pork shoulder, it can take eight hours or more for it to penetrate, and the salt concentration will be much lower. A pork butt needs much more salt than ribs because it contains a lot more meat for the salt to penetrate.

Salt is two simple atoms, sodium and chloride (NaCl), and when they get wet, they ionize, get electrically charged, and slowly penetrate deep. But table sugar is 45 atoms

$(C_{12}H_{22}O_{11})$; piperine, the action part of black pepper, has 40 atoms $(C_{17}H_{19}NO_3)$; and goodness knows how many there are in garlic, oregano, thyme, ginger, etc. They are just too large to get beyond the pores and cracks in the surface of most meats and produce (fish and porous veggies like zucchini and eggplants are an exception).

Because salt penetrates deep and herbs and spices cannot, the Meathead Method says to apply salt based upon the weight of the meat and apply herbs and spices based on the surface area. Here's how:

Dry brine with about ½ teaspoon of Morton Coarse Kosher Salt per pound of meat hours before cooking. For herbs and spices use about ½ teaspoon per 24 square inches. That's the size of a 4 × 6-inch postcard or a typical pork chop. You don't want it on so thick you can't see the meat.

There are numerous good reasons why my rub recipes don't have salt:

1. Applying the salt in advance, dry brining, is an important technique because it takes time for salt to work its way to the center of the meat. Adding spices in advance does little to benefit the meat below the surface. So it is a good practice to salt well in advance, and if your rub has salt as well as herbs and spices, gahead, the salt will penetrate while the herbs and spices sit there waiting for you to fire up.

2. You do not need *any* salt at all on cured meats like a ham, bacon, or corned beef because they are heavily salted at the processor. But you might want sugar, savory, spices, and heat. So a commercial rub will likely make the meat too salty.

3. Nowadays almost all turkeys and many ribs, pork loins, chickens, and other meats are injected with a salt solution at the processor to improve flavor and water retention. It also adds weight that increases

Do This Experiment

Here is an assignment: Buy a piece of pork loin about 6 inches long (not a tenderloin). Cut it in two 3-inch lengths. Season one half with a variety of spices 24 hours in advance but no salt. Put nothing on the other half. Cook both until they reach 140°F. Cut a ¾-inch slice from the center of each half. With a wet paper towel, wipe the cut surfaces to make sure that the knife has not pushed spices from the outer surface onto the cut surface. Now cut core samples from the center. Taste them and serve them to a friend or two. Ask if there is a significant difference between the cubes cut from each half. I'm betting they taste the same.

profit. Meat that is labeled "enhanced" or "flavor enhanced" or "self-basting" or "basted" has been injected with a brine at the packing plant. Kosher meat has also been treated with salt at the plant. You don't need to salt them, and a commercial rub might make them too salty.

4. Some people like salt more than others. By keeping salt separate you can tailor saltiness to your taste. On more than one occasion I have been sent rubs to taste and after applying them discovered, to my dismay, that the meat was way too salty.

5. Some people are on salt-restricted diets, although when you do the math, ½ teaspoon of Morton Coarse Kosher Salt per pound is way below their limit. The recommended daily allowance is 2,300 mg. If a serving is 6 ounces of meat, then you are getting only 355 mg.

6. Leaving salt out of the mix gives you room to add a finishing salt just before serving. A sprinkle of large-grain salt on a steak as soon as it hits the table gives it real pop. In some cases, such as Memphis-style "dry" ribs, which are served without sauce, you might want to apply a layer of rub after cooking as they do at the famous Rendezvous restaurant in Memphis. If there is no salt in the rub there is no risk of oversalting.

7. It is a good idea to salt foods before sous vide cooking, because the salt has plenty of time to penetrate, but if you add a salty rub before sous vide, the herbs and spices will mostly wash off so you will need to add them after sous vide. If there is salt in the rub, the food will end up too salty.

8. Sometimes you want to put oil on the meat to help with browning and moisture retention. Salt will not dissolve in oil. If you apply salt first, the water in the meat will pull it in. Then the oil won't interfere.

So salt, rub, and sauce are like antifreeze, oil, and gas. They all go into the engine, but don't mix them. Apply them in the right proportion at the right time. You want to apply the salt a little heavier on thick sections of meat so the fat end of a turkey breast will get more salt than the thin.

You can still use commercial rubs with salt as a dry brine; just sprinkle the rub on well in advance to give the salt time to penetrate. Salt appears first in the ingredients list because law requires ingredients be listed by weight and salt weighs a lot more than pepper, paprika, thyme, and all other spices.

HOW TO USE RUBS

First, make sure to store rubs in an airtight jar in a cool dark place. Believe it or not, most plastic bags breathe and rubs contain oils that can oxidize.

You can put a rub right on bare meat, or you can help it stick by moistening the meat with a little water, mustard or ketchup (which are mostly water), or oil or mayo (which is mostly oil). Many pitmasters put down a slather of mustard before the rub thinking it will add flavor. Bottled mustard is a mix of powdered mustard with water, and/or vinegar, and/or white wine. The amount of mustard powder is so small that by the time the water steams off and drips away, the mustard powder remaining is minuscule. If you want a mustard flavor, you will do much better by simply sprinkling mustard powder on the meat (be careful, it is strong).

I wondered about the solubility of the herbs and spices in a rub. So I put my two most popular rub recipes in oil and water: Meathead's Memphis Dust, which is mostly spices, and Simon & Garfunkel Rub, which is mostly herbs. As you can see, they dissolved better in water. Think of how well tea leaves infuse water.

Most of the time I just wet my hands in the sink and pat the meat to moisten it and then sprinkle on the rub. I hope that the spices and herbs will melt a bit, make a nice flavorful slurry that will become a major part of the desirable flavorful and lovable bark when it is heated and dries out.

To prevent cross-contamination, one hand sprinkles on the rub and the other hand does the rubbing. Don't put the hand that is rubbing into the powder or onto the bottle.

After salting, place the food on a wire rack set over a pan, no wrap. There is nothing about plastic wrap that forces salt or rub molecules into the meat. It is not some sort of vacuum or pressure system. Plastic wrap just gets stuck to the rub and pulls it off when you remove the plastic. Liquid also accumulates in the plastic and washes away some of the rub. Plastic wrap does not prevent some aromas from entering

1 ounce vegetable oil · 1 ounce tap water
½ teaspoon Meathead's Memphis Dust 30 minutes
½ teaspoon Simon & Garfunkel Rub 30 minutes

and exiting, so if you're in Satriale's Pork Store and one of Tony Soprano's dead friends is hanging in your walk-in, the smell can penetrate most of your plastic-wrapped meats. If you don't believe me, pour some ammonia in a zipper bag and sniff.

If you are dry brining, you want airflow around the meat to help desiccate the skin. In the restaurant world you are required by law to cover or wrap the meat so juices won't contaminate other foods like veggies. This makes sense, so be very careful if you leave raw meat uncovered in the fridge.

MAKE YOUR OWN PEPPER FLAKES, POWDERS, AND SMOKED PAPRIKA

Stop spending money on flavorless paprika, smoked paprika that is stale, hot chile pepper flakes that are mostly seeds, and chipotle without a bite. It is a snap to make your own.

The method is simple. Split the fresh peppers in half and remove the seeds and stems. Depending on how much heat you want, leave or remove the ribs. Put them in the smoker, oven, or dehydrator at about 140°F for about 6 hours or more depending on the thickness of the flesh, and crush or grind them in a spice grinder, blender, or food processor. You can use whatever chiles you want, any color. Wear gloves and a mask if you are fiddling with the hot stuff. I make these every year:

CAYENNE FLAKES AND POWDER. Far superior to grocery-store pepper flakes and no seeds. If you have a garden, cayennes are easy to grow.

CARMENS FOR PAPRIKA. Carmen peppers are a tapered thin-skinned sweet pepper that get ripe and red in about 75 days in our cool-climate garden west of Chicago. You can buy seeds and grow them. If not, good old sweet red bell peppers will do. I powder them to make mild paprika that tastes better than anything from Hungary or Spain.

SMOKED PAPRIKA. I also smoke Carmens and grind them for smoked paprika. Smoked paprika is an easy way to add character to a dish and bring the outdoors indoors on cold days. It is also an excellent addition to a spice rub. I use it on everything from baked potatoes to pasta to pizza to chicken paprikash.

JALAPEÑOS. I split, seed, and smoke red jalapeños, dry them, and grind them, and I have chipotle. My first choice for pizza and pasta.

POBLANOS. When you dry these meaty green peppers, they are called anchos, and they have slight chocolaty/pruney flavor. Anchos are the cornerstone of chili powders and mole sauces and when added to a stew they bring a richness and depth. Their heat is barely noticeable.

MEATHEAD'S MEMPHIS DUST, AN ALL-PURPOSE RUB

This is the only recipe in this book that appeared in my last book and on my website, but I simply could not leave it out. It's that good. Meathead's Memphis Dust began as a pork rub back in 2001 and many competition teams have won big bucks with it on their ribs and pulled pork. Readers and I have been using it for everything from poultry to smoked salmon, with celery stuffed with cream cheese, on the rim of Bloody Marys, and even on popcorn. Therefore, I hereby christen it an All-Purpose Rub and it is used in Championship Pork Ribs (page 245), Championship Chicken (page 266), and on salmon.

MAKES *About 2½ cups*

TAKES *15 minutes*

- ¾ cup firmly packed dark brown sugar
- ¾ cup granulated white sugar
- ½ cup mild paprika
- ¼ cup garlic powder
- 2 tablespoons medium-grind black pepper
- 2 tablespoons ground ginger
- 2 tablespoons onion powder
- 2 teaspoons ground rosemary

ABOUT THE ROSEMARY. Several readers tell me they hate rosemary and leave it out. Trust me, it hides in the background and you will never know it is there. It is subtle and important in this blend. Substitute dried thyme or oregano if you must, but I think rosemary is the best choice. Just grind the rosemary leaves in a mortar and pestle or in a coffee grinder or a blender. It will take 2 to 3 tablespoons of leaves to make 2 teaspoons of powder.

SMOKE IT. Amp it up by using smoked paprika, and instead of garlic powder, smoke and dehydrate garlic cloves, then grind them. Go nuts and smoke the ginger and rosemary, too!

Blend all the ingredients together in a bowl. Store in a dark place.

TO USE. Dry brine your meat at least 1 hour in advance, more if it is thick. Pat the surface with wet hands or a wet paper towel to help the rub stick. Apply about 1 teaspoon per 4 × 6 inches of surface.

RED MEAT RUB

In Texas, beef brisket is king and the old standby is "Dalmatian rub," just salt and pepper, black and white. But I like to amp it up a bit, and if you are going to take a crack at competition with the Championship Brisket (page 212), you need more than S&P. This rub works just fine on all red meats—beef, lamb, even duck breasts. It can handle high heat. Use fresh ingredients. I made a batch with old ingredients and it sucked. I know baking soda seems weird, but it raises the pH and contributes to the crust.

MAKES *About ¼ cup, enough for about 8 large steaks*
TAKES *5 minutes*

- 1 tablespoon dried rosemary leaves
- 3 tablespoons medium-grind black pepper
- 1 tablespoon smoked mild paprika
- 2 teaspoons ground ancho chiles
- 1 teaspoon ground chipotle chiles
- 1 teaspoon garlic powder
- 1 teaspoon onion powder
- 1 teaspoon instant coffee
- 1 teaspoon dark brown sugar
- ½ teaspoon baking soda

Grind the rosemary into a coarse powder. In a bowl, stir together all the ingredients. Store in a dark place.

TO USE. Dry brine your meat at least 1 hour in advance. Pat the surface with wet hands or a wet paper towel to help the rub stick. Apply ½ teaspoon per 4 x 6 inch surface. Reverse sear.

CAJUN SEASONING

Lotta stuff in this rub, but hey, in Louisiana they learned to cook from the kings of complexity, the French. This can also be used as seasoning for jambalaya, gumbo, dirty rice, and blackened fish, and as the seasoning for andouille sausage. It is used in my Pastalaya (page 274) and Crawfish on Dirty Rice (page 334).

MAKES *⅓ cup*
TAKES *15 minutes*

SPECIAL TOOLS *Blender*

- ½ teaspoon celery seeds
- ½ teaspoon caraway seeds
- 4 dried bay leaves
- 1½ teaspoons dried rosemary leaves
- ½ teaspoon dried oregano
- 1½ teaspoons red pepper flakes
- 1½ teaspoons medium grind black pepper
- 4 teaspoons mild paprika
- 2 teaspoons ancho chile powder

1 teaspoon garlic powder

1 teaspoon onion powder

1 teaspoon granulated white sugar

½ teaspoon mustard powder

⅛ teaspoon ground mace

⅛ teaspoon ground allspice

In a grinder or blender, combine the celery seeds, caraway seeds, bay leaves, rosemary, oregano, and pepper flakes and pulse until finely ground. Combine with the remaining ingredients. Store in a dark place.

TO USE. Dry brine your meat at least 1 hour in advance. Pat the surface with a wet paper towel to help the rub stick. Apply about ½ teaspoon per 4 × 6 inches of surface.

FRENCH RUB

This is an easy all-purpose herbcentric mix for white meats, chicken, turkey, pork, veal, and many veggies. I have also used it in bread stuffings and meatloaf. They sell Poultry Seasoning in jars, but you can make it yourself easily, and modify the ingredients to your taste. Each manufacturer has its own proprietary mix, but sage is the main ingredient in most of them. It is used in my Pastalaya (page 274), Paris Chicken (page 271), Real Fried Chicken on a Gas Grill (page 278), and Mary's Meatballs (page 294).

MAKES ½ cup, enough for 4 chickens
TAKES 15 minutes

3 tablespoons rubbed sage

1 tablespoon dried rosemary

1 teaspoon celery seeds

2 tablespoons dried thyme

1 tablespoon dried marjoram

1 tablespoon medium-grind black pepper

¼ teaspoon ground cloves

If you can only get whole sage leaves, finely crumble them. Crush the rosemary and celery seeds fine. Mix everything together and store in a tightly sealed glass bottle.

TO USE. Dry brine your meat at least 1 hour in advance. Pat the surface with a wet paper towel or wet hands to help the rub stick. Apply about ½ teaspoon per 4 × 6 inches of surface.

MARINADES, BRINERADES, VINAIGRETTES, INJECTIONS

"Knowledge is knowing a tomato is a fruit. Wisdom is not putting it in a fruit salad."

—ANONYMOUS

It helps to think of marinades as wet rubs. One can even think of a rub as a dry marinade. Many marinades are simply salad dressings, aka vinaigrettes. Brinerades are marinades that have enough salt to do double duty as both a marinade and a brine, and in my humble opinion all marinades should have enough salt to be brinerades.

As with rubs, only salt in a marinade penetrates deep. Flavorings do not penetrate meat more than ⅛ inch, usually less. Marinades can penetrate a bit deeper in some soft meats and veggies such as fish, shrimp, zucchini, eggplant, and mushrooms. The bottom line is that most marinades should be viewed as surface treatments. Oh, and maybe you've heard buzz about using marinades after cooking. Excuse me, I think that is called a sauce.

SAF: SALT, ACID, FLAVOR

The best marinades/brinerades usually contain three working components: Salt, acid, and flavor. How long? That depends on thickness. As a rule of thumb, let's say one to two hours per inch.

S IS FOR SALT. Shoot for about 6% salt by weight. Soy sauce, miso, and pickle juice make good brinerades.

A IS FOR ACID. Acid can denature protein on the surface and make the surface of the meat tender, but too much can make it mushy, so use it judiciously, no more than one-eighth of the blend. Citrus is a good acid because it also has sugar and flavor.

F IS FOR FLAVOR. Typical flavorings include herbs and spices. It's a good idea to add some umami and some sugar. This is a good opportunity to find flavors that are not on the spice rack. So look for liquids such as juices, soft drinks, liqueurs, jams, tamarind, miso, soup bases, pickle juice, vinegars, extracts. Sugar aids in browning, but go easy. Too much will burn.

NO OIL. Notice there is no oil in this formula, but practically every marinade recipe in the world contains oil of some sort. Remember, meat is 70+% water, and oil and water don't mix. Oil may help extract flavor from some herbs and spices, but it's not necessary.

BEWARE OF BOOZE. Wine, beer, and spirits have wonderful flavors, but they contain a powerful solvent: Alcohol. When used in marinades they can significantly alter the structure of proteins and other compounds, and rarely for the better. It is a good idea to boil adult beverages for 20 to 30 minutes to drive off most of the alcohol and concentrate the flavors before adding them to a marinade or sauce.

Basting adds flavor

BUSTED. Apple juice, beer, and wine are mostly water and they don't have many flavor molecules. Most of the liquid just runs off when you baste, taking your rub with them, in effect *removing* flavor. Worse, they moisten the surface, preventing browning. On the plus side, basting cools the food and slows cooking, and that tenderizes. Also, wet surfaces hold smoke, and sugar aids in browning.

PROS AND CONS OF MARINADES/BRINERADES

PRO. Salt in marinades/brinerades penetrates and improves moisture retention and amplifies flavor.

CON. They leave the surface wet, so before you can get browning you have to cook long enough to dry out the surface. By then the meat may overcook. You can rinse or pat the surface, but then you lose much of the flavor.

PRO. They can have flavors that are not on your spice rack such as fruit juices, vinegars, soft drinks, yogurt, soy sauce, miso, tamarind, etc.

CON. They take up more space since you need a container to hold all that liquid.

PRO. They often have sugar in them, which will help with browning.

CON. They often have sugar in them, which can burn.

PRO. If the brinerade has some acid, it will tenderize the surface a bit. If marinades contain enzymes, such as bromelain or papain, they can also tenderize the surface. But you have to be careful, long soaks can make meats mushy.

CON. Many cooks use salad dressings for marinades, but they usually are 50% or more oil. I don't recommend them because the oil can drip and cause flare-ups and soot.

PRO. They are best on thin things. Think of Korean kalbi, very thin slices of beef, perhaps ¼ inch, so penetration of ⅛ inch on each side can be 100% penetration. Marinating skirt steak is good for the same reason.

USING THEM AS SAUCES

Once you put meat in a marinade, the marinade is contaminated, so you can't use it as a sauce. Boiling it will probably make it safe, but some spores can survive boiling, and boiling may not destroy toxins left behind by bacteria. You can baste with marinades but you should stop basting 10 minutes or so before you serve so the heat can kill any bad guys in the baste or on the brush. If you want to use that tasty marinade as a sauce, and that is often a good idea, make twice as much and set half aside. Sauces are usually thicker than marinades but there are several ways to convert them into a sauce.

COOK IT DOWN. By simmering or boiling a marinade you can reduce the amount of water. Beware, if you reduce a marinade that has salt and sugar in it, the water will diminish but you might end up with something too sweet or salty.

Toothpick Depth Gauge

When a recipe says to reduce the sauce by half or another fraction, pour it into a pan and stick a toothpick into the pan and mark the depth of the liquid on the wood. Then lift it out and mark the fractional point. Boil the liquid and occasionally turn off the heat and insert the toothpick. When the liquid is down to the mark, take the pan off the heat.

ADD FAT. Fats are large molecular chains and they move much less rapidly than water, so emulsifying water with something like melted butter, oil, cream, animal fat, or even yogurt will make it more viscous and thicker.

ADD STARCH. When starches are mixed with hot water they swell and form gels, trapping water in complex networks. As the sauce cools it thickens even more. To thicken 1 cup of sauce, soup, or stock, just mix 1 tablespoon cornstarch thoroughly into 2 tablespoons water, add it to the warm sauce, bring it almost to a boil, and stir until it thickens. Other thickeners are arrowroot and tapioca starch.

ADD BOTH FAT AND STARCH. Take equal parts of softened butter and flour and mix them to form a paste and you have a beurre manié. Just stir it into a sauce and watch it thicken. To make a roux, mix equal amounts of butter and flour in a pan over medium heat and stir until it turns blond. You can keep going until it turns brown if you wish. Brown roux has more flavor but less thickening power.

ADD A PUREE. Puree cooked vegetables, potatoes, tomatoes, rice, or legumes and stir them in.

ADD EGG YOLK. This makes an emulsion, but using yolks is tricky because they can cook and coagulate quickly if the liquid is too hot.

MOUNT WITH BUTTER. Just before serving, swirl a knob of butter into the warm sauce (which is called mounting), so it melts and forms a simple emulsion.

LEMONGRETTE

Vinaigrettes are most commonly used as salad dressings, but they make fine brinerades. They get their name from vinegar, but one can often substitute lemon juice, lime juice, very tart wine, or a blend of them all. When lemon is the acid, I call it a Lemongrette. Traditionally the oil is a quality extra-virgin olive oil, but you can change the flavor profile with a different oil. The fun begins when you amp it up with fresh herbs and other goodies, and this recipe calls for fresh, not dried, herbs. If time permits, make it 12 to 24 hours in advance to allow the flavors to marry.

MAKES *About 2½ cups*
TAKES *30 minutes to make, at least 1 hour to marry*

SPECIAL TOOLS *Blender or food processor*

- 1 ripe to overripe tomato
- 3 medium lemons
- 1 large shallot or small red onion
- 1 garlic clove
- 1 black garlic clove
- 2 tablespoons fresh lavender, tarragon, thyme, or other herb leaves
- ½ cup high-quality olive oil
- 3 tablespoons Dijon-style mustard
- 1 tablespoon honey
- ½ teaspoon Morton Coarse Kosher Salt
- ¼ teaspoon fine-grind black pepper

ABOUT BLACK GARLIC. Black garlic is regular garlic that has been fermented at a controlled temperature and humidity. It is sweet and reminiscent of prunes or balsamic vinegar in flavor. It tastes nothing like regular garlic and you cannot substitute for it. It is expensive but its rich, raisin character is very special. If you've never tried one, here's your chance for something mind-blowing.

ABOUT THE LAVENDER. For this recipe I like lavender but you can use tarragon, thyme, oregano, whatever is available. I prefer fresh herbs, but you can use dried. If you use dried herbs, cut the quantity by half.

1 PEEL THE TOMATO. There are two good methods. (1) Cut an X in the tomato skin at the bottom and dunk it in boiling water for 60 seconds, no more. Then dunk it into a bowl of ice water. After a minute or so it should peel easily. (2) Stick a fork in the tomato and hold it over the flame on a gas stove. The skin will start to crack and peel. Turn it around a bit until you have several cracks. Let it cool and peel away.

2 SQUEEZE THE LEMONS. Squeeze all the juice from the lemons and discard the seeds. Pour it into a measuring cup. You need 4½ fluid ounces (9 tablespoons). If you don't have that amount, add a little water. Pour 2 ounces (¼ cup) into a blender or food processor and set the rest aside.

3 CHOP. Peel and mince the shallot. Take about 4 tablespoons and add it to the blender. Peel and mince or press the white garlic clove. Peel and coarsely chop the black garlic clove. Everything goes into the machine.

4 BLEND. Add the lavender, oil, mustard, honey, salt, and pepper and take it for a spin for about 30 seconds, until all the big chunks are gone. Tasting a fresh vinaigrette nekkid is not very helpful since it will be significantly altered as it sits and sucks the flavors from the solid ingredients. Pour it into a bottle or jar, and let it sit in the fridge for at least 1 hour or overnight to allow the flavors to marry.

5 TASTE AGAIN. The salad also alters the flavors, so dip a leaf of lettuce in it to taste it. You can then adjust the ingredients to your taste. After it sits it will separate a bit, so give it a good shake before using it.

SMOKED TOMATO VINAIGRETTE

In August, one of the two greatest sensual pleasures of the year is plucking a fat, red, ripe tomato, still warm, and chomping down right there in the garden. The other is buying an ear of sweet corn at the farmers' market,

one of the new sugar-enhanced varieties, and chomping down, no salt, no butter, no nothing. This recipe takes full advantage of ripe tomatoes and uses them in Corn Salad (page 351), but it can be used on just about any salad.

MAKES *About 1½ cups*
TAKES *45 minutes to 1 hour 15 minutes*

SPECIAL TOOLS *Grill topper, blender or food processor*

- 1¼ pounds ripe meaty tomatoes
- 3 garlic cloves
- 2 tablespoons champagne vinegar or white wine vinegar
- 1 teaspoon salad-grade balsamic vinegar
- ½ cup high-quality olive oil
- 1½ teaspoons roughly chopped fresh dill
- ½ teaspoon Morton Coarse Kosher Salt
- ½ teaspoon medium-grind black pepper

1 PREP. Quarter and core the tomatoes. Peel the garlic and mince or press the cloves.

2 FIRE UP. Fire up your smoker and aim for 225°F. Or set up a grill for 2-zone cooking, aim for about 225°F on the indirect side, and get some smoke going. Put a grill topper in indirect heat.

3 SMOKE. Smoke the tomatoes skin side down on the grill topper until they are slightly shriveled, perhaps 45 minutes. Don't worry if they blacken a bit.

4 PEEL THE TOMATOES. See Step 1 in the Lemongrette recipe above. Remove from the heat, and when they are cool enough to handle, you can pinch the skins and they should slip right off.

5 MAKE THE VINAIGRETTE. Set aside 6 of the tomato quarters. Transfer the rest to a blender or food processor. Add the garlic, both vinegars, olive oil, dill, salt, and pepper and give them a spin until smooth. Cut the reserved tomatoes into ¼-inch dice and add them for texture.

MISO VINAIGRETTE

Miso is a used in many Japanese recipes and should be used more often in American cuisine. It is an umami-laden paste made by fermenting soybeans with salt and kōji (see page 111) and sometimes other ingredients. Most common are white miso and red miso. Red is the stronger of the two. I use this sauce/dressing on Asparagus with Miso Vinaigrette (page 354).

MAKES *About ½ cup*
TAKES *5 minutes*

- 3 tablespoons unseasoned rice vinegar
- 2 tablespoons white miso
- 2 tablespoons vegetable oil
- ½ teaspoon fish sauce or Worcestershire sauce
- ¼ teaspoon ground chipotle, or more to taste

In a small bowl, combine the ingredients and whisk together until blended. Refrigerate until you are ready to use it as either a marinade or dressing.

SESAME-LIME DRESSING

This zesty dressing is used in the recipe for Smoked Tuna with Sesame-Lime Dressing (page 314).

MAKES *About ½ cup*
TAKES *About 5 minutes*

- 2 garlic cloves
- 1 large lime
- ¼ cup vegetable oil
- 2 teaspoons toasted (brown) sesame oil
- 1½ teaspoons dark soy sauce
- 1 teaspoon sesame seeds

1 teaspoon whole-grain mustard

Fine-grind black pepper to taste

Peel and mince or press the garlic. Scrub the lime, zest it, and squeeze out 2 tablespoons of juice. In a screw-top jar, combine all the ingredients and shake vigorously.

TERIYAKI BRINERADE AND SAUCE

There are numerous formulae for this classic sauce. Mine has a few more ingredients than most but it makes both a sterling brinerade and sauce. Try the sauce on the Beef Back Ribs (page 220) and the Mushroom Sandwiches with Teriyaki Sauce (page 356).

MAKES *A bit less than 3 cups*

TAKES *45 minutes*

1-inch finger of fresh ginger

4 garlic cloves

1 teaspoon cornstarch

1 cup soy sauce

¾ cup mirin

¼ cup orange juice

¼ cup unseasoned rice vinegar

½ cup packed brown sugar

1 PREP. Peel and finely grate the ginger. Peel and press the garlic. In a small cup, stir the cornstarch into 2 tablespoons water.

2 COOK. In a saucepan, combine the ginger, garlic, soy sauce, mirin, orange juice, vinegar, brown sugar, and cornstarch slurry. Bring it to a boil. Reduce to a simmer and cook until it reduces to the thickness of cream, about 15 minutes.

3 STORE. Remove from the heat and let it cool. Strain out the solids and store in the fridge in an airtight jar.

YOGURT MARINADE

Yogurt, buttermilk, and milk are used often in marinades. In Indian cuisines many meats are marinated in whole-milk plain yogurt like my Tandoori Chicken (page 277). Mediterranean cultures use buttermilk or yogurt on pork, goat, and mutton. In the American South, biscuits, corn bread, pancakes, hush puppies, and hotcakes often contain buttermilk, and of course, many recipes for fried chicken call for marinating in buttermilk. Fishermen soak fishy-smelling fish in milk, and hunters soak boar, venison, and other game in milk.

What's the point if most of the molecules in these dairy products are too large to penetrate? It's mostly about the acid. Yogurt and buttermilk have lactic acid and that can slightly tenderize the surface proteins and penetrate a tiny bit below the surface, adding a brightness to meat's taste. This lowering of the pH can tenderize the surface, help it retain

Marinades will penetrate faster under a vacuum

BUSTED. There are several gadgets that create a vacuum during marination and they are sold with the promise that they will get the flavor in deep. Well, this idea sucks. This idea fails with just logic, but food scientists have proven it doesn't work. Why? If you suck the air out from around meat, you are not creating a vacuum within the meat, you are creating a vacuum outside the meat and you will suck juices out of the meat, not in. Doh!

moisture, and help it form a better crust. It may also stimulate enzymes, and enzymes in dairy may also aid with tenderizing.

Buttermilk and yogurt can contain live microbes. Before refrigeration their bacteria can outcompete harmful and spoilage bacteria. The lactic acid is also unfriendly to pathogens. Fat in dairy is also good at trapping unpleasant aromas from game and stale seafood. Lactose, a sugar, can aid the Maillard reaction and browning.

My favorite Indian restaurant, Kama in La Grange, Illinois, slips a little papaya into their yogurt marinades because the enzyme papain in fresh papaya is a tenderizer. Yogurt is great for holding spices onto the surface, and of course so much of the charm of Indian cuisine is in their complex spice blends.

When meats come out of the dairy marinades, they are coated with fat and other compounds that stick to the surface during cooking. Because whole-milk yogurt is about 3% milk fat, 8% other solids, and the rest mostly water, in an Indian tandoor, which is a lot like a kamado, the sugars, fats, and proteins in yogurt brown a bit, even burn, but the water prevents the meat from searing and developing Maillard flavors. The yogurt coating can stick to the grates, so hanging yogurt-marinated foods in a kamado as they do in a tandoor, or spearing them on a rotisserie, is a good way to cook them.

As Indian dishes tend to, this marinade has a lot of spices. They bring complexity and depth. The marinade is also a transformer because it can be used as a sauce for lamburgers, hamburgers, spiedies, salmon, and grilled chicken. Yes, I said hamburgers. You'd be surprised. This recipe is used in my Tandoori Chicken (page 277).

MAKES *About 1 cup*
TAKES *10 minutes*

> 4 garlic cloves
> 2-inch finger fresh ginger
> 3 tablespoons vegetable oil
> 1 cup whole-milk plain thick yogurt
> ¼ cup fresh lemon juice
> 4 teaspoons garam masala
> 1 tablespoon dried oregano

2 teaspoons Morton Coarse Kosher Salt

2 teaspoons Tabasco Chipotle Sauce

1 teaspoon smoked mild paprika

½ teaspoon ground turmeric

Press or mince the garlic. Grate the ginger until you have 2 tablespoons, skin and all. In a bowl, stir together all of the ingredients and store in the refrigerator.

TO USE: For use as a marinade, coat the meat and leave it in the refrigerator for at least 3 hours. For use as a sauce, double the yogurt.

BASIC INJECTION

Because I like the unadulterated taste of most meats, I don't inject often, and when I do, I usually keep it to either saline solutions, which I use for only very lean meats like pork loin, or butter, which I use for turkey breasts (see Buttered-Up Turkey Breast, page 285). There are also commercial injection blends and many of them contain compounds that help with capturing and retaining water.

We know the magic of salt but want no more than 1% of the weight of the meat in salt, so make a brine about 3% by weight.

MAKES *1 quart 3% brine*

TAKES *15 minutes*

1 ounce (28.4 g) any salt

1 quart (946 g) warm water or low-sodium stock

OPTIONAL. 1 ounce (28.4 g) sugar

Mix the salt and sugar (if using) into the warm water and stir until it all dissolves.

TO USE: Inject every inch or so into lean meats like pork loins, chicken breasts, and turkey breasts.

BARBECUE SAUCES

"A good upbringing means not that you won't spill sauce on the tablecloth, but that you won't notice it when someone else does."
—ANTON CHEKOV

When I say "barbecue sauce," I know what you are thinking: Thick, red, sweet, a sauce that has become known as Kansas City–style. But not everyone has the same vision. In parts of South Carolina, barbecue sauce is yellow because it is mostly mustard (see Carolina Gold for Grownups, page 182). In other parts of the Carolinas and parts of Georgia, it is mostly vinegar (see Lexington Dip #2, page 184). There's a white sauce from Alabama (see Alabama White Sauce, page 183), in Western Kentucky, Worcestershire dominates, and old-timers in Texas will tell you about a thin tomato-based sauce laced with chiles. I delved into most of them on my website.

Sauces should go on the food just before the food comes off the heat. Especially if they are sweet like the Kansas City Red (page 180) or Carolina Gold (page 182). Sweet sauces can burn quickly. Paint a single layer or two of sauce on and let it bake in low indirect heat, about 225°F, for 15 minutes or so. Keep an eye on it. My favorite trick is to then move it to a scorching-hot gas grill, lid open, and let the sauce sizzle and caramelize for just 1 to 2 minutes per side.

Show restraint. When I judge competitions, it is disconcerting when there is so much sauce I can't taste the meat. A 20-ounce bottle of Kansas City–style sauce should be enough for 3 to 4 slabs of ribs.

Be conscious of cross-contamination. Pour what you need into a coffee cup, spoon it on the meat with a clean spoon, then brush it around, but don't put the brush into the cup. When you're done, any left in the cup can go back into the bottle.

As with rubs, it is often cheaper and more fun to make your own. If that's too much work, a lot of folks like to buy a commercial sauce and doctor it by adding heat, sweet, and flavor.

Lola Longo. Photo by Natalie Longo

You can then pour it into a jelly jar and put your own label on it, and when people ask for the recipe use the old line "I'd tell you but then then I'd have to kill you."

Some recipes call for sautéing onions and garlic and adding herbs. Sometimes it is nice to leave the sauce chunky and rustic, but if you want a smooth sauce, run it through a fine-mesh sieve. Keep in mind, if you use animal fats like butter or bacon fat, they can go rancid with time, so they must be kept in the fridge and you would also be limiting shelf life. If you know how to do sterile canning, boil the jars for 35 minutes. If you don't know how to do this properly, *The Ball Blue Book* is considered the bible for home canners, and Ball also has an informative website. I am especially fond of the book *The Preservation Kitchen* by Paul Virant, chef and owner of several wonderful restaurants in Illinois: Petite Vie in Western Springs, Vistro Prime in Hinsdale, and Gaijin in Chicago.

KANSAS CITY RED

Here's a fun version of the classic Kansas City sweet red barbecue sauce with an exotic undertone. You'll know the raspberries are in there but can offer a prize to the guest who guesses the secret ingredient. And the best part of this sauce is that it is quick and easy.

Liquid smoke is optional here, but it adds a nice dimension. It is a dominant player in the flavor profiles of the most popular commercial BBQ sauces. Without liquid smoke, most people feel a barbecue sauce is missing something. Many commercial "smoked" meats, such as bacon, hams, jerkies, and sausages are created with liquid smoke. It appears on labels as "natural smoke flavor." Purists consider it an evil adulterant. But it

is simply smoke that has been gathered in a condenser and aged in barrels. When smoke hits cold meat, it condenses on the cold meat in the same manner, so what's the big deal? The process for liquid smoke is pretty much the same as the way they make whiskeys. So if you know someone who derides liquid smoke, I authorize you to take away his bourbon.

Try this sauce on the Championship Pork Ribs (page 245), BBQ Beans (page 345), beef jerky, grilled chicken, pork chops, French fries, whatever.

MAKES *About 2 cups*
TAKES *30 minutes*

> ½ cup bourbon
>
> 1 cup ketchup
>
> ½ cup Smucker's Seedless Red Raspberry Jam
>
> ¼ cup honey
>
> ¼ cup apple cider vinegar
>
> ½ teaspoon Tabasco Chipotle Sauce
>
> ½ teaspoon garlic powder
>
> ½ teaspoon ground ginger
>
> ½ teaspoon Morton Coarse Kosher Salt
>
> 2 tablespoons liquid smoke

ABOUT THE JAM. I normally don't specify brands lest someone think I'm on the take. I call for the Smucker's because it has a great natural flavor and it should be easy to find. Use it to replicate my recipe, but feel free to use another jam of your choice (blueberry is nice). Go for something without skins and seeds. Beware that it may be sweeter or less sweet

than my choice, so you may want to start with less and add more to taste.

ABOUT THE HOT SAUCE. I use Tabasco Chipotle. Use whatever you like, but chipotle has an affinity for raspberry.

1 COOK OFF THE ALCOHOL. In a saucepan, bring the bourbon to a boil. Continue to boil until you have only 2 to 3 tablespoons left. The alcohol will be mostly gone so it is child safe, but the elixir that remains will have a rich, woody, smoky flavor. If you don't have bourbon or don't want to do this step, use 2 more tablespoons liquid smoke.

2 SIMMER THE SAUCE. Add all the remaining ingredients to the saucepan and whisk until smooth. Bring to a simmer over low heat and cook for 15 minutes. Sample it and adjust the ingredients to your taste. Want more heat, reach for the hot sauce. Want it sweeter, add sugar or more jam. Simmer a few more minutes.

3 STORE. Pour the finished sauce into a very clean bottle and store in the fridge for months.

CAROLINA GOLD FOR GROWNUPS

South Carolina is known for its BBQ sauces made with mustard, especially in the belt between Columbia and Charleston. Most are simply made with yellow ballpark-style mustard, cider vinegar, hot peppers, and sugar. Others are simple variations on the honey-mustard theme. I love these classic South Carolina sauces, especially on pulled pork, and there's a winning recipe in my last book, but for this book I wanted something a bit more interesting and complex.

Savory herb flavors are great with pork and mustard, so I started with the classic SC barbecue mustard recipe and added layers of complexity by adding rosemary and other more subtle flavors. If the classic SC mustard sauces are trumpet solos, this is a full orchestra. There's a lotta stuff in this recipe, but try not to leave anything out. When I served it to my neighbor Keith Miller, he said, "Wow, this is a mustard sauce for grownups!" And it had a name.

This sauce is especially good on pulled pork, pork chops, hot dogs, and other sausages. I love it on baked potatoes, for dipping pretzels, or as a mustard substitute in many recipes.

MAKES *About 2 cups*
TAKES *5 minutes to prep, 30 minutes to cook*

- 1 medium onion
- 1 small red bell pepper
- 2 garlic cloves
- ½ teaspoon whole dried rosemary leaves
- 2 tablespoons vegetable oil
- 1 teaspoon medium-grind black pepper
- 1 teaspoon dried thyme
- 1 cup Dijon-style mustard
- ⅔ cup packed dark brown sugar

¼ cup fresh lemon juice

¼ cup apple cider vinegar

2 teaspoons tomato paste

2 teaspoons Worcestershire sauce

1 teaspoon gochujang or your favorite hot pepper sauce

1 teaspoon chicken bouillon granules, chicken base, or 2 bouillon cubes dissolved in 2 tablespoons water

1 PREP. Peel and mince the onion and save ½ cup for the recipe. Mince the bell pepper and save 3 tablespoons for the recipe. Peel and mince or crush the garlic. Crush the rosemary leaves into bits.

2 COOK THE AROMATICS. In a 1-quart saucepan, warm the oil over medium-low heat. Add the onion and bell pepper and sweat them until the onions are limp, 5 to 6 minutes. Add the black pepper, rosemary, and thyme and cook, stirring, for about 3 minutes to develop and extract their flavors. Add the garlic and cook it for about a minute.

3 FINISH. Add the mustard, brown sugar, lemon juice, vinegar, tomato paste, Worcestershire sauce, gochujang, and chicken bouillon and mix thoroughly. Bring to a low boil for 3 minutes, stirring frequently to keep the sugar from burning or sticking to the bottom. Turn down the heat and simmer on low for another 15 minutes.

4 FINISH. Let it come to room temperature and puree it in a blender. A few lumps are OK. Taste, adjust, and chill.

ALABAMA WHITE SAUCE

This is not what most folks think of when they think of barbecue sauce. It is white and it is not sweet. It was designed for chicken by Big Bob Gibson in Decatur, Alabama, probably in the 1920s.

There are now two Big Bob Gibson Bar-B-Q restaurants in Decatur and they have achieved national fame and acclaim on the shoulders of my fellow Barbecue Hall of Famer Chef Chris Lilly and his competition team. The original location on 6th Avenue is destination dining.

Lilly's team has won practically every major competition on the circuit and their white sauce is now widely imitated. The actual recipe is, of course, secret, but Chris tells me mine

Ken Hess, former pitmaster at Big Bob Gibson's.

is close. The chefs butterfly the chickens and dunk them in a vat of the sauce. My version of their chicken recipe can be found with a link here: AmazingRibs.com/mm.

MAKES *1 quart, enough for 4 or more chickens*
TAKES *10 minutes to prep, 2 hours to marry flavors*

- ¾ cup full-fat mayonnaise
- ⅓ cup apple cider vinegar
- ¼ cup fresh lemon juice
- ¼ cup apple juice
- 1 tablespoon garlic powder
- 1 tablespoon prepared white horseradish in vinegar
- 1 tablespoon coarse-grind black pepper
- 1 teaspoon mustard powder
- ¼ teaspoon Morton Coarse Kosher Salt
- ½ teaspoon cayenne pepper

In a large bowl, whisk together all the ingredients. Refrigerate in a jar for at least 2 hours to allow the flavors to meld.

LEXINGTON DIP #2

There are so many flavor profiles on the BBQ sauce menu, and I have covered them in depth in my last book and on my website. But I feel the need to offer a low-sugar recipe in this book.

In the Piedmont (aka Hill Country or the Foothills) of the western Carolinas around Lexington, North Carolina, they call BBQ sauce "dip" and they apply it to pork shoulder most of the time. It is apple cider vinegar–based and the result is thin and penetrating. The acidity is the perfect counterpoint to fatty meats. To the chagrin of the rest of the Carolinas, where ketchup is eschewed, in the Piedmont they let a little of the red stuff in.

MAKES *2 cups*
TAKES *30 minutes*

- 1 cup apple cider vinegar
- ¼ cup apple juice
- ¼ cup ketchup
- 3 tablespoons light brown sugar
- 1½ teaspoons Morton Coarse Kosher Salt
- 1 teaspoon adobo sauce from canned chipotle in adobo sauce
- 1 teaspoon medium-grind black pepper

ABOUT CHIPOTLES IN ADOBO SAUCE. This is chipotles (smoked and dried jalapeños) swimming in a vinegary tomato sauce with onion, garlic, and a kiss of sugar in small cans. Just use the sauce from the can, not the chiles for this recipe. Traditional Carolina vinegar sauces use hot pepper flakes. They take a day or two to blend in. This version is ready to go.

In a bowl, whisk together all the ingredients.

TO USE: Divide the sauce in two and use half for frequent basting, the other half for serving.

BUTTERS AND DAIRY-BASED SAUCES

"Age is not important unless you're a cheese."

—ANONYMOUS

Dairy provides a rich and luxurious texture to a wide variety of dishes. Velvety and decadent sauces made with cheese, butter, and cream can elevate simple ingredients into an indulgent experience.

COMPOUND BUTTERS

Compound butters are just flavored butter. Once you have mastered the basic method, feel free to experiment with your own flavor combinations and try it on your favorite foods.

BLEND. Simply cut the butter into chunks and let it come to room temperature for, say, 1 hour. Smush in the other ingredients with a fork. Start with herbs, spices, honey, maple syrup, hot sauce, or try some of my rubs. Below are three examples that are used in other recipes in this book.

WRAP AND ROLL. Lay the flavored butter on a sheet of plastic wrap, waxed paper, or parchment paper and roll it around to form a log about 1 inch thick. Twist the ends like

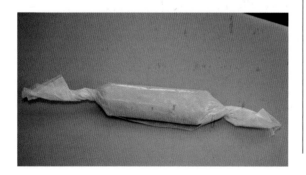

a Tootsie Roll. If you use plastic wrap, wrap it again tightly with foil, because plastic wrap can breathe. Or put the butter in a clean used margarine tub.

STORE. You can store it in the fridge for a week, but if you plan to keep it longer, freeze it, because some of the add-ins may have mold spores.

BLACK GARLIC BUTTER

Black garlic is an umami-bomb that was once rare and only seen in Asian kitchens, but is now the darling of restaurant chefs and adventurous home cooks. It is raw garlic that has been fermented and aged at a controlled warm temperature. When you chop it, it is sticky, almost tar-like, and full of slightly sweet, earthy, and savory flavor, at once reminiscent of figs and fine balsamic. No resemblance to garlic. Once you try it, you'll fall in love.

This mix of black garlic and butter made my knees buckle when I used it with scallops (page 317). Spread it on toast or English muffins, add it to a baked potato, use it when you fry eggs, make a beurre blanc (page 190) with it, top a steak or a fillet of fish, or melt it and pour it over grilled seafood. The flavor of black garlic is delicate and complex, so don't waste it in strong sauces.

MAKES *About ¾ cup*
TAKES *5 minutes*

> 6 tablespoons (3 ounces) unsalted butter
> 8 black garlic cloves
> ½ teaspoon grated lemon zest
> ¼ teaspoon Morton Coarse Kosher Salt

1 PREP. Take the butter out of the fridge, cut it into 3 chunks, let it come to room temp in a bowl. Peel the black garlic and mince it as much as you can. A little oil on the knife might help keeping it from sticking.

2 BLEND. Mash the softened butter, black garlic, lemon zest, and salt together with a fork. Roll it and wrap it as described above.

3 STORE. Because it is made from garlic, which grows underground and is friendly to botulinum bacteria, for safety, don't keep it in the fridge for more than a day or two.

YUZU BUTTER

Yuzu is a yellow citrus fruit from Japan. It has flavors reminiscent of grapefruit, lemon, and orange, and if you can't find a yuzu, use one of them. I use yuzu butter in Panko Pebbles (page 374) and they go on the grilled Broccoli with Panko Pebbles and Yuzu Butter (page 352).

MAKES *About 5 ounces*
TAKES *About 5 minutes, 10 minutes standing time*

> 8 tablespoons (4 ounces) unsalted butter
> 1 tablespoon fresh yuzu or citrus juice
> 2 teaspoons dark soy sauce

In a small saucepan, melt the butter over medium heat. Add the yuzu juice and soy sauce and stir. If you won't be using it immediately, let it solidify, then wrap and roll (see page 186) and refrigerate.

MISO BUTTER

This is a lovely salty blend that I use on shellfish.

MAKES *5 ounces*

TAKES *20 minutes*

> 1 lemon
>
> 8 tablespoons (4 ounces) unsalted butter
>
> 1½ teaspoons white miso

Scrub the lemon and zest the surface enough to get about ½ teaspoon. Melt the butter in a pan or microwave. Whisk together the zest, melted butter, and miso until well blended. The miso will want to settle out because it doesn't blend well with fats and if it cools the butter will solidify. No problem, you can always warm it just before serving and give it another stir. Let it cool, then wrap and roll (see page 186), and refrigerate.

SMOKED BUTTER

Simply cut a stick of butter into 4 parts and freeze it because cold surfaces attract more smoke. Then put it in a shallow pan or bowl on the smoker or on the indirect side of your grill, generate heavy white smoke, and smoke at about 225°F for 1 hour. Take it out of the heat and let it sit at room temperature for about 1 hour or until it can be worked like clay. Scrape it all together and roll it into a log about 1 inch thick, wrap and roll (see page 186), and refrigerate. Have some fun and use it in a hollandaise sauce or Béarnaise sauce (recipes on my website).

CLARIFIED BUTTER, DRAWN BUTTER, GHEE

Most butter is made by skimming the fatty cream from the top of fresh cow's milk, leaving behind skimmed milk. The cream is heated to pasteurize it, then beaten (churned) until the fat molecules clump and they separate into butter and buttermilk. After rinsing, the butter that emerges is an emulsion of about 82% fat, about 16% water, and 2% milk proteins. European butters tend to be slightly higher in fat. Salted butter has 1.5% to 2% salt added (all my recipes call for unsalted butter). The exact proportions vary from creamery to creamery.

Because of the water and protein, butter behaves differently in cooking than oils that are 100% fat. Butter melts at about 90°F and then it goes up to the boiling point of water, 212°F, and doesn't go much higher until the water foams and boils off. Most frying is done between 350° and 375°F. Because of its water and protein-laden milk solids, butter is not a good candidate for deep-frying. Remove the water and solids and you have clarified butter or drawn butter. Ghee is clarified butter that has been allowed to brown slightly (but not always). These clarified butters can go up to 450°F and fry just about anything and still impart a buttery taste. I call for clarified butter in Scallops with Black Garlic Butter Sauce (page 317) and Shrimp with Grand Marnier

Butter (page 331). You can buy both clarified butter and ghee in stores and online. Or make it yourself. It's easy. I like to do a pound or more because it is easier to manage and the end product can keep for months in the fridge.

MAKES *1¾ cups*
TAKES *20 minutes*

4 sticks (1 pound) unsalted butter

1 CLARIFY THE BUTTER. In a saucepan, melt the butter over medium-low heat. Gently turn up the heat until it starts to boil at 212°F. It will bubble away at this temp for a few minutes, foam will accumulate, until all the water boils off, then the foam will subside and the temperature will rise. Let it go up to about 230°F. The proteins in the foam will settle to the bottom. That's clarified butter on top.

2 GHEE. Cook it a little longer, up to about 250°F, and the solids will begin to brown, giving the clarified butter the slightly nutty taste of browned butter. That's Indian-style ghee.

3 STORE. Pour it into a largish measuring cup or water glass and let it sit at room temperature for about 30 minutes. It will separate into three distinct layers. A thin layer of milk solids at the top, a large layer of yellow clarified butter fat below it, and a layer of milk solids at the bottom. Scoop the top layer off with a spoon, carefully pour the clarified butter into a jar, and discard the milk solids at the bottom. Seal the jar and store in the fridge for months.

BROWN BUTTER

This method transforms butter from a joyful rhyme to a Pulitzer poem. You can use it in just about any recipe that calls for butter. Bakers love its nutty caramel flavors (make your next batch of chocolate chip cookies with brown butter). It is excellent on most fish and many veggies. Gnocchi with BB blows me away. How's it work? First the emulsion breaks and the water separates from the fat and the milk solids, which are loaded with amino acids and sugars. Then the water evaporates and we discover that the Maillard reaction is not just for meats. Most folks use frying pans, but I use a saucepan to contain the spatters. Cook a fish fillet in this.

MAKES *3 ounces brown butter*
TAKES *20 minutes*

SPECIAL TOOLS *Small light-colored (such as stainless steel) saucepan*

2 sticks (8 ounces) unsalted butter

Cut the butter into chunks. In a small light-colored saucepan (a stainless steel pan is perfect), melt the butter over medium heat. Don't use a dark-colored pan or you won't see the transformation. Stir with a silicone or wooden spoon. Stay right at the stove and watch carefully, it can burn in a hurry. It will foam and sizzle if there is water left in the butter and then go silent after about 4 minutes. The color will move from yellow to golden to tan to amber to brown. You should smell a nutty scent (hence the French name *beurre noisette*, "hazelnut butter") and the milk solids should settle along the bottom. Remove from the heat and pour into a bowl or jar, solids and all. Let it separate and pour off the clear liquid, leaving the precipitate behind, or not. The solids could burn when you use it at high heat, but at a medium heat they add complexity. Store in the fridge or freezer.

HERBED LEMON BROWN BUTTER

This tangy brown butter sauce is the perfect way to crown a fillet of whitefish, lobster, or a mess of shrimp. This recipe is used in Tea-Smoked Whitefish (page 310).

MAKES *About 3 ounces, enough for 2 fish fillets or lobsters*

TAKES *About 10 minutes*

> 1 lemon
>
> 8 tablespoons (4 ounces) unsalted butter
>
> 1 teaspoon fresh thyme leaves, or
>   ½ teaspoon dried thyme
>
> ¼ teaspoon red pepper flakes
>
> ¼ teaspoon Morton Coarse Kosher Salt

VARIATIONS. Try this with lemongrass, rosemary, oregano, sage, or your favorite fresh herb. You can also do it with lime, grapefruit, or oranges.

1 PREP. Scrub the lemon well and scrape off the zest. Then cut it in half and squeeze the juice from one half into a small bowl. Get rid of the seeds. Save the other half for another dish.

2 COOK. Start the process as in the recipe for Brown Butter (above), but when melting the butter add the thyme, zest, pepper flakes, and salt. As soon as the butter turns a medium-dark tan, remove the pan from the heat, stand back, and slowly pour in the lemon juice. It will spatter. Use it or store it.

BEURRE BLANC

Beurre blanc, which means "white butter," is a classic French sauce, rich and versatile, and the natch match for seafood. It also works beautifully on veggies, even potatoes, wherever you would want butter. Despite its historic importance, there are many variations on the recipe out there. It is an emulsion, a blend of water (the wine, vinegar, and water in the butter) and oil (in the butter). We convince them to mate by using the milk solids in the butter as the emulsifier. You can make a batch of this and freeze it. Temperature control is important. I use it in Lobster with Beurre Blanc (page 328).

MAKES *About ½ cup, enough for 2 fish fillets*

TAKES *About 25 minutes*

6 tablespoons (3 ounces) unsalted butter

1 large shallot or small onion (size of a golf ball)

⅔ cup good-quality dry white wine

3 tablespoons white wine vinegar, champagne vinegar, or fresh lemon juice

¼ teaspoon Morton Coarse Kosher Salt

¼ teaspoon ground white pepper

VARIATION. To make a Beurre Rouge for red meat, use red wine. Gahead, also throw in ¼ teaspoon ground chipotle.

1 PREP. Cut the butter into 6 pieces. Peel and mince the shallot as fine as you can get it.

2 REDUCE. In a small saucepan, combine everything except the butter. Bring to a boil, then immediately reduce the heat to medium and cook until the liquid is reduced by half. Reduce the heat to low and simmer gently until reduced to about 2 tablespoons of liquid surrounding the shallots. Monitor it carefully to be sure the liquid does not completely evaporate—this can happen quickly when the quantity gets low. Remove from the heat. The mixture may be set aside at room temperature for an hour or so or even refrigerated for days.

3 FINISH. When ready to finish the sauce, warm the shallot mix over very low heat. When it is warm, add all the butter at once and whisk the sauce constantly until the butter melts and blends thoroughly. Beware, if it gets hotter than about 135°F, some of the proteins that help form the emulsion might break down and the sauce separates. If it cools too much,

the butterfat can clump. Taste and add more salt, pepper, or vinegar if you wish. If you want a smooth sauce, pour it through a fine-mesh sieve and press the liquid out of the shallots. But I love the solids and I leave them in. Keep it warm.

TO USE. Give it a good whisking, spoon it onto the plate, and place the fish or veggies or potatoes on top.

BUTTERY MIGNONETTE

Typically used on clams, oysters, mussels, and scallops, the classic French mignonette couldn't be simpler, just shallots, wine vinegar, and a pinch of salt. But I wanted something a bit more interesting and I require butter on my bivalves, so . . .

MAKES *About ¾ cup*
TAKES *30 minutes*

1 small fresh hot red chile

1 large shallot or small onion (size of a golf ball)

1 tablespoon fresh tarragon

1 small orange or tangerine

8 tablespoons (4 ounces) unsalted butter

¼ cup semi-dry white wine, such as a Riesling Kabinett or Chenin Blanc

2 tablespoons white wine vinegar or champagne vinegar

2 pinches Morton Coarse Kosher Salt

1 **PREP.** Mince the chile and set aside
1 teaspoon for this recipe. Peel and mince the
shallot. Mince the tarragon. Scrub and zest the
orange. Peel and eat the orange.

2 **COOK.** In a saucepan, melt the butter. Add
the wine, vinegar, shallot, tarragon, chile,
zest, and salt and cook over low heat for about
10 minutes to extract all the flavors.

TO USE. Put the bivalves on the grill and spoon
some mignonette on top as in Smoke Catcher
Clambake (page 321).

DUXELLES AND MUSHROOM CREAM SAUCE

The problem with chicken breasts, turkey
breasts, and most pork chops is that they
have so little fat that they get dry easily, even if
you don't overcook them—and they are bland.
So they really benefit from a sauce. This sauce
works on all of them and even veggies like
asparagus.

I found this while I was on a "research"
trip to France, and by including it here I think
the trip is tax deductible, right, Phil? My wife
and I were drawn to a tiny bistro on a narrow
side street in the Latin Quarter near the
Luxembourg Gardens, named Les Racines.
Alas, I think it closed during COVID. The menu
in the window caught my eye because it was so
old-fashioned French, the kind you rarely find
in Paris where *nouvelle* cuisine is already *vielle*.
It was not on any of the lists of top eateries,
and the internet had several complaints
of rude service lodged by tourists who
don't understand that small restaurants in
France only seat people for full meals. Don't go
there asking for just soup, salad, appetizer, or
dessert. You will be turned away brusquely.
Their precious few tables are for people
having a full meal, and that's just the way it
is. If you want something light, go to a café or
a brasserie. This is not anti-Americanism—it is
the way some French restaurants work.

Judging by the meal my wife and I ate at Les
Racines, sauces were their specialty. She had
veal kidneys in an incredible rich brown sauce,
and I had a chicken breast in a silky cream and
morel mushroom sauce that was over the top.
My attempt to replicate it comes close, and no,
it bears no resemblance to normal cream of
mushroom things typical in the US. And if you

work for the IRS, try this recipe and then just give me the deduction.

Duxelles is a French term for finely chopped mushrooms and other goodies. They are most famously used on top of beef tenderloin wrapped in puff pastry to make Beef Wellington. But duxelles are also used as a stuffing for chicken breasts or pork loins, as a spread on grilled bread, and in sauces. I also use duxelles in this book for a kick-ass mushroom sauce used in Kōji Filet Mignon (page 225) as well as Paris Chicken (page 271). You can use this sauce on pork chops, turkey, even baked potatoes. If you want, make a batch and freeze it.

MAKES *About ¾ cup duxelles or 1½ cups sauce, enough for 4 chicken breasts or pork chops*
TAKES *45 minutes to make the duxelles plus 15 minutes to make the sauce*

DUXELLES

3 ounces shallots (2 to 3 medium shallots)

8 ounces fresh mushrooms

4 tablespoons (2 ounces) unsalted butter

½ teaspoon Morton Coarse Kosher Salt

½ teaspoon coarse-grind black pepper

2 garlic cloves

1 teaspoon fresh thyme, or ½ teaspoon dried

MUSHROOM CREAM SAUCE

¾ cup Duxelles (the recipe above)

1 tablespoon white wine vinegar

½ cup plus 2 tablespoons cream sherry

¾ cup low-sodium chicken stock or broth

½ cup half-and-half

2 tablespoons cream cheese

Morton Coarse Kosher Salt

ABOUT THE MUSHROOMS. This is time to splurge. You can use plain old button mushrooms if you wish, or you can put on your tux and go for morels, porcinis, chanterelles, or another fancy 'shroom. See more about mushrooms on page 147.

ABOUT THE SHERRY. This recipe calls for a cream sherry, which is sweet. You can use a drier sherry like an Amontillado, but I like the hint of sweetness that the cream sherry brings to the party. Sweet Marsala will also work great. Splurge here, too.

1 PREP. Peel and mince the shallots. Clean the mushrooms. Set aside 4 whole mushrooms and mince the rest.

2 COOK THE MUSHROOMS. Place a pan on the direct heat side of a grill or on a burner. Add the butter and as soon as it starts to melt, add the shallots, all the mushrooms, salt, and pepper. Cook, stirring often, until some of the pieces begin to brown, about 10 minutes. You want to drive off most of the water to concentrate the mushroom flavor.

3 FINISH THE DUXELLES. Peel and mince, mash, or press the garlic and add it along with the thyme and cook for a minute or two until the garlic is translucent. Taste and adjust the S&P to your taste. You now have duxelles if you want to make Wellington or stuff a chicken breast. If you wish, go on and make a sauce with them.

4 MAKE THE MUSHROOM CREAM SAUCE. Add the vinegar, sherry, and stock to the pan with the duxelles and boil it until about half has evaporated. Stir in the half-and-half and cream cheese in that order. Taste and adjust the salt to your taste. You can serve it or refrigerate it or even freeze it.

POBLANO-BASIL CREAM SAUCE

This versatile refined sauce can be used on many meats and I especially like it on fish, potatoes, and grilled veggies. There is one thing to know: While poblanos have practically no heat, occasionally one can surprise you, so taste it as you go. The recipe below produces a sauce with timid heat that is mitigated by the cream and the food you put it on—unless you get a rogue chile. If you want more heat, add a green jalapeño. You can see this sauce in action in Close Proximity Smoked Fish (page 308).

MAKES *About 1 cup*
TAKES *35 minutes*

SPECIAL TOOLS *Grill topper, blender or food processor, fine-mesh sieve (optional), dried herbs, wood chips, sawdust, or wood pellets*

> 1 large poblano pepper (about 3 ounces)
> 2 garlic cloves
> 1 large shallot or small onion (size of a golf ball)
> 1 teaspoon high-quality olive oil
> ¼ cup sour cream
> 6 large fresh basil leaves
> ½ cup half-and-half
> 2 tablespoons low-sodium chicken broth
> ¼ teaspoon ground cumin
> ¼ teaspoon Morton Coarse Kosher Salt
> ¼ teaspoon fine-grind black pepper

1 PREP. Slice the poblano in half lengthwise and remove the stems and seeds. Taste it to make sure it is not very hot. Peel the garlic cloves. Peel and chop the shallot into chunks about the same size as the garlic cloves.

2 FIRE UP. Set up the grill for 2-zone cooking and aim for Warp 10 on the direct heat side.

Place the grill topper over the direct heat and preheat it. Toss some dried herbs or wood chips or sawdust or wood pellets on the flame.

3 GRILL THE VEG. Lightly coat the shallots and garlic cloves with the oil and when the smoke starts to roll, spread them out on the grill topper. Grill, lid down, tossing every 1 to 2 minutes until softened and slightly golden, about 5 minutes. Remove them and the topper. Grill the poblano lid down, skin-side down, until blackened in spots but not thoroughly charred. Flip over and grill for 1 to 2 minutes more until tender.

4 PEEL THE POBLANO. Toss the poblano halves into a bowl and cover it with a plate. When they are cool enough to handle, scrape the skins off the peppers with a serrated knife and cut the poblano into ½-inch chunks.

5 SIMMER. In a blender or food processor, combine the garlic, shallot, poblano, sour cream, and basil leaves but don't blend yet. In a small saucepan, combine the half-and-half, broth, cumin, salt, and black pepper, bring to a high simmer, and cook to reduce by about one-quarter, stirring occasionally.

6 BLEND. Carefully pour the hot mixture into the blender/processor and whup everything up until completely smooth. This takes a while, but a bit of patience results in a smoother and more refined sauce. If you wish, push it through a fine-mesh sieve.

7 TO SERVE. When the meal is ready you can gently heat the sauce in a pan or microwave.

MELTY CHEESES

The meltiness of a cheese depends on how much water is in the cheese, how well the emulsifiers do their job of holding the fat and water together, and the age of the cheese. Young cheeses melt more readily because there is more water in them. If you want a melty cheese for grilled cheese sandwiches, cheeseburgers, or quesadillas, try these: American, Asiago, Brie, Camembert, Emmenthaler, Fontina, Gouda, Gruyère, Havarti, Jack, Jarlsberg, Manchego, mozzarella, Muenster, provolone, raclette, Reblochon, Swiss, Taleggio, and young cheddar. Some cheeses, like Gouda, melt well one day and other times, not so much. This is often a by-product of how they are stored. Cheeses coated in wax tend to have problems melting. It's not the wax, it's the nature of the cheese.

And yes, you can grill grilled cheese sandwiches on a grill. Doh! Add apple or pear slices and thin slices of ham. Use medium heat and put a upside down pan over them to capture the heat from above.

GRILLING CHEESES

Yes, you can actually grill some dry cheeses. Halloumi, paneer, and queso seco can be grilled over medium heat and retain their integrity. Serve them with a drizzle of honey and some fresh herbs. The folks at Milk Street have a recipe for fried Halloumi and peach sandwiches with honey.

Here's a fun one. Parmigiano-Reggiano wheels are not coated in wax. The rinds are hard because they are dehydrated. After you grate a wedge, toss the hard rind on your grill over direct heat and it will soften like taffy in a few minutes. Plop it on some garlic bread for a treat.

SMOKED CHEESES

You can smoke almost any cheese in a smoker. Start with cold (but not frozen) cheese, keeping the cooker's temperature under 90°F, and put the cheese in a pan that will contain it if it melts. Sit the pan with the cheese on top of another pan with ice. Or use a smoke generator such as a smoking gun and cover the pan with foil or plastic wrap. I've even done it in a grill with a disposable pan filled with sawdust. Wait until there is an amber hue to the cheese, as long as an hour or two. Try it with mozzarella, Gouda, cheddar, and even cream cheese.

I guarantee you'll love smoked cream cheese. Take a brick and score it with crosshatches every inch or so, sprinkle it with your favorite rub, place it in a pan, and smoke it at about 80°F until golden. Spread it on crackers, bagel chips, or garlic bread, or fill celery sticks with it.

OTHER FUN SAUCES AND COATINGS

"I grill, therefore I am."

—ALTON BROWN

BALSAMIC SYRUP

You can amp up cheap grocery-store balsamic by simmering it until it is reduced by half and about the thickness of maple syrup. This is by no means a substitute for high-end aceto balsamico tradizionale or condimento balsamico, but it does concentrate the flavor, richness, and sweetness, and it makes a lovely syrup that is sweet and tart, suitable for drizzling. Try it on the Warm Fruit Salad (page 376), Torched Figs (page 379), fresh strawberries, grilled peaches, and tomato salads like a caprese salad.

Just be careful and keep an eye on the boiling. There's sugar in balsamic and it can burn. In fact, some cheap balsamics have so much sugar they can turn to hard candy when reduced (which is kinda fun).

ORANGE TUSCAN-STYLE BOARD SAUCE

Adam Perry Lang is a classically trained chef and a fellow member of the Barbecue Hall of Fame. He has worked in such hallowed kitchens as Le Cirque and Daniel in New York City, as well as Restaurant Guy Savoy in France.

In 2012 he published an excellent book, *Charred & Scruffed*. In it he describes a technique I have fallen in love with: "Board dressings" or "board sauces." This is a clever idea that works superbly on beef, lamb, chicken, shrimp, lobster, and who knows what else. It is a good substitute for chimichurri, and I have even put it on sandwiches like Clint's Chuck Roast Sandwiches (page 226). Nothing salvages a lean or overcooked steak like a board sauce.

Here's how the concept works. Take a handful of fresh herbs and toss them on a cutting board. Then pour some olive oil on the herbs, mince them together, lay hot grilled meat on the mixture, carve the meat, and roll the meat around in the "board sauce," enriching the sauce with the meat juices and adding fat to the meat. The board sauce keeps the meat juicy and brings intriguing flavors to the cut surfaces of the meat. Shockingly, fresh herbs do not mask the meat.

Best of all, there is no set recipe and you don't have to measure; but if you are nervous, I'll share some guidelines and then a recipe.

The core is *fresh* herbs, then garlic, chiles, shallots and/or scallions, and oil. If you want, mix in some balsamic vinegar, citrus juices, and zest. If you wish, you can add salt and pepper, but be careful that you don't overdose if salt is already on the meat.

I often chop the herbs while the meat is cooking and put them in a bowl. Then I pour in enough high-quality olive oil to cover them so the oil has a chance to extract more flavor from the herbs. You cannot make it up hours

in advance because the anaerobic (oxygen-free) environment in the oil is friendly to the botulinum microbe. Even in the fridge. If you make it in a cup just before cooking, you can spoon some sauce on each guest's plate, sit a steak or chop on top, then slice the meat and roll it around in the sauce. We use this in Double-Wide Lamb Rib Chops (page 238). It also works well on seafood. Here's a recipe, but feel free to get creative.

MAKES *About 1½ cups, enough for 4 large steaks*
TAKES *10 minutes*

- 1 large orange
- 1 cup brine-cured French or Italian green olives, such as picholine or Cerignola
- 2 garlic cloves
- 2 large shallots or small red onions (size of golf balls)
- 1 tablespoon capers
- 2 anchovy fillets
- 1 small red bell pepper
- ⅓ cup fresh flat-leaf parsley leaves
- ⅓ cup high-quality olive oil
- 2 teaspoons Balsamic Syrup (page 198)
- ¼ teaspoon large-grain finishing salt, such as Maldon
- Medium-grind black pepper

Scrub the orange well and scrape off all the zest. Pit and mince the green olives until you have ½ cup. Peel and mince or press the garlic. Peel and mince the shallots. Drain and mince the capers and anchovies. Mince the red pepper and scoop 3 heaping tablespoons into a medium bowl. Add all the other ingredients to the bowl and mix. Pour onto the cutting board just before you place the meat on it. Slide the meat on top and roll it around to coat.

THAI SWEET CHILE SAUCE (NAM JIM KAI)

This tangy hot, sweet, sauce is common in Thai restaurants and I put it on everything there except the tea. So I had to reverse engineer it. Came out darn good! Then I met Leela Punyaratabandhu online, bought her excellent book *Simple Thai Food,* and there it is on page 187! Nailed it!

Leela says *nam jim kai* means "dipping sauce for chicken." That may have been the original plan, but you have our permission to use it anywhere you feel the need for sweet heat. I read once that the singer Trisha Yearwood puts it on waffles with a sunny-side egg on top! I think it really sings on fried foods. Sometimes I make a coleslaw and then sprinkle it on and pile the slaw on pulled pork: OMG.

Restaurants use red Thai bird's eye chiles, which are hard to find and even harder to eat, right up there with habaneros and much hotter than jalapeños. For this recipe I have substituted red jalapeños, which are more in line with my capacity, and easier to find. I have even made it with green jalapeños. Splash this on all fried chicken and onion rings.

MAKES *About 2 cups*

TAKES *30 minutes*

SPECIAL TOOLS *Food processor or blender*

6 ounces fresh red jalapeños (3 or 4)

3 garlic cloves

¾ cup sugar, plus more to taste

½ cup unseasoned rice vinegar, plus more to taste

2 teaspoons cornstarch

OPTIONAL. Red pepper flakes

ABOUT THE PEPPERS. The heat of the final sauce depends on how much of the chiles' ribs you leave in. This sauce is not meant to be 3-alarm, but you can add more chiles if you wish or you can use serranos, or cayennes, or other hotter peppers. Just beware that in a handful of peppers, some can be hotter than others and that they can also vary in heat depending on where they were grown or when they were grown. I made a batch with habaneros once and it was delicious, but hard for me to handle.

Simmering mellows the chile heat and after a day in the fridge it loses a bit of its bite. Also, putting it on food tempers it, especially fatty foods. After you make it, if you feel you want more heat, you can add red pepper flakes or dried ground chiles. Most store-bought pepper flakes are loaded with seeds, so I strongly recommend you remove them. Then give them a day to soften and do their stuff.

ABOUT THE VINEGAR. If you don't want to buy unseasoned rice vinegar, you can use distilled white vinegar or apple cider vinegar, but increase the water content by about 4 tablespoons since they are more acidic.

ABOUT THE CORNSTARCH. This slightly thickens the sauce but also keeps it from separating.

1 PREP. Wearing gloves, remove the stems and seeds from the peppers. Chop the peppers into large chunks. Peel and chop the garlic into quarters.

2 BLEND. In a food processor or blender, combine 1 cup water, the chiles, garlic, sugar, vinegar, and cornstarch and pulse it a little.

Leave it a bit chunky so there are some pretty flecks floating around.

3 SIMMER. Pour the blend into a saucepan, bring to a simmer, and cook for 5 minutes, then remove from the heat. The simmering kills any bacteria on the ingredients. Don't let it bubble more than a gentle simmer or the sugar can foam up, overflow, and burn on the sides of the pot.

4 AGE IT A DAY. Pour into a very clean jar. I think it is best after a day of aging. It will keep forever in the fridge although it may lose some of its kick and the natural pectins will make it thicker. Although the cornstarch keeps the chile bits floating magically, shake before using.

5 TASTE. After it has aged a day, taste and then, if you wish, you can add red pepper flakes, more sugar, or more vinegar to your taste.

ZHUG

This hot and garlicy green sauce from Yemen is kin to all the fantastic green sauces of the global village, like pesto, salsa verde, and chimichurri. Serve this on Lamb Kebabs Are Better Without Skewers (page 240) and Vigneron Method for Flank Steak Subs (page 231).

MAKES *About 1 cup*
TAKES *10 minutes*

SPECIAL TOOLS *Food processor or blender*

- 6 garlic cloves
- 2 large green jalapeños, plus more to taste
- 1 cup (about 1 ounce) fresh cilantro leaves and tender stems
- 1 cup (about 1 ounce) fresh flat-leaf parsley leaves
- 30 coriander seeds, or ¼ teaspoon ground coriander
- ½ teaspoon cumin seeds, or ¼ teaspoon ground cumin
- ¼ teaspoon cardamom seeds
- ½ cup high-quality olive oil
- 2 tablespoons plain yogurt
- 1 teaspoon Morton Coarse Kosher Salt
- 2 teaspoons fresh lemon juice

ABOUT THE CARDAMOM. The seeds are little black pellets smaller than peppercorns. They come housed in a cardboardy pod like an oversized grapefruit seed. We want the seeds, so if you have pods, crack them open with your fingernails or by putting them on the counter and pressing down on them with the flat part of a sturdy knife or even a frying pan.

1 PREP. Peel and mince or press the garlic. Don some gloves and stem, seed, and roughly chop the chiles. Roughly chop the cilantro and the parsley. In a mortar and pestle or spice grinder, powder the coriander, cumin, and cardamom seeds.

2 BLEND. In a food processor or blender, combine the chiles, chopped herbs, ground spices, olive oil, garlic, yogurt, and salt. Pulse until more or less smooth. A little chunkiness is OK. Taste and add more salt or jalapeño if you want.

3 CHILL. Refrigerate, covered, until serving time. Stir in the lemon juice just before serving.

TARE SAUCE

Yakitori is one of the most beloved foods in Japan with numerous restaurants and street vendors serving it. *Yaki* (grilled) *tori* (chicken) is usually coated with tare sauce. It is a thick, rich, sweet soy sauce–based elixir.

After work "salary men" often stop at yakitori joints for a snack and a drink. I first tasted yakitori when I held up two fingers at a baseball stadium in Tokyo as a vendor with a hot box came up the aisle. I thought I would get hot dogs, but I got skewers of grilled chicken livers glazed in this marvelous sauce.

In Japan, chefs grill skewers of meat over hot binchōtan charcoal just a few inches from the food on a long trough-type grill called a *konro*. Then they dip the skewers in the tare glaze and cook it for a minute or two more over medium heat, turning the skewers often. I do this on a cheap American portable charcoal grill with briquets at least two layers deep.

Tare sauce keeps for weeks in the fridge and it is terrific painted on Not Grannie's Meatloaf (page 296), Torched Beef Short Ribs (page 230), or grilled chicken, squid, salmon, turkey, pork, dumplings, and many veggies, especially onions. Even chicken livers. My last book had an interpretation of the recipe. Here is a different version.

MAKES *2 generous cups, enough for 2 small chickens, a whole mess of chicken livers, or 2 slabs of ribs*
TAKES *About 45 minutes*

SPECIAL TOOL *Fine-mesh sieve*

- 2 scallions
- 2 garlic cloves
- 1 ounce fresh ginger
- 1 cup reduced-sodium soy sauce
- ½ cup sake
- ½ cup low-sodium chicken stock
- 2 tablespoons dark brown sugar
- 1 tablespoon white miso
- 1 teaspoon toasted (brown) sesame oil
- ½ teaspoon of your favorite hot sauce

ABOUT THE LOW-SODIUM STUFF. I call for low-sodium chicken broth and reduced-sodium soy sauce. This is because we will be cooking the sauce down and concentrating it and that will make it too salty if you use full-salt ingredients.

1 PREP. Clean the scallions, peel off any dead skin, chop off the roots, and coarsely chop the whites. Save the green tops for use as a garnish. Peel, then mince or press the garlic. Grate the ginger on a Microplane or the small holes of a box grater and save the juices.

2 SIMMER. In a 2-quart nonreactive saucepan, combine the ginger (and juice), scallions, garlic, soy sauce, sake, stock, brown sugar, miso, sesame oil, and hot sauce and gently simmer over medium-low heat for about 30 minutes.

3 STRAIN. Pour the sauce through a fine-mesh sieve into another saucepan. By now, the chunky stuff has given its all and it's time to discard it like letters from ex-lovers. With a ladle or spoon, gently press the mush left in the sieve to release all those good juices.

4 THICKEN. Put the saucepan back over medium-low heat and simmer until it's almost as thick as motor oil, 10 to 15 minutes. Taste the sauce and adjust the sugar or hot sauce if you wish. Now paint it on everything except the walls.

BEEF, BISON, LAMB

"I take a vitamin every day—it's called a steak."
—COACH JIM HARBAUGH

Beef and bison are genetically related and their meats are similar. They have slightly different flavors because bison is usually leaner and tougher, but recipes for beef work just fine on bison with little modification.

The muscles that have the most marbling—the most flecks of fat mixed in with the muscle fibers—like ribeye, are the juiciest and most flavorful. The muscles that don't work a lot, like tenderloin, are the most tender. Often the harder-working muscles, like flank, are more flavorful, but they are tougher and less juicy. You can combat the low fat and chewiness by proper cooking, slicing, or adding a sauce or a compound butter.

Most calves suckle for 6 to 10 months, then they eat grass and hay until they are 14 to 16 months old. Then they are sold to feed lots where they stay for about 4 months and are fed grain. The grain adds weight and

flavorful marbling. At 20 to 24 months and about 1,350 pounds, the animals are killed, butchered, and shipped.

Important Beef Sections

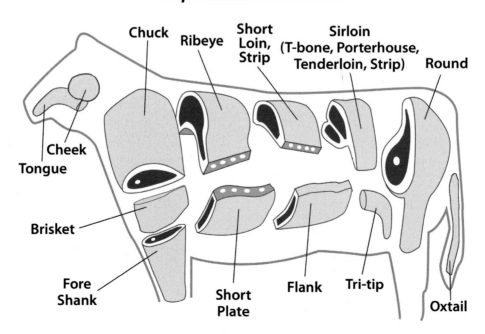

There are more than fifty standard cuts of beef fabricated in the US, and some go by multiple confusing names. Some have been confused for so long that nobody can say for sure what exactly a Delmonico or London Broil really is. Names and shapes in other countries can be different.

BEEF GRADES

USDA inspectors grade beef primarily on the age of the animal and the amount of marbling as measured between the twelfth and thirteenth ribs. Here are some boneless ribeye steaks that show the marbling differences.

Harold McGee, the eminent food scientist, has written that much more than marbling influences quality. "Despite the prestige of Prime beef, the current consensus among meat scientists is that fat marbling accounts for no more than a third of the variation in the overall tenderness, juiciness, and flavor of cooked beef. The other important factors include breed, exercise and feed, animal age, conditions

USDA Choice

USDA Prime

American Wagyu

Wagyu From Japan

The bones flavor the meat

BUSTED. The exterior of the bone is calcium, which is flavorless. There is lots of flavor within the marrow, but it can't get out when you cook with dry heat in an oven or grill or smoker. You can get it out by submerging the meat in liquid as in a stew or braise. Worse, bones block heat penetration on a grill or in an oven, so the meat next to bone is usually less cooked than meat far away from the bone. So bone-in ribeyes, for example, are often not evenly cooked.

during slaughter, extent of post-slaughter aging, and storage conditions before sale."

USDA UTILITY, CUTTER, CANNER, STANDARD, COMMERCIAL, SELECT. These lowest grades of beef are best for soups, stews, Hamburger Helper, chili, sloppy joes, etc. You will not likely see them sold as steaks in a grocery. *About 2% to 4% fat.*

USDA CHOICE. About half of all beef is marked USDA Choice and it is a good option. Walmart "Choice Premium" is equivalent to USDA Choice. *About 4% to 7% fat.*

USDA TOP CHOICE. Top Choice is USDA Choice with a bit more marbling than normal, just below USDA Prime. *About 7% to 9% fat.*

USDA PRIME. My all-round favorite. Often from younger cattle. Prime is almost always better tasting and more tender than Choice. Only about 3% of all beef is Prime and most is reserved for the restaurant trade. It can be pricey, but the good news is that there is more of it than ever before. *Normally about 10% to 13% fat; a dry aged Prime steak can be 15% to 18% fat.*

ANGUS. Angus cattle are an especially flavorful breed. Alas, it is almost impossible to know if what you are buying really is Angus since nobody does genetic testing and few herds have documented their heritage. Farmers and USDA just look for black hides, a strong indicator that there are Angus genes in the animal, but not a guarantee. Most black-hided animals are Angus, but not all. The Certified Angus Beef (CAB) label is a trademarked brand that is reliably above average, because the carcass must be USDA Choice or USDA Prime and pass other quality-control standards. But they are "certified" by hide color, not genetics. CAB costs a bit more because the association charges farmers a fee to "certify" the cattle and markups take place on down the chain of distribution.

AMERICAN WAGYU BEEF. Do not die without having tasted true American wagyu steaks. These cattle are said to have Japanese bloodlines, but are often crossed with Angus (sometimes called Wangus) and there is no proof that they are actually of Japanese descent. True wagyu meat is expensive because it is remarkable, shot through with thin whisps of buttery marbling. I am addicted to wagyu flank steak. American wagyu is more marbled than USDA Prime but not as much as Japanese wagyu. It can also be about twice the price of USDA Prime. *Ribeye steaks can run up to 30% fat.*

KOBE, TAJIMA, and other **JAPANESE WAGYU.** Japan is famous for highly marbled fabulously expensive beef, most notably Kobe beef, a trademarked name. The Kobe brand oversees producers, slaughterhouses, distributors, retailers, and restaurants who use the name. The steer must be born, raised, and slaughtered in Hyogo prefecture (a prefecture is like a state), and fed a specified diet. There are only about four thousand cattle slaughtered each year for Kobe beef, a pittance, and 90% of all Kobe is consumed in Japan. They have a high ratio of fat to muscle and are revered for flavor, tenderness, and most of all, the richness that comes from the chemistry of the fat, which practically melts in your mouth. In Japan, beef is graded A1 to A5, with A5 having the most marbling. A 12-ounce ribeye can cost $300 in a US restaurant. You will not likely find it in a store.

Despite what you may have read, they are not fed beer. Nor are they routinely massaged for tenderness, although a few may be hugged by loving owners who are paying for their childrens' college.

There are several other bloodlines of extraordinary-tasting rare Japanese beef such as Tajima that are similar to Kobe. I have been able to get Tajima online and it is mind-blowing. Beware of fraudsters.

OTHER BEEF TERMS

GRASS-FED BEEF. A misleading term because all beef are fed grass and hay most of their lives.

GRAIN- or **CORN-FED BEEF.** Many cattle spend their last few months being fed corn and other grains to add weight and marbling. They should be called grain or corn *finished*.

GRASS-FINISHED BEEF. Cattle that are fed grass or hay their entire lives. More and more cattle are being finished on grass rather than on grain. Grass-finished beef is popular because people believe it to be more humane and natural, but in taste tests, grain-finished usually wins because it tends to be tastier and more tender. Grass-finished also tends to be more expensive because they grow more slowly, so they need more food and care. Grass-finished animals are killed at about 150 pounds lighter, so there is less meat per animal and it takes longer for them to reach slaughter age, so they make more methane, CO_2, and manure.

ORGANIC BEEF. USDA rules state that certified organic beef must be produced according to strict rules and have been verified with an elaborate paper trail on every animal, including its breed, feed, and medical history. To be certified organic, it must eat only organic grasses and grains, have unrestricted outdoor access, must not be given antibiotics or hormones, and must be treated humanely. It is hard to verify all of this. Organic beef is more expensive.

NATURAL BEEF. Natural beef must not be given antibiotics or hormones, but they can be grown, fed, and handled in the same way as other common cattle.

KOSHER BEEF and **HALAL BEEF.** These cattle are grown and slaughtered according to Jewish law (kosher) or Muslim law (halal).

Their requirements are similar. Both require that the animal be slaughtered by slitting the animal's neck veins and drained of practically all blood. Some experts believe this method is painful and inhumane.

BLADE-TENDERIZED BEEF. Beware. Some meat suppliers try to tenderize beef by using a device called a blade tenderizer or jaccard. It is a series of thin sharp blades or needles that stab the meat and cut through tough fibers and connective tissues. Pathogens are common on the surface of meat, but they are killed almost instantly when cooked. However, if the meat has been blade-tenderized, these pathogens can be pushed down into the center of the meat, which often is not heated enough to kill them. That makes the process risky. USDA requires raw meats tenderized this way to be labeled "Mechanically Tenderized," or "Needle Tenderized," or "Blade Tenderized." Use them only for stews.

Buy only rosy colored beef

BUSTED. If that beef in the store is looking rosy and delicious, that's because it's fresh, right? Well, maybe not. It seems that packers have perfected a process for sealing meat in an air-tight package with a carbon monoxide atmosphere, and that keeps the meat from oxidizing and turning brown even if it is stored improperly for a long time. Check the sell-by date. Of course, health inspectors can tell you tales about unscrupulous butchers changing meat labels to extend the sell-by date. The solution? Get to know your butcher and stay on your toes.

AGING BEEF

After cattle are slaughtered, chemical changes called rigor mortis makes the meat tough. The carcass must be chilled rapidly but not frozen, and it takes a couple of days for enzymes to relax the muscles enough to be sold. After that it can be further aged to increase tenderness. Aging must be done under carefully controlled temperature and humidity constraints. Enzymes and oxygen begin to work on the meat during the aging process, but too much age can spoil the meat.

WET-AGING. Most meat is shipped from slaughterhouses in large wholesale cuts packed in plastic vacuum bags in boxes. If kept this way for about 30 days, enzymes tenderize the meat, but the flavor is not changed as much as it is in dry-aging.

DRY-AGING. If you haven't tried it, you need to taste dry-aged beef. Start saving now: It ain't cheap. You can occasionally buy dry-aged beef from specialty butchers, or you can do it my favorite way: Get some friends and head to a restaurant that specializes in dry-aged beef and do an expensive comparison tasting with several ages.

Dry-aging is an expensive process for tenderizing beef and concentrating its flavor. Dry-aged beef is noticeably different tasting than fresh beef because the chemistry of the fat changes drastically. Some describe it as earthy, nutty, gamey, leathery, or even mushroomy. Some people are addicted, some just plain don't like it.

Large hunks of meat, usually the best cuts, such as the rib primal, are held in a room at 34° to 38°F and 70% to 80% humidity, with brisk airflow, typically for 30 to 75 days. Dry-aging is sometimes called controlled rotting because the exterior of the muscle gets dark, and mold sometimes grows on the outside of the meat. The interior remains rosy.

Dry-aged bone-in rib primals at Allen Brothers Meats in Chicago.

During aging, natural enzymes break down connective tissue and tenderize the meat while moisture evaporates, shrinking the meat as much as 20%. The outside crust is trimmed off before the meat is sliced into steaks and cooked, so another 15% is lost.

You can dry-age large cuts of beef at home, such as rib primals. Attempting to dry-age individual steaks doesn't work well. The simplest and best way is to buy one of the commercial dry-aging boxes on the market by SteakAger or Steak Locker. They come in a variety of sizes and can talk to your smartphone. If you are a do-it-yourselfer, you can adapt a spare fridge for the job.

Another option that I recommend is the UMAi Dry system. You simply wash the meat and place it in a vacuum bag made from a semipermeable membrane that allows for both oxygen and moisture exchange. Place the bagged meat in your refrigerator and come back in 28 or more days. It produces high-quality meat and it is fairly inexpensive.

CHAMPIONSHIP BRISKET
AND BURNT ENDS

Anyone who has traipsed the Texas barbecue trails knows beef brisket is king. Brisket is one of the toughest cuts on the animal and mastering it is climbing the Mt. Everest of barbecue. But when you nail it, the sun shines on you wherever you walk.

Brisket is two overlapping pectoral muscles that bear more than half the weight of the animal, because cattle lack collar bones. Traditional Texas brisket is USDA Choice, trimmed, seasoned with salt and pepper, and smoked for 12 hours or more until tender. No sauce. Brisket needs sauce like cattle need wolves.

Brisket turn-in box by Chris Grove.

My last book and website teach you how to make traditional Texas brisket. Now let's cook Championship Brisket. I turned to several top competitors for tips in creating this recipe, especially Clint Cantwell, president of AmazingRibs.com and winner of the Travel Channel's "American Grilled" nationwide cooking competition series and named one of the "10 Faces of Memphis Barbecue" by *Memphis Magazine*. Also Travis Clark of Clark Crew BBQ restaurant in Oklahoma City. He is one of the hottest cooks on the circuit, winner of the massive American Royal Barbecue Invitational, a World Championship title in brisket, and his team has been named Team of the Year in brisket. I have also looked over the shoulder of dozens of other top pitmasters.

I cannot stress this enough: When shopping for brisket, go for the highest grade you can afford, and handpick the slab with the most marbling. Pay no attention to the fat cap. You will trim most of that away. Please don't write to me and say you can't figure out why your brisket was tough if you did not buy USDA Choice or better. Competition BBQ teams often cook expensive wagyu briskets because the extra internal fat makes for more flavorful, tender, and juicy bites.

Anatomy of a Packer's Cut Brisket

Point muscle (*pectoralis superficialis*)

Fat Layer

Fat Cap

Flat muscle (*pectoralis profundi*)

Clark starts at the top, with an 18- to 20-pound Snake River Farms American Wagyu brisket that sells for about $250. "Once you try one of these it'll be impossible to go back to anything else."

That's a lot of meat. After trimming and shrinkage you still have 13 to 15 pounds, enough for at least 26 servings of 8 ounces each. You can buy a whole "packer" brisket and cut it into partial portions, but here is the way it is cooked in competition.

Traditionally flat and point (the two parts of the brisket) have been cooked together, but it is easier to get them cooked properly if you separate them. That's because, when together, one side is a lot thicker than the other, so it cooks unevenly. If you separate them, the flat is pretty uniform in thickness. It makes a beautiful matched card deck of slices, perfect for sandwiches. When you separate point and flat you can get bark and smoke ring on all sides of both muscles making for a prettier presentation and more bark flavor.

The point muscle, sometimes called the deckle, has more marbling, is juicier and more tender, and is usually designated for scrumptious cubes called "burnt ends." Originally burnt ends were simply edges and ends that were overcooked, perhaps burnt, and munched on by the kitchen staff. Today there's nothing burnt about burnt ends.

In 1970, in his marvelous book *American Fried*, humorist Calvin Trillin wrote the following about Arthur Bryant's restaurant in Kansas City: "The main course at Bryant's, as far as I'm concerned, is something that is given away for free—the burned edges of the brisket. The counterman just pushes them over to the side as he slices the beef, and anyone who wants them helps himself.

I dream of those burned edges. Sometimes, when I'm in some awful overpriced restaurant in some strange town—all of my restaurant-finding techniques having failed, so that I'm left to choke down something that costs seven dollars and tastes like a medium-rare sponge—a blank look comes over my face: I have just realized that at that very moment someone in Kansas City is being given those burned edges free."

Many champions use commercial rubs, injections, and sauces to get the flavor profiles they want out of brisket. Clark uses four rubs, for example! I have made substitutions for them with recipes you can make yourself.

MAKES *13 to 15 pounds of meat, enough for 26 to 30 servings*

TAKES *45 minutes to prep, 8 to 12 hours to cook, and 1 to 2 hours to hold*

SPECIAL TOOLS *Meat injector, large aluminum pans, fine-mesh sieve*

MEAT

1 whole brisket (18 to 20 pounds)

INJECTION

2½ cups low-sodium beef broth

1 tablespoon Morton Coarse Kosher Salt

1 teaspoon ground white pepper

RUB, MOP, AND SAUCE

1½ cups Red Meat Rub (page 166)

1 cup low-sodium beef broth

6 tablespoons apple cider vinegar

1 cup Kansas City Red (page 180)

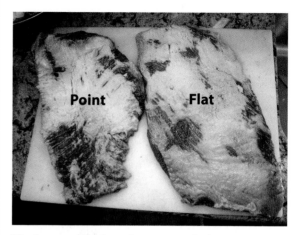

The two muscles have been separated.

1 TRIM THE MEAT. Separate the point and flat from each other. It's easy to follow the fat layer between them with your knife and once you get a cut started you can often just tear them apart with your hands. Remove as much of the surface fat as you can and all the silverskin, because you want your rub on the muscle, not on the fat. As I have explained, fat cannot enter the water-laden meat. Besides, no one wants a big mouthful of fat. And if the fat is too thick, you will not get a good smoke ring. Notice which way the grain is running and cut a chunk from a corner across the grain so that when the meat is finished you will be able to know how to slice since by then you will no longer be able to see the grain.

2 MAKE THE INJECTION. Clark uses Kosmos Q Reserve Brisket Injection, but for a simple injection, in a bowl, stir together the beef

How Are Competitions Scored?

The Kansas City Barbeque Society (KCBS) runs the most competitions in the world. Entries are judged on appearance, taste, and tenderness with taste weighted twice as much as tenderness and about 4 times appearance.

broth, salt, and white pepper until the salt dissolves completely. Place the flat and point in two pans. Inject as much liquid as possible using the method described in Injecting (page 32). Let the brisket rest for 30 minutes.

3 APPLY THE RUB TO THE MEAT. Season the meat with a generous coat of the Red Meat Rub.

4 FIRE UP. Prepare a smoker or a grill for indirect smoking and shoot for about 400°F. Remove the meat from the pans and place them on the smoker or on the indirect side of the grill for 30 minutes. Clark and several other cooks like to start hot. Clark says, "I find this does something really special to a wagyu brisket." This high heat gives bark formation a good running start and it shrinks the fibers on the surface but not within so the meat gets plump.

5 SMOKE. Reduce the temperature to 275°F and allow the brisket to cook until it

has a nice mahogany color and the beginning of a nice bark, about 160° to 170°F internal temperature. Let color be your primary guide.

6 MEANWHILE, MAKE THE MOP. Stir together the beef broth and cider.

7 CRUTCH. Wrap the flat in a double layer of aluminum foil along with ¾ cup of the mop and crimp it tight so no steam will escape. Wrap the point in a double layer of foil along with ½ cup of the mop. Return the meats to the smoker or indirect side of the grill and continue cooking until they are jiggly tender (called "wubba-wubba" tender), about 203° to 206°F for the flat and 210°F for the point. Let tenderness be your guide. This can take 8 to 12 hours depending on thickness of the meat, oven temp, and other variables.

8 HOLD. Place the meat, still in foil, in an insulated holding box, such as a beer cooler, for at least 1 hour. The temperature will drop slowly to 145° to 160°F.

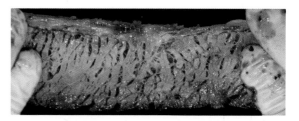

9 BURNT ENDS. If you are competing, prepare the turn-in box by lining it with whatever greens are allowed. Remove the point from the foil and save the liquid. With a very sharp knife cut the point into ¾-inch to 1-inch cubes. Pour the Kansas City Red into an aluminum pan big enough to hold all the cubes in one layer with some space between them, add the cubes, and stir until they are covered in sauce and pour off the excess. Place the pan back on the smoker or the indirect side of the grill for 10 minutes to set the sauce. Some cooks like to go longer or even roll them around on a hot grill for a few minutes to crisp the edges.

10 SLICES. Remove the flat from the foil and add the liquid in the foil to the liquid from the foil the point cooked in. Slice the flat across the grain. Many teams use an electric knife for this job. The slices should be about the thickness of a pencil, about ¼ inch, and when tugged should be elastic and separate slightly with the grain, but not tear easily. Pour all the liquid from the foil through a fine-mesh sieve to remove any solids. Taste it to make sure it is not too salty. If you wish, you can add a wee bit more Kansas City Red but you don't want to make your slices sweet. Brush both sides of each slice with the liquid from the foil. Line up a minimum of 6 identical slices in the turn-in box, one for each judge, and add at least 6 burnt ends. Turn-in time is 1:30 sharp.

LEFTOVERS. There are always leftovers. The next day I lay the slices out on a platter in a single layer, drip some jus or plain water on both sides, and microwave. Burnt ends only need to be nuked. Better still, make Pho with Leftover Brisket and Smoked Bone Broth (next recipe).

PHO WITH LEFTOVER BRISKET AND SMOKED BONE BROTH

Pho (pronounced *fuh*) is a rich, complex Vietnamese broth usually loaded with meat and veggies. This one contains leftover brisket and bone broth from smoked bones. It provides an awesome payoff in an exotic sweetness.

MAKES *6 servings*

TAKES *About 20 minutes to prep, about 5 hours to make the broth*

SMOKED BONE BROTH

- 4 carrots
- 3 medium onions
- 4 garlic cloves
- 4 cremini or button mushrooms
- 2 celery stalks with leafy tops
- 5 pounds beef femur bones split lengthwise
- ¾ cup dry white wine
- 4 whole star anise pods
- 2 teaspoons Morton Coarse Kosher Salt
- 2 teaspoons fish sauce

PHO

- 1½ pounds leftover tender beef brisket
- 12 ounces cremini or button mushrooms
- 4 red radishes
- 6 ounces medium rice stick noodles
- 6 ounces greens, such as collards, kale, mustard, or spinach, or a blend
- Sriracha sauce
- **OPTIONAL.** Thinly sliced scallion greens and small cilantro leaves, for garnish

ABOUT THE MARROW BONES. Marrow bones are beef leg bones, or femurs. They are widely available but are often cut crosswise into 2- to 3-inch lengths. Ask your butcher for 6-inch bones cut in half lengthwise to expose far more of that flavorful marrow so it can be distributed throughout your tasty broth.

ABOUT THE MUSHROOMS. I recommend plain cremini or button mushrooms, but if you want to take it up a notch, go for chanterelles or something more exotic.

ABOUT THE NOODLES. Rice stick noodles vary in size and method of preparation, so consult the package instructions for soaking method. If you can't find them, you can use ramen, glass noodles, vermicelli, or even soba.

1 PREP. Peel the carrots and cut them into 3-inch chunks. Peel the onion and cut into quarters. Peel and mince or press the garlic. Rinse and cut off the bottom of the mushroom stems. Rough chop the rest of the mushrooms as well as the celery.

2 FIRE UP. Set up your grill or smoker for smoking and aim for 225°F and get some smoke rolling.

3 SMOKE. Smoke the bone pieces marrow-side up for an hour or two.

4 SIMMER. Transfer the bones, veggies, wine, and star anise to a large stockpot and add 1½ gallons of water or whatever is needed to cover everything. Bring to a boil, then immediately reduce the heat to a gentle simmer. If there is scum, scoop it off. Partly cover the pot and simmer until the liquid has reduced by about half, 4 to 5 hours.

5 STRAIN AND SEASON. Cool the broth slightly then pour through a large sieve or colander into a bowl or another pot. Discard the solids. Let the broth settle for 5 minutes, then using a fat separator or a large flat spoon, skim off and discard most of the fat and any scum (check out the OXO fat separator that is a large measuring cup with a hole in the bottom). Stir in the salt and fish sauce and taste. You can use it now or freeze it.

6 START THE PHO. Cut the leftover brisket into thick slices. Rinse the mushrooms clean, trim the ends, and slice thinly. Scrub and thinly slice the radishes.

7 BRISK IT. Warm the brisket in the microwave or wrap it in foil and warm in a low oven.

8 NOODLING. Soften the rice noodles in water, according to the package directions. Bring the broth to a simmer and add the noodles to the broth. Cook until tender.

9 GO GREEN. Sturdier greens like collards, kale, or mustard greens will need a quick 2-minute blanch in boiling water, while more tender spinach leaves do not. Frozen greens are fine, too, but be sure to thaw completely and squeeze out the excess water.

10 SERVE. Divide the hot noodles, brisket, greens, and mushrooms among six bowls. Ladle the hot broth over all the ingredients. Let the soup stand for 5 minutes to wilt the greens. Scatter the radishes on top and add a few squirts of Sriracha sauce. If desired, garnish with scallion greens and cilantro leaves.

THE ULTIMATE PRIME RIB ROAST

This is what I serve for Christmas dinner. The method is perfect because prime rib takes time and dinner is over the river and through the woods.

Well before Christmas. I shoot for 8 ounces cooked meat per person after trim and shrinkage, plus extra to send home. So, for

14 people, about a month in advance I call and order an 18-pound, 7-bone, USDA Choice, 28-day wet-aged in Cryovac, bone-in prime rib, for about $30 a pound for pickup Christmas Eve. Seven bones is a full rib primal, the entire hunk of meat from the rib section. Because it is one of the large primal cuts, it is called prime rib, but please note, that does not mean it is USDA Prime. You can get USDA Prime for a primo price, but this is such a tender cut that USDA Choice is plenty good and a good deal cheaper. You can splurge for dry-aged, but not everybody likes that taste.

8:00 A.M. CHRISTMAS MORNING I get up early, pour a glass of eggnog, and start the butchering. Meat comes out of the fridge at 38°F internal. I run a boning knife or fillet knife along the top of the bones to remove them from the roast. I set them aside for another meal (see next recipe).

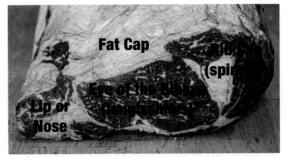

These seven bones are enough to feed 2 to 3 people, because there is plenty of meat between these back ribs.

I then cut off the thin strip of meat called the lip or nose right along the strip of fat between it and the big muscle, the eye of the ribeye, the longissimus dorsi. After trimming excess fat from the nose, I usually have enough for a nice stew or stir-fry for two. I cut and freeze it.

Then I find the seam of fat between the two muscles, the curved rib cap (spinalis dorsi) and the round eye of the ribeye (longissimus dorsi). I can remove it most of the way by simply pulling the two muscles apart, but a knife helps. The rib cap is about the same size as a flank steak, enough to serve 3 or 4 people, and it is absolutely the richest most succulent muscle on the animal. I set it aside, too, and cook it the same way I cook flank, hot and fast (page 231). The image on the next page is half a USDA Prime rib cap.

Now I remove almost all the surface fat from the eye. I put the fat in a Dutch oven in the oven at about 200°F for about 8 hours and it melts into about 1 cup of golden tallow and some scrap to be discarded. I put aside about ½ cup for later and the rest goes into a jar in the fridge for Yorkshire pudding or frying potatoes.

Partial rib cap.

When I am done trimming, I have four meals—the eye of the ribeye for Christmas dinner, the rib cap for a New Year's Eve feast, the slab of bones for gnawing during the football playoffs, and stir-fry meat for Lunar New Year—and tallow for frying.

Here's how it breaks down.

18 pounds bone-in rib primal to start contains:

> 3 pounds bones
>
> 2 pounds fat
>
> 2 pounds trim for stir-fry or stew
>
> 3 pounds rib cap
>
> 1 pound water, which drips and evaporates during cooking

That yields about 7 pounds eye of ribeye meat, enough for 14 servings of 8 ounces each.

9:00 A.M. Dry brine and then smoke on pellet smoker at 165°F for about 90 minutes until about 60°F internal.

10:30 A.M. Meat goes into sous vide bags and into a beer cooler filled with water and a sous vide immersion circulator set for 131°F. The salt will migrate deep in the warm water.

4:30 P.M. I take the beer cooler with the meat, the sous vide immersion circulator, and half the tallow in a jar over the river and through the woods to my niece's house.

5:00 P.M. Arrival. Water temperature has dropped to only 125°F. I turn on the sous vide machine to 131°F and preheat her gas grill to Warp 10.

5:30 P.M. Remove the meat from the bags (total sous vide time about 7 hours), let it cool for about 15 minutes, pat it dry, paint with rendered beef fat, apply my Red Meat Rub (page 166), start to sear on grill, lid up, for 5 to 10 minutes, give it a quarter-turn, repeat on all sides until dark mahogany all over. Be careful that the meat doesn't go above 135°F in the center.

6:30 P.M. Carve.

6:45 P.M. Serve and take a bow.

BEEF BACK RIBS

In my last book I talked about beef short ribs, which come from the side of the steer. Less well known are back ribs. These are the bones attached to the prime rib roast and ribeye steaks, so they are rarely sold as slabs. Because the bones contribute nothing to the flavor of roasts or steaks (see the myth busted on page 208) I remove them. The roast and steaks will cook more evenly and you'll have more seasoned crust without the bones. Now you have a slab of bones with some spectacular meat between them.

By slow-roasting them in a smoker or grill, the copious marbling practically fries the meat, making shards crispy, while other pieces

are succulent and tender like brisket. It's good by itself or with Teriyaki Brinerade and Sauce (page 175) or Tare Sauce (page 203). I know, that sounds like heresy. Just try it.

MAKES *2 servings*

TAKES *2 hours to dry brine, 5 to 6 hours to smoke*

    1 teaspoon Morton Coarse Kosher Salt
    1 (6- to 7-bone) rack of beef back ribs
    2 teaspoons Red Meat Rub (page 166)

1 PREP. Rain salt onto both sides of the rack. Add the Red Meat Rub. Make a sauce if you are going to use it, but it isn't necessary.

2 FIRE UP. Set up your grill or smoker for smoking and aim for 225°F with lots of smoke.

3 COOK. Add the slab, bone side down. Cooking a whole slab like this takes time, but cutting into individual ribs before cooking will result in dry meat, so don't do it. You may want to put a large drip pan directly underneath the grate because a lot of fat will be rendered. Roast until the temperature in the meat between the bones reaches 190°F to 203°F. This will take about 5 to 6 hours. The meat will shrink quite a lot. That's normal.

4 SERVE. Carve these ribs differently than normal. Run the knife along the bone on one side, so one side of the bone has no meat and the other side has twice as much than if you had run the knife down the middle. We call that Cadillac cut. Have lots of napkins on hand.

SMOKED SOUS-VIDE STEAK

The internet is flooded with "The Best Way to Cook a Steak." So let me jump on the bandwagon. It always starts with the thickness of the steak. If you prefer crust to the rest of the meat, then cook ¾- to 1-inch steaks. If you prefer juicy rosy meat over crust, then go with 2-inch steaks. For me, ideal thickness is about 1½ inches. This size also makes it easier to get the exterior good and dark without overcooking the interior.

Doneness is another issue. Meat scientists can measure tenderness and juiciness with machines, and in general they agree with most steak lovers, medium-rare, 130° to 135°F is best. There is also debate over what temperature the meat should be before cooking. Taking the meat out of the fridge and letting it "temper" at room temp for 30 minutes or so has been shown to improve it if you are cooking indoors. But aside from the bacterial risk of letting it sit around that long, we know that cold surfaces attract more smoke than warm surfaces, so I advocate for going from the fridge to the grill, no stops in between.

There is little doubt that dry brining amps up flavor and helps denature proteins so they hold on to more moisture. So salt is required.

This recipe may seem convoluted, but every step makes supreme sense as it marches toward an unparalleled taste experience. It uses sous vide even though I am on record saying I prefer reverse sear. In head-to-head competition, sous vide is slightly juicier and more tender, but reverse sear has more flavor that comes from being exposed to smoke and flame longer. With this recipe I give you the best of both worlds. We salt, sous vide, smoke, and sear. Effin killer.

MAKES 1 *steak*

TAKES *2 hours to sous vide, 1 hour to chill, 30 minutes to smoke, 10 minutes to sear*

SPECIAL TOOLS *Sous vide machine, sous vide bag*

1 ribeye or strip steak, about 1½ inches thick

½ teaspoon Morton Coarse Kosher Salt per pound of meat

Red Meat Rub (page 166)

1 tablespoon Smoked Butter (page 188)

ABOUT THE SMOKED BUTTER. If you don't have the smoked butter on hand, you can make it when you smoke the steak in step 4.

1 PREP. Trim excess fat from the exterior of the steak and save it in the fridge. Salt the meat. You don't have to let it sit around in the fridge waiting for the salt to penetrate because it will move in the sous vide bag.

2 SOUS VIDE. Preheat a sous vide bath to 131°F. Stick the meat into a sous vide bag and get out the air (see Sous-Vide-Que, page 107). Leave it in the bath for 2 hours. In a hurry? You can pull it after an hour. Have to pick up a kid at school, it can go for 3 hours.

3 CHILL. Take the bagged meat from the warm water bath and submerge it in a bowl of ice water. Leave it there for 30 minutes to thoroughly chill the core, then pat the bag dry

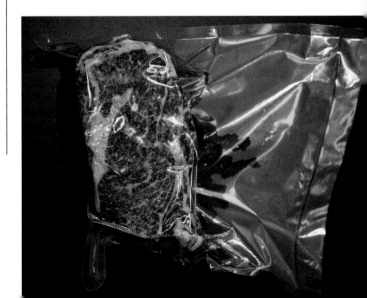

and put it in the fridge until the day you wish to serve it. You can keep it there for 4 days.

4 SMOKE. Take the meat out of the bag but don't pat it dry. Cold wet meat attracts and holds smoke. Because of the magic of thermophoresis, you only need to smoke the meat for 30 minutes to get a nice smoky taste and to begin the process of re-therming the meat.

5 SEAR. You have four choices below:

5A SEAR IN A PAN. Take the meat out of the smoker and get a cast-iron skillet or griddle as hot as possible. You can do this in the smoker. Put the fat you trimmed and reserved into the pan and melt it. You just need enough to coat the bottom. If there's not enough, add some high-smoke-point oil like ghee. Then plop the steak in the pan and press it down so it is making maximum contact with the metal. After 3 minutes or so it should be getting a nice dark crust. If it is, flip and repeat.

5B SEAR WITH THE "AFTERBURNER." Use the afterburner (see Afterburner Fajitas, page 227). Load a charcoal chimney at least halfway and fire up. When the coals are white, put a wire grate on top, and put the steak on the grate. Flip every minute until it is seared beautifully.

5C SEAR IN A DEEP-FRY. Yup. You can submerge the steak in oil at 350°F until it has the crust you want.

5D SEAR UNDER THE BROILER OR IN A PIZZA OVEN. A good pizza oven is like a broiler with a hot flame just above the meat.

Weather's bad? Bring the steak inside and pop it under the broiler.

6 SERVE. Rapidly move the meat to the dinner plates and top it with the Smoked Butter while it is hot. No need to rest it (see Resting Meat, page 55).

JAPANESE BEEF

While it is highly unlikely you can find true Japanese Kobe in a store or even online, occasionally you may be able to buy Tajima beef, which comes from nearby prefectures. If you can afford it, do it. The fat content and texture of Tajima beef is close to pork belly or bacon. It is far too rich to just choke down a whole ribeye. Tajima steaks are rarely more than 1 inch thick. I figure about 4 ounces per person. I like to serve it after a vegetarian course, such as the big salad. After tasting Tajima or Kobe, eating more meat would be pointless.

My favorite method is to cut it into strips 3 to 4 inches long, 1 inch wide, and 1 inch thick, and I let my guests sear them on a raging hot salt block in the middle of the table.

Salt block purveyors say you should heat them on a stovetop gradually. They say salt blocks shouldn't be heated in gas ovens or a closed grill because the gas produces too much moisture and as the moisture condenses on the slab, it dissolves the salt, forms steam,

Get the best sear by putting the meat right on the coals

BUSTED. I know you want to try this caveman gimmick, and a lot of macho cookbooks love it. They say that once you get your charcoal down to glowing embers take a piece of cardboard and fan the coals so any loose ash is blown away. Take a steak about 1 inch thick, pat it dry, salt it, and lay it right on the hot coals. Surprisingly, when you turn it in about 3 minutes, there will be very little ash stuck to the meat, and it produces an all-over sear in a hurry. But every time I've tried it, small amounts of ash and even whole coals have stuck to the surface and there have been scorched dry spots. I can't recommend this method. It is a much better technique to put a cooking grate right above the coals, as close as you can get, but keep the meat off the coals. You'll get a much more even sear. And no ash.

and the steam can split the block. Of course, I disobeyed and nothing happened. I think that's because gas grills are so well vented. They also say you want to heat them gradually or they could split. I've never seen this happen either. I place the block on the indirect side of my gas grill and crank up the heat for about 45 minutes, lid down. I then place it on a trivet or cutting board on top of a thick stack of cloth placemats or kitchen towels in the center of the dining table. For this recipe we are using beef, but you can use it for calamari, shrimp, strips of other meat, and vegetables. I've used mine for rare Ibérico pork from Spain. Your guests can then plop the meat on the salt and it will sizzle away, rapidly cooking. Amazingly it will not be too salty. In fact, it usually is perfectly seasoned. As the block cools, food cooks more slowly and absorbs more salt.

If you don't have a salt block, dry brine and cut up the steaks. Heat a heavy pan, a cast-iron skillet is a good choice, as hot as possible. You do not need to oil the pan; the melting fat will be enough. Toss the meat on and cook it long enough to get a dark brown surface, perhaps 3 to 4 minutes per side. Serve with a sprinkling of large-grain salt like Maldon. Sit down to eat so you don't collapse in a dead faint.

KŌJI FILET MIGNON

The classic French method is to sear meat in a pan, finish it in the oven, and build a sauce with what's left in the pan. The juices, the rendered fat, and the browned bits of meat that stick to the bottom of the pan are called fondly the *fond* by the French and it is loaded with umami. Fond means "bottom." By moving the process outdoors, you can amp it up by bringing smoke flavors to the party. By adding kōji, you can amp up the umami even more.

The secret here is to use a thick piece of meat like a filet mignon; and the filet is perfect because it is lean and mild flavored, so the sauce can really amp it up to 11. Use a stainless steel, aluminum, or a cast-iron skillet with an ovenproof handle. Do not use a nonstick pan, because fond cannot develop well on a nonstick surface and the coating might not be able to handle the high heat on the grill.

Filets are often wrapped with a strip of bacon because they are so lean and the bacon brings fat, flavor, salt, and even more umami to the party.

MAKES *2 servings*

TAKES *1 hour*

SPECIAL TOOLS *Toothpicks and a skillet with an ovenproof handle*

> 2 filets mignon (8 to 10 ounces each), 1½ to 2 inches thick
>
> ¼ cup Hanamaruki Liquid Shio Kōji (see page 112)

> 2 slices bacon
>
> 1½ cups Duxelles and Mushroom Cream Sauce (page 192)
>
> Tallow (beef fat), bacon fat, clarified butter, or ghee

SERVE WITH. Something to soak up extra sauce like mashed potatoes, rice, couscous, or pasta.

1 MARINADE. Put each filet in a zipper bag and pour in 2 tablespoons of Hanamaruki Liquid Shio Kōji. Zip the bag and tumble it around so all surfaces get wet. Let the meat marinate for 6 hours or more in the fridge. No need to salt the meat; the shio kōji has salt in it (*shio* means "salt" in Japanese).

2 WRAP. Wrap a bacon slice around each filet and fasten in place with toothpicks. Snip the toothpicks so they don't prevent the meat from being intimate with the pan. Set them in the fridge until you are ready to cook.

8 FINISH. Add the meat to the pan. Check the meat temperature. Shoot for medium-rare, 130° to 135°F. If it is not there yet, put it over direct heat. Spoon some of the sauce and mushrooms onto each dinner plate and drop your filets right into the center of the pools of sauce.

3 MAKE THE MUSHROOM SAUCE: Make the Duxelles and Mushroom Cream Sauce, taste it, and adjust the salt if necessary.

4 FIRE UP. Set up your grill in a 2-zone configuration. On the direct heat side, heat a frying pan and add just enough fat for a thin coat on the bottom of the pan.

5 SEAR. Put wood on the hot side and as soon as it starts smoking, pat the top and bottom of the filets dry with a paper towel and place in the hot pan. Close the lid and sear them, perhaps 3 minutes, flip them and place them in a different location on the pan because the spot where they just sat has cooled a bit. Keep flipping until they are dark, but don't let the interior go above 120°F.

6 KEEP WARM. Remove the meat from the pan and place it on the indirect side of the grill to keep warm and to pick up some more smoke while you are finishing the sauce.

7 'SHROOMS. Pour the Duxelles and Mushroom Cream Sauce into the hot skillet and scrape the bottom of the pan well with a wood or silicone spatula to get up all those bits of fond and incorporate the flavor into the sauce.

CLINT'S CHUCK ROAST SANDWICHES

Clint Cantwell turned me on to the wonders of beef chuck roasts. Taken from the "shoulder clod," a large primal cut that includes a mass of intertwined muscles (think of how complex your shoulder is), chuck offers as much flavor as brisket but it is smaller and often less expensive. There is also more marbling, so it is juicier and more tender. He approaches it like he would a brisket.

Photo by Clint Cantwell

MAKES *About 2¼ pounds of meat after trimming and shrinkage, enough for 6 sandwiches*
TAKES *5 minutes to prep, 6 hours to cook, 1 hour to hold*

- **3 pounds boneless beef chuck roast**
- **1½ teaspoons Morton Coarse Kosher Salt**
- **1½ teaspoons coarse-grind black pepper**
- **12 sandwich-size slices Smoke-Roasted Garlic Bread (page 367)**
- **6 tablespoons Orange Tuscan-Style Board Sauce (page 198)**

1 PREP. Trim off excess surface fat from the chuck roast. Season the roast with the salt and pepper at least 2 hours before cooking. Prepare the garlic bread but don't grill it yet. Make the Board Sauce.

2 FIRE UP. Prepare a smoker or grill for smoking at 225°F and get some smoke rolling.

3 COOK. Put the meat in the smoker or on the grill. When it hits 150°F, wrap it tightly in aluminum foil.

4 HOLD. When it reaches an internal temperature of 180° to 190°F, take it off, wrap it, still in foil, in a clean towel and let it sit for 1 hour in a beer cooler.

5 SERVE. Grill the garlic bread. Unwrap the meat, slice it across the grain in ¼-inch slices and build your sandwiches topped with the Orange Tuscan-Style Board Sauce.

AFTERBURNER FAJITAS

So, I was cooking some ¾-inch ribeyes one night. I started some charcoal in a chimney to toss on my trusty Weber Kettle because I wanted max heat for that great whiskey-colored crust.

It was getting dark and when I looked at the chimney, I noticed it looked like the afterburner of a fighter jet. Long blue and red flames, hardly visible. So I put a wire grate right on top of the chimney, and tossed the meat on.

Perfect sear, deep mahogany brown, in less than 3 minutes per side and cooked perfectly to medium-rare in the center! In the image on the next page, you can see me cooking some ¾-inch ribeyes on three afterburners at the conference of the International Sous Vide Association with scores of hungry chefs in attendance. Fortunately, I had scores of USDA Prime ribeyes that had been sous vided to 131°F and needed a good sear. Needless to say, the steaks were a huge hit, as was the show, and now everybody is doing it. There are even small grates designed for the chimney being sold.

The technique works superbly on sous vide meats and it can even be used on raw meats ½ to 1 inch thick. It will burn anything thicker before the center is done. It is ideal for outside skirt steaks, a long tough muscle rarely more than ¾ inch thick, the traditional meat for fajitas. The secret is that it puts massive amounts of heat on one surface at a time and cooks it so quickly that the interior doesn't get too warm. At regular grill temps the heat progresses through the surface to the interior, and by the time you have a good dark sear on the outside, the inside is overcooked. That's the problem with fajitas. You have such a tasty piece of meat in the skirt steak, but the center is almost always gray. Nevermore.

MAKES *6 fajitas*
TAKES *20 minutes to prep, 3 hours to marinate, 20 minutes to cook and slice*

SPECIAL TOOLS *Blender or food processor, charcoal chimney and briquets, wire grate to sit on top, cast-iron frying pan, aluminum foil*

MARINATED MEAT

- 2 oranges
- 2 limes
- 3 garlic cloves
- 3 canned chipotle chiles in adobo sauce
- 3 tablespoons fresh cilantro leaves
- ½ teaspoon ground cumin
- 1 teaspoon Morton Coarse Kosher Salt

½ teaspoon fine-grind black pepper

3 tablespoons vegetable oil

1½ pounds outside skirt steak

FAJITAS

2 medium bell peppers, any color

1 large onion

2 medium tomatoes

1 tablespoon vegetable oil, or more as needed

6 tortillas, your choice of flour or corn

1 avocado

ABOUT SKIRT STEAKS. Go for the outside skirt. It comes from the diaphragm between the sixth and twelfth ribs. It is thicker, more tender, and more uniform than the inside skirt. You may have to order it from your butcher because most go to restaurants. It may come with a membrane attached that is easy to remove.

The white cuts are parallel to the grain about 6 inches apart and you make them before cooking so the sections will fit on the chimney. The black lines are how you slice the sections across the grain after cooking to make it easier to chew.

1 MAKE THE MARINADE. Squeeze the oranges and pour ¾ cup of the juice into a large bowl. Squeeze the limes and add 3 tablespoons of the juice into the bowl. Peel and press or mince the garlic. Finely chop the chipotles and the cilantro and add them along with the cumin, salt, black pepper, and oil. Puree all this in a blender or food processor. It doesn't have to be perfectly smooth. Measure out ¼ cup and put in a small bowl in the fridge for use as a sauce. Return the rest to the large bowl.

2 MARINATE THE MEAT. Cut the meat with the grain into 6-inch lengths (the white lines in the image below left). Put on some gloves, add the meat to the big bowl with the marinade, and massage it in. Let them get to know each other in the fridge for 1 to 3 hours.

3 PREP. Slice the peppers in half, rip out the stems and seeds, and cut what's left into ¼-inch slices. Cut the top off the onion, peel the onion, cut it in half pole to pole, and then slice it into ¼-inch half-moons. Cut the tomatoes in half and, over the trash, squeeze out the seeds and gel. Chop what is left into ¼-inch chunks and put them in a bowl. (Don't cut the avocado yet or it will turn brown.)

4 FIRE UP. Put the charcoal chimney on top of the cooking grate on your grill. Fill it halfway with briquets and light it.

5 COOK THE FAJITA VEGETABLES. When the coals are white and flame is shooting out of the top, put the cast-iron skillet on the chimney, add the oil, spread it around, and add the bell peppers and onion. Cook just until they soften a bit, but leave some crunch, about

4 minutes. If they scorch a bit, that's OK. Pour them into a bowl.

6 WARM THE TORTILLAS. With a paper towel wipe the oil from the pan being careful not to burn yourself. Working quickly place one tortilla at a time in the pan and heat them until they toast slightly on one side. Take them out and stack them on a plate and cover with foil to keep them warm.

7 COOK THE MEAT. Remove the pan and put a wire grate on the chimney. Remove the meat from the marinade, wipe the excess marinade, put the meat on a plate, and head for the chimney. Grill one chunk at a time. Sear one side quickly, 1 minute is all, and then flip and sear the other side until it has good color and the interior is 125° to 130°F. Move the cooked meat to a cutting board. Cut across grain into ¼-inch strips as shown by the black lines in the photo on page 229. Put them in a bowl.

8 AVOCADO. Carefully run a knife around the avocado cutting it in half from pole to pole. Twist the halves apart and with a spoon pop out the seed. Cut the halves again so you have 4 quarters. Scoop the meat of the fruit out of the skins, cut it into ¼-inch slivers, and put them in a bowl.

9 SERVE. Now set all the bowls on the table, meat, onions and peppers, tomatoes, and avocados—and don't forget the sauce in the fridge. Let your guests assemble their fajitas as they wish.

TORCHED BEEF SHORT RIBS YAKITORI, SORTA

Here's an idea I stole from my favorite dessert, crème brûlée. The custard is made in a bowl in advance and kept in the fridge. Just before serving you sprinkle sugar on top and blast it with a handheld torch to make a hard crunchy candy cap.

So I took a 1-gallon propane tank and nozzle I had for soldering, bought an attachment called a Searzall that spreads the flame, tossed a thin inside skirt steak into a cast-iron skillet, and hit the surface with the torch. Within minutes I had a richly Maillarded slab that remained rosy rare in the center. For this recipe I have chosen a thinner cut, flanken-cut short ribs, which are cut into ½-inch-thick strips perpendicular to the rib bones.

A propane torch with a Searzall spreader.

MAKES 2 *large servings*

TAKES *20 minutes to prep, 3 to 12 hours to marinate, 10 minutes to cook*

SPECIAL TOOLS *Butane or propane torch, Searzall spreader attachment (nice but optional)*

> 3 pounds flanken-cut short ribs
>
> 2 kiwis
>
> ½ cup Tare Sauce (page 203)
>
> 1 cup basmati rice

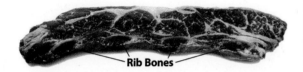

Rib Bones

ABOUT THE RIBS. I use beef short ribs because they are heavily marbled for flavor, and tough. But the flanken cut is thin and cut cross-grain, so that makes them easier to chew. It is common in Mexican groceries.

ABOUT THE KIWI. Kiwis have an enzyme, *actinidin*, that tenderizes meat. If you can't find them, use fresh pineapple, not canned. They have a tenderizing enzyme, *bromelain*.

1 PREP THE MEAT. Remove the bones and the membrane and freeze them for making stock. Lotta good marrow in those bones.

2 TENDERIZE. Cut the kiwis in half, scoop out the green interior, and puree it in a blender. Marinate the meat in the kiwi juice in a zipper bag for 2 hours, stirring occasionally.

3 MORE PREP. Make the Tare Sauce and get the rice started.

4 FIRE UP. Take the meat from the marinade, rinse it, and blot the surface really dry with paper towels. Place the meat on a griddle or grill topper or cast-iron skillet and hit it with your torch until you get some nice browning on the surface. Move the flame closer or farther away until you find the sweet spot. Flip it and torch it until the red is gone and it turns tan, but don't try to sear the second side. Paint the meat with the Tare Sauce and hit with the torch until it sizzles; flip the meat and repeat.

5 SERVE. Scoop some rice onto each plate. Top it with the meat and pour a couple of tablespoons of the Tare Sauce over the meat and rice. Sake anyone?

VIGNERON METHOD FOR FLANK STEAK SUBS

When I was a boy, I would watch Dad grill flank steaks and that's where I learned to love outdoor cooking. Every bite of flank steak today sends echoes through my mouth of those nights and memories of Dad.

I learned this method when visiting wineries in Bordeaux, the French region that makes wine perfectly designed for steaks. Vines grow there like, well, vines. In the

winter, vineyard owners prune away most of the branches, called canes, that will not bear fruit. They then have huge piles of grapevine wood, most about the thickness of a pencil. During the fall harvest season, the vignerons will take a big stack of the dried canes and set them on fire. They quickly burn down to a glowing mound, and the workers grill meats over the scorching hot embers. The flavor is exquisite. My hosts called this method *sarment* (pronounced *sar-MOHN*) which means "vine shoot." I call it the "Vigneron Method."

You can get grapewood from a winery, or in many states, grapevines abound wild in the woods and grow on fences along the roadside. In a pinch you can use sticks cut from fruit or hardwood trees. To ensure a steady flow of grapewood, I planted five vines. I've also done this method with twigs from my neighbor's cherry tree.

In Bordeaux they cooked small butterflied quail, but at home I completely destroyed some Cornish game hens, so I steer away from poultry. Too hot. I now use this method primarily for flank steaks because they are about the right thickness, under 1 inch, and beef loves heat and smoke. Lately I have been splurging on wagyu flanks and they are perfect. Much juicier and more tender than USDA Choice but loaded with beefy goodness. I find wagyu ribeyes a bit unctuous, but flanks, which are lean, really benefit from the extra marbling. It is the only wagyu I buy. There is a video of me cooking flank steak this way. I link to it on AmazingRibs.com/mm.

MAKES *4 big sandwiches*
TAKES *2 to 3 hours to dry brine, 30 minutes to fire up and cook*

SPECIAL TOOLS *Charcoal grill, newspaper, and lots of dry grapevine or other thin hardwood or fruitwood cuttings about pencil thick, 12 charcoal briquets*

> 1 flank steak (about 2 pounds)
> Morton Coarse Kosher Salt
> Medium-grind black pepper
> 4 rolls
> Good olive oil
> Zhug (page 202)

ABOUT THE ROLLS. Use your favorite roll. Choose from big Italian rolls, Kaiser rolls, or smaller hoagie rolls, adjusting the amount of meat and sauce per sandwich accordingly.

1 PREP. If the steak has a fat end, pound it with your fist so it is more even in thickness. Pat the surface of the meat with wet hands. Sprinkle the steak with salt, using ½ teaspoon per pound of meat, and dry brine for 2 to 3 hours in the refrigerator.

2 SET UP. To start, I take out the bottom grate from my Weber Kettle and open the lower vents. I crumple two sheets of newspaper and put them in the bottom of the bowl (I hope you still read a newspaper, says the former journalism major). Then I stuff as many dried vine canes as I can fit on top of the paper, all the way to the level of the upper grate. I then put about 12 charcoal briquets on top so I will still have energy after the wood fire dies down if I need it. On goes the top grate.

3 FIRE UP. I light the paper through the bottom vent holes, and the whole thing goes poof in about 2 minutes with very impressive 5-foot flames. Make sure nothing flammable is overhead. (I once came close to melting the television cable.) Within a few minutes I have glowing white hot embers. I wait until most of the flames have died down, then I quickly clean the grate and on goes the meat. I've

got a window of about 20 minutes before the embers die out. That's Warp 15, Spock.

4 COOK. Flip the meat often or it will burn. You should not need the lid. Temp the meat often in both the thick and thin end. If the thin end is done but the thick end is not, I let the thin end hang over the edge of the grill where it can't cook anymore. Now add the black pepper. If you add it earlier it can burn. If the embers are dying and the meat is below 120°F, put the lid on.

5 TOAST. Cut the buns open and coat the cut sides of the rolls lightly with olive oil. Go all the way to the edges with the oil or else they will burn. As the fire dies, toast the rolls cut sides down, lid open, for 1 minute or so, until golden. Stand right there because they can burn faster than a barn full of hay.

6 SERVE. Slice the meat across the grain (this is crucial for easy chewing) about ¼ inch thick. Top the base of the toasted rolls with a few slices of steak, then spoon some Zhug on top.

BEEF FRIED RICE

I f you need to make a larger portion of this recipe, consider doing it in two batches, as you never want to crowd a wok.

MAKES *4 entrée servings, 6 side dish or appetizer servings*

TAKES *About 3 hours to prep (most of it marinating time), 20 minutes to cook*

SPECIAL TOOLS *Charcoal chimney with holes drilled or cut just below the top to allow air in under the wok (as described on page 91), a slotted spatula, many bowls, and a dump bucket for rinse water*

> 2 cups cooked rice, preferably basmati from India, aged at least 1 day in the fridge
>
> ½ teaspoon baking soda
>
> ½ pound lean beef
>
> 3 tablespoons oyster sauce
>
> 3 tablespoons dark soy sauce
>
> 1 teaspoon toasted (brown) sesame oil
>
> 1 teaspoon your favorite hot sauce
>
> 1 teaspoon unseasoned rice vinegar
>
> ½ cup vegetable oil
>
> 1 large onion
>
> ½ red bell pepper
>
> 3 ounces mushrooms
>
> 1 medium carrot
>
> 3 garlic cloves
>
> 1-inch finger of fresh ginger
>
> 24 sugar snap peas
>
> 3 scallions
>
> 3 large eggs

ABOUT THE RICE. Plain old white rice is fine, but I use basmati from India. It resists clumping and it smells and tastes lovely. To get rice grains that aren't gummy and clumpy you need to rinse them thoroughly, cook them, and

let them air-dry for a day or two in the fridge. Freshly cooked rice is too wet and the results will stick to the wok and be mushy. Often, when I cook a rice dish, I double the amount of rice I need and I put the leftovers in the fridge for a day or two and then use it in a stir-fry. I also make fried rice from the extra rice I get when I order takeout.

ABOUT THE BEEF. This recipe uses lean beef, but you can use pork, chicken, shrimp, city ham, even leftover grilled meats.

ABOUT THE OIL. Peanut oil is the standard, but there are people allergic to peanuts, so I use another high-smoke-point oil such as a blended vegetable oil.

ABOUT THE MUSHROOMS. Try for shiitakes if you can find them, but plain old button mushrooms will work just fine.

ABOUT THE SAUCES. I prefer dark soy sauce for this recipe, but if you only have reduced-sodium, feel free to use it. And if you don't have oyster sauce, you can substitute more soy, but it is not the same.

ABOUT THE SUGAR SNAP PEAS. You can also use snow peas, fresh peas, even frozen peas. If you are using frozen peas, add them to the wok when you cook the onions, mushrooms, carrots, etc.

1 READ TWICE. Read this procedure twice. Once you get started, things move fast. You need to be ready. When it comes time to cook, it is vital to have everything chopped, sliced, and diced and ready to go. Clean the wok, dry it, and coat the inside with a thin layer of oil.

2 THE RICE. If the rice is still wet, with your hands, break up the clumps into a single layer of discrete grains on a sheet pan. Then pop it in the oven on the lowest temp for about 15 minutes.

3 TENDERIZE THE BEEF. Here's the trick for tenderizing the beef that I learned from J. Kenji López-Alt's book *The Wok*. In a small bowl, stir the baking soda into 1 cup water and then mix it with the meat. Coat it thoroughly and work the meat roughly with your hands, massaging it in, and let it sit for at least 15 minutes but an hour or two in the fridge is better.

4 WHEN READY TO COOK. Rinse the excess baking soda from the meat with plenty of cool water, squeeze out excess water, and then pat the meat dry because too much baking soda can taste bitter or metallic.

5 MAKE THE MARINADE/SAUCE. In a large bowl, mix 2 tablespoons water, the oyster sauce, soy sauce, sesame oil, hot sauce, rice vinegar, and vegetable oil. Add the meat and squish it so the meat is covered and so the marinade gets into the pores.

6 OTHER BOWLS. Peel and dice the onion and put it in the bowl. Chop the bell pepper into ¼-inch chunks and add them to the same bowl. Wash and slice the mushrooms and chop into bite-size chunks. You need 1 cup chopped mushrooms for this recipe. Freeze the rest. Bowl them. Peel and slice the carrot lengthwise into 4 sticks and then cut across them into small dices. You know where they go. Peel and mince or press the garlic cloves and put them in a different small bowl. Mince the ginger until you have 2 teaspoons (you don't have to peel it) and put it in the bowl with the garlic. Cut the sugar snap peas in half lengthwise and put them in a different bowl. Clean the scallions, peel off any dead skin, chop off the roots, and slice them on the bias. Add them to the bowl with the peas. Break the eggs into yet another bowl and beat them lightly with a fork.

7 FIRE UP. Start a chimney full of charcoal and wait until they are covered in ash.

8 HEAD OUT. Head outside and bring with you all the bowls of ingredients, the wok, the wok spatula, a strainer, a cup of water, the vegetable oil, a roll of paper towels, a dump bucket or bowl for used rinse water, several clean bowls, and a table knife. They need to be all ready to go on a moment's notice.

9 START FRYING. Put the cold wok on top of the chimney. Leave it there for 30 seconds. Add about 2 tablespoons oil and roll it around, coating the sides all the way up. Toss in the onions, peppers, carrots, and mushrooms, stirring and tossing often. When they are a bit limp and gaining brown on the edges, but still a bit crunchy, add the garlic and ginger and fry them for only 1 minute. Push everything to one side of the wok. Pour in the eggs and keep moving them around until you have scrambled eggs. When they are approaching firmness, chop them into bite-size chunks with your spatula, mix them with the veggies, and then slide everything into a clean bowl.

10 CLEAN THE WOK. If the wok gets caked with food bits and you fear they will burn, pour in a bit of water and clean the bowl with the paper towels. Let it dry thoroughly. You can clean the wok like this whenever you need to.

11 COOK THE MEAT. With a strainer, separate the meat from the sauce, reserving the sauce. Add 2 tablespoons vegetable oil to the wok and fry the meat, keeping it moving, until it is cooked through with a little browning on the surface. It should only take a couple of minutes. Slide it into a big clean bowl (not the same bowl that held the raw meat).

12 FRY THE RICE. Add 3 tablespoons oil to the wok and roll it around. When it is hot, in about 30 seconds, add the rice and stir-fry until warm and some of it is lightly browned and the rice grains start jumping around on the metal. They may stick to the wok. Keep frying and scraping the bottom of the wok and the stuck gunk will eventually let go.

13 SAUCE. Pour the marinade/sauce down the sides of the wok so it heats rapidly. Mix it into the rice until everything is tan. Because the sauce is cooked with such a high heat, I make an exception to the rule that you should not use marinades as sauces for fear of contamination.

14 COMBINE. Dump the veggies, eggs, and meat into the wok and mix everything together. Add the scallions and peas. Mix until everything is warm, about 2 minutes. Get your chopsticks out and pour your wine.

LAMB

If you say you don't like lamb, I'm guessing you've never had it cooked properly.

Since lambs are so small, there are fewer cuts of interest to the backyard cook. They are also leaner and more tender than pigs and cattle. I like beef in the medium-rare range, 130° to 135°F, but I prefer lamb in the low-medium range, 135° to 140°F.

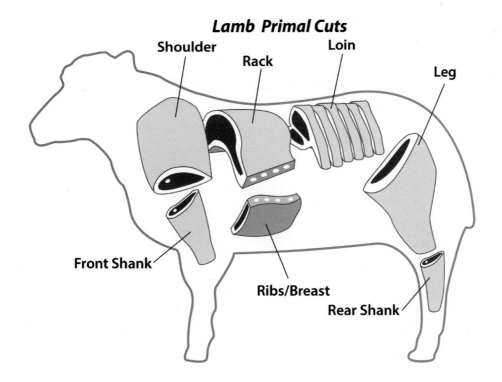

Lamb Primal Cuts

Shoulder · Rack · Loin · Leg · Front Shank · Ribs/Breast · Rear Shank

SPRING LAMB. Sheep less than 3 months old. Very tender and very mild flavor.

LAMB. A sheep younger than 1 year. Very tender and beef-like.

MUTTON. Sheep older than 1 year, often as old as 3 years. Strong flavor, but still tender.

DOUBLE-WIDE LAMB RIB CHOPS WITH BOARD SAUCE

I have tried to share creative recipes in this book, most of which you've never seen before, but variations on this one are as old as the hills and tried and true. This much we know about lamb since the first one was sacrificed in the Bible: Lamb goes with garlic and rosemary like chocolate goes with milk. And yes, there's a lotta garlic in this one, but remember, cooking defangs it.

Lamb rack is not a cheap cut, but if you love beef, you will love this. It is a section of eight ribs with the loin meat attached, about 2 pounds, the same cut as bone-in prime rib of beef or pork crown roast, just a lot smaller. After you trim the fat, you might have only 1 pound of gorgeous meat, enough for 2 people to ride to heaven.

MAKES *One 8-rib rack of lamb, serves 2*

TAKES *30 minutes to trim and prep, 2 hours to dry brine, 30 minutes or so to cook (depending on the temperature you get in the indirect zone)*

> One 8-rib rack of lamb
>
> 1 teaspoon Morton Coarse Kosher Salt
>
> ½ cup Panko Pebbles (page 374)
>
> ½ cup Orange Tuscan-Style Board Sauce (page 198)
>
> 3 garlic cloves
>
> 2 tablespoons coarsely chopped fresh rosemary or 2 tablespoons dried
>
> 1 lemon
>
> 1 teaspoon medium-grind black pepper
>
> 2 tablespoons whole-grain Dijon-style mustard

1 PREP. Many racks of lamb come in vacuum-sealed bags. Whenever they are opened, the smell can be off-putting. It dissipates rapidly. If the rib bones are connected at their base, called the chine, use your kitchen shears and snip the connection so you can slice the rack more easily when it is done. Get as much fat off as the lamb possible. While surface fat on beef and pork are tasty, lamb, not so much. Carve all the fat out from between the bones, a method called Frenching. There's a tiny bit of meat in there, but it is not worth fighting for it. Sprinkle with salt and dry brine for about 2 hours if possible.

2 MAKE THE SAUCE AND PANKO. Make the Panko Pebbles and the Orange Tuscan-Style Board Sauce. Peel and press or mince the garlic. Coarsely chop the rosemary. Scrub and zest the

lemon. In a bowl, mix the Panko Pebbles, garlic, zest, rosemary, and black pepper.

3 COAT THE RACK. Slather the mustard all over the meat but not the bones. Sprinkle on the bread crumb mix, trying to distribute it evenly and press it into the mustard.

4 FIRE UP. We are going to cook this almost all the way in the indirect zone, so get it as hot as you can over there, say 400°F. If you have a thermometer with a probe, you can leave in the meat, insert it from dead center in the side. When it hits 130°F move it over on direct heat and toast it on both sides. Watch it like a hawk so the bread crumbs don't burn. It's dinnertime when the lamb hits 135° to 140°F.

5 SERVE. Pour the board sauce onto the cutting board, spread it out a bit, place the rack on top of the sauce, and slice between every second bone while it is hot. Get some of the board sauce on the cut surfaces and serve. Sometimes I snap or snip off a bone from each cut so it looks like a lollipop.

LAMB KEBABS ARE BETTER WITHOUT SKEWERS

This dish can be made with boneless leg or shoulder. Lamb hind legs are a cone-shaped twisted mass of muscles, sinew, and fat. The leg can be roasted whole, with or without smoke, and you get a whole range of doneness from well-done at the tapered end by the shank, to medium-rare at the hip end. When you remove the bone from a lamb leg you get a huge bumpy slab of meat. Some butchers will sell you a boneless leg rolled up in a cookable mesh.

The shoulder is another group of complex muscles that can be cut into arm chops and blade chops, but is better roasted whole or cut into chunks for grilling. Some butchers sell boneless shoulders wrapped in mesh.

Lamb is especially popular in the Middle East, where it is often paired with ras el hanout, a spice blend, and served on couscous. You can buy ras el hanout in stores or online, or make it yourself from a recipe on AmazingRibs.com/mm. Kebabs are typically cooked on skewers, but I prefer to cook largish chunks on a grill topper so I can remove each at optimum color and temperature.

MAKES *4 servings*
TAKES *2 hours to dry brine, 10 minutes to make the couscous, about 8 minutes for the lamb*

2½ pounds boneless shoulder or leg of lamb

1½ teaspoons Morton Coarse Kosher Salt

1 teaspoon medium-grind black pepper

2 tablespoons ras el hanout

1 small onion

1 tablespoon vegetable oil

2 (8.8-ounce) packages Israeli couscous (about 3 cups)

4 cups low-sodium chicken broth

½ teaspoon white miso

Olive oil, for coating

2 sweet bell peppers, any color

OPTIONAL. Serve this with Zhug (page 202) drizzled on top.

ABOUT THE LAMB. You will probably need to buy 2½ to 3 pounds of lamb to get 1¾ pounds of meat after trimming. More if there is bone.

ABOUT THE COUSCOUS. Israeli couscous are small balls of toasted pasta developed in Israel in the 1950s when rice was hard to get. It is also called pearl couscous.

1 PREP. Trim the fat and gristle from the lamb and end up with 1¾ pounds. Cut it into 2-inch cubes. We want large cubes so the surfaces can brown properly without overcooking the center of the meat. Most kebabs are cut so small that the meat is badly overcooked. Make the Zhug if you are going to use it. I recommend you do.

2 DRY BRINE. In a large bowl, toss the lamb cubes with 1 teaspoon of the salt. Add the black pepper and ras el hanout and make sure all the surfaces are evenly coated with the salt and spice mixture. Refrigerate for 2 hours.

3 FIRE UP. Set up your grill for 2-zone cooking and aim for a little less than Warp 10 at the grate level on the direct heat side. Preheat a grill topper over direct heat.

4 MAKE THE COUSCOUS. Peel and finely chop the onion. In a saucepan, warm the vegetable oil over medium heat. Add the onion and cook, stirring, until softened, about 5 minutes. Add the couscous and cook and stir until the couscous has just begun to pick up a little brown, about 6 minutes. Add the chicken broth and miso. Bring to a boil over high heat. Immediately reduce the heat to a simmer, cover the pan, and simmer until tender and all the liquid is absorbed, about 8 minutes.

5 GRILL THE PEPPERS AND LAMB. While the couscous is cooking, coat the lamb cubes lightly with olive oil. Cut the bell peppers in half and remove the stems and seeds. Slice them into strips. Grill the peppers first on the grill topper. Set them aside in a bowl. Now grill the lamb chunks on the grill topper, lid up, flipping with tongs every 2 to 4 minutes, until crusty brown on the outside and pink within, 135° to 140°F.

6 SERVE. Transfer the couscous to a platter and top with cubes of lamb and the peppers. If desired, drizzle Zhug over all.

PORK

"All animals are equal, but some animals are more equal than others."

—NAPOLEON, LEADER OF THE PIGS, FROM *ANIMAL FARM* BY GEORGE ORWELL

nterestingly, pork is not graded by the USDA as is beef. Beef gets its grades depending on the amount of fat marbling within the meat. When buying pork, look for meat that is pale pink with white fat, but the best muscles are rosy because they have more myoglobin and thus more flavor. Notice that loin and tenderloin are very different cuts although they are often confused.

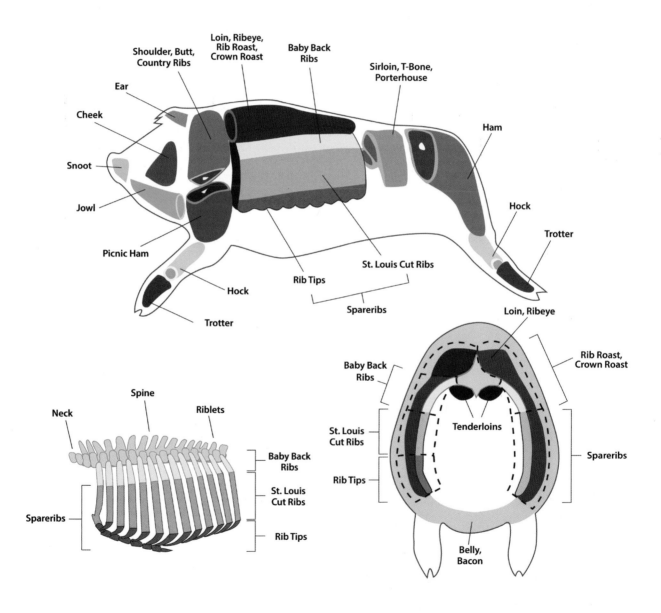

CHAMPIONSHIP PORK RIBS

It is impossible to be melancholy listening to a banjo or eating pork ribs. Nothing is happier than pig on a stick. They come in three cuts.

BABY BACKS are curved and attach to the spine. Their meat is leaner and on top of the bones.

SPARE RIBS are flatter, thicker, and attach to the back ribs and descend down the side of the animal. Most of the meat is between the bones and slightly more fatty and therefore tastier. Spare ribs have both rib tips and center cut ribs.

Turn-in box of championship ribs by Darren and Sherry Warth.

RIB TIPS are a flap of cartilage and fatty meat at the lower end of the spares. They can be chewy, but tasty.

ST. LOUIS OR CENTER CUT. If you cut the tips off spare ribs, you have a rectangular slab from the middle of the rib cage called St. Louis cut or center cut. This is my favorite cut. The meat is mostly between the bones rather than on top of them and it is fattier and juicier.

COUNTRY RIBS are not really a rib cut, although some have rib bones in them. They are pork chops from the shoulder and should be cooked much differently than ribs to a much lower temperature.

In my last book I gave you the recipe for how I cook ribs at home. Competition cooks do a few things differently. There are more than one thousand barbecue competitions around the nation. Many are informal events held by the local chamber of commerce but there are at least four hundred formal events supervised and sanctioned by the Kansas City Barbeque Society (KCBS) and prize money can be in the thousands. Some regions prefer baby backs, some St. Louis cut. Other cuts are eschewed. If you are competing, ask the local teams which cut scores best. This recipe is for 1 slab, but nearly all competitors cook at least 3 in order to get 8 tip-top ribs that are about the same size.

Judges taste only one or two bites, so they better be impressive. If you plan to compete, start here, because this recipe is the way champions cook ribs in competition, although most of them do a simpler recipe at home, like Last Meal Ribs in my first book or on my website. Note: Most competitions do not allow gas cookers, only charcoal, logs, or pellets.

MAKES *1 slab, enough for 2 to 3 servings*
TAKES *30 minutes to prep, 1 to 2 hours to dry brine, 4 to 5 hours to cook at 225°F*

SPECIAL TOOLS *Spray bottle, heavy-duty aluminum foil*

- 1 slab pork ribs, St. Louis cut or baby backs
- ½ teaspoon Morton Coarse Kosher Salt per pound of meat
- 2 tablespoons yellow ballpark mustard
- 4 tablespoons Meathead's Memphis Dust (page 165)
- 6 ounces liquid margarine from a squeeze bottle
- ¼ cup brown sugar
- ¼ cup honey
- 3 tablespoons apple juice
- ¼ cup Kansas City Red (page 180)

ABOUT THE MUSTARD. The mustard helps the rub stick to the meat without significantly altering the flavor of the finished meat.

ABOUT THE MARGARINE. I know you are clutching your chest and asking "Why not butter?" Many cooks prefer the convenience of a liquid in a squeeze bottle. It's easier to spread.

ABOUT THE APPLE JUICE. I've noticed some top cooks using citrusy soft drinks like 7 Up lately instead of apple juice.

ABOUT THE SAUCE. Don't stray far from Kansas City–style here. Mustard sauces and Asian-accented sauces don't stand a chance. Too bad. Also, judges like shiny ribs. Corn syrup makes shiny sauce. You might consider adding some. Many cooks buy a commercial sauce and doctor it to their tastes. Clint adds liquid margarine to his sauce for extra richness.

1 PREP. Many cooks do this step at home before they leave for the event. Rinse the ribs in cool water to remove any bone bits from the butchering. If you are cooking more than one slab, make sure the bones are all the same length so they will look pretty in the turn-in box and that they are not too large for the box. There is a membrane, the pleura, on the concave side of the slab. Sometimes it gets rubbery when cooked, sometimes it gets leathery. Get rid of it. The best way is to wiggle a table knife between the membrane and one of the bones in the center of the slab. Then wiggle a finger in there and lift. The membrane will start to pull free. If you're lucky, it will all come off in one pull. Sometimes it takes a couple more pulls. A paper towel helps you get a grip on the membrane. Remove excess fat and the flap of meat on one corner of the back side of St. Louis–cut ribs. Your meat will be inspected soon after you arrive at the contest, so do not season or inject your ribs until after you get the all-clear.

2 AMPLIFY. Dry brine and give the salt at least 1 to 2 hours to be absorbed. Some champs even inject special compounds such as Butcher BBQ

Pork Injection between each rib. It helps the meat retain moisture and boosts flavor. If you inject, skip the dry brine.

3   FIRE UP. You can do this on a charcoal grill set up in 2 zones, even a gas grill at home, but you will need a smoker to win. Get the temperature to 225°F and stable. Add some wood and make sure it is blowing blue smoke. Small hot fire with lots of air.

4   RUB. Coat both sides of the meat with the mustard, then sprinkle on Meathead's Memphis Dust generously. Keep in mind, much will drip off.

5   COOK. Place the slab meat side up on the cooking grate as far away from the heat source as possible in about 225°F air. Close the lid. Allow the ribs to smoke until the meat begins to shrink back from the ends of the bones and when the meat color gets dark, about 2 hours. Some cooks spritz it with a mist of apple juice to keep it wet and cool. This slows the cook, attracts smoke, and replenishes a small amount of the moisture that has dripped off and evaporated.

6   TEXAS CRUTCH. Lay out two layers of 18-inch-wide heavy-duty aluminum foil at least 8 inches longer than the slab on each end. Squirt a few stripes of margarine the length of the slab on the foil, then scatter half the brown sugar and honey on top. Lay the ribs meat side down on the butter, brown sugar, and honey. On the bone side, squirt some more margarine, and the rest of the brown sugar and honey. Pull up the sides of the foil to create a boat, pour in

the apple juice, and tightly crimp the foil all around the slab. No air leaks.

7   COOK SOME MORE. Place the packet on the smoker and cook another 3 hours for St. Louis cut, 2 hours for baby backs.

8   PREPARE THE TURN-IN BOX. Make sure you know what is allowed in the Styrofoam turn-in boxes. Line the box artfully. Many teams prefer parsley because it is bright green and does not easily wilt.

9   FINISH. Carefully open the foil. That escaping steam can peel your nose. Check the temperature between the bones. It should be between 195° and 205°F. For home cooking I use the bounce test. I pick up the slab in the center with tongs and bounce. If the surface cracks it is done. But for competition you can't do this for fear of defacing a few ribs. Cut off the end bone and taste the second bone in. The meat should pull nicely off the bone leaving bare bone, but it should not fall off the bone. If they are done, remove the ribs from the foil and set them back on the smoker meat side up. Close the lid for about 15 minutes to firm up the bark.

10   SET THE SAUCE. Now brush Kansas City Red on both sides of the ribs and turn them meat side up on the smoker or grill. One coat is enough. The judges want to taste the meat. Close the lid and cook until the sauce sets and becomes tacky. Some cooks like to hit them with high heat on a grill to caramelize the sauce. Be careful it does not burn.

11 CUT AND BOX. Turn the slab meat side down on a cutting board and slice the ribs carefully midway between the bones. It is easiest to cut bone side up. Make sure your knife is wicked sharp. You want clean cuts, and be certain that no two ribs are attached to each other or you can be disqualified. You have now messed up the finish on the meaty side during cutting, so line up the bones in the board meat side up and touch up their finish with more sauce, top and bottom only, never the sides. Some cooks give them a spritz with corn syrup to make them shiny. Select the 8 prettiest ribs that are similar in size. Now carefully place the ribs in the turn-in box. Greenery should frame them. If you got sauce on the inside of the box, wipe it off. Remember, the first thing judges do is score the appearance in the box.

12 WALK. Slip the turn-in box into an insulated bag like those pizza deliveries use to keep the food warm, and carry it carefully to the turn-in. For KCBS contests, turn-in time is 12:30.

MYTH

Boil ribs to make them tender

BUSTED. Don't do it! You wouldn't boil a steak, would you? Water is a solvent and it pulls much of the flavor out of the meat and bones and makes the meat mushy. When you boil meat and bones, the water turns tan because you are making a flavorful stock. Ribs are most flavorful when roasted slowly at low temps in the presence of smoke. If you are in a hurry, you are better off steaming or microwaving ribs and then finishing them on the grill or under the broiler. Just don't boil 'em. When you boil ribs, the terrorists win!

TUFFY STONE'S CHAMPIONSHIP PORK BUTT

Tuffy "The Professor" Stone is a fellow BBQ Hall of Famer and owner of the Q Barbecue restaurants in Glen Allen and Midlothian, Virginia. He is a frequent judge on the TV show *BBQ Pitmasters* and is the author of the book *Cool Smoke: The Art of Great Barbecue*. We have a number of his instructional videos on AmazingRibs.com. He also makes a line of rubs and sauces and teaches classes.

His Cool Smoke BBQ Team has won more grand championships than he can count, including the big ones: The American Royal Open in Kansas City, the Kingsford Invitational, and the Jack Daniel's Invitational (twice). His pork butt has won first place in the Memphis in May World Championship Barbecue Cooking Contest. We asked him for tips on how to cook it for competition. This recipe is inspired by his method. It is more complex than the method I use at home and described in my last book. There's a lotta stuff going on here, but that's what it takes to win the big bucks.

There is a simpler method in my last book and on AmazingRibs.com.

MAKES *An 8- to 10-pound pork butt, 5 to 7 pounds after trimming and cooking, enough for 10 to 16 servings*

TAKES *30 minutes to prep, up to 12 hours after applying rub and injection before cooking, 8 hours to cook*

SPECIAL TOOLS *Meat injector, spray bottle, heavy-duty aluminum foil. Fist-size wood chunks if you are cooking with charcoal or gas.*

COOL SMOKE CHILI POWDER (MAKES A GENEROUS 2 CUPS)

- 1 cup smoked mild paprika
- 5 tablespoons plus 1 teaspoon dried Greek oregano
- 3 tablespoons plus 1 teaspoon ground cumin
- 3 tablespoons plus 1 teaspoon granulated garlic

Pork turn-in box by Darren & Sherry Warth.

- 3 tablespoons plus 1 teaspoon powdered cayenne pepper
- 2 tablespoons granulated onion

COOL SMOKE RUB (MAKES A GENEROUS 1½ CUPS)

- ¼ cup Cool Smoke Chili Powder (above)
- ½ cup turbinado sugar
- ¼ cup plus 2 tablespoons Morton Coarse Kosher Salt
- 2 tablespoons plus 2 teaspoons ground cumin
- 1 teaspoon cayenne pepper
- 1 tablespoon plus 1 teaspoon cracked black pepper
- 1 tablespoon plus 1 teaspoon granulated garlic
- 1 tablespoon plus 1 teaspoon granulated onion

COOL SMOKE INJECTION (MAKES A GENEROUS 2 CUPS)

1½ cups apple juice

3 tablespoons dark brown sugar

⅓ cup Butcher BBQ Pork Injection

¼ teaspoon xanthan gum

COOL SMOKE BARBECUE SAUCE (MAKES 1 QUART)

3 cups ketchup

1 cup packed dark brown sugar

¾ cup distilled white vinegar

¾ cup water

¼ cup molasses

¼ cup apple cider vinegar

3 tablespoons Worcestershire sauce

2 tablespoons smoked mild paprika

1 tablespoon Cool Smoke Chili Powder (above)

2 teaspoons ground cumin

2 teaspoons granulated onion

2 teaspoons granulated garlic

2 teaspoons cayenne pepper flakes

1½ teaspoons fine-grind black pepper

½ teaspoon Morton Coarse Kosher Salt

COOL SMOKE PORK BUTT

1 (8- to 10-pound) bone-in pork butt

2 cups Cool Smoke Injection (above)

1 cup Cool Smoke Rub (above)

2 cups apple juice (in a spray bottle)

3 cups Cool Smoke Barbecue Sauce (above)

ABOUT GRANULATED GARLIC AND ONION. They are similar to the powdered garlic and onion I use in most of my recipes only the grains are larger, closer to the size of sand. If you substitute garlic or onion powder, use about two-thirds as much.

ABOUT THE BUTCHER'S BBQ PORK INJECTION. Tuffy, like many top competitors, injects his meat with a special commercial product from Butcher BBQ to amplify flavor and retain juice. You can buy it online or substitute chicken stock. It helps but leaving it out will not ruin the meat.

ABOUT THE XANTHAN GUM. This is a soluble fiber powder widely used as a thickener or stabilizer. It's made by fermenting sugar with the bacteria *Xanthomonas campestris*.

IN ADVANCE. Many cooks will prepare their rubs and sauces in advance and even trim their meat. When they get to the competition site, the meats are inspected, they are issued turn-in boxes, and they will arrange the garnish in advance and chill the box.

1 COOL SMOKE CHILI POWDER. Mix all the ingredients in a bowl. You can make this ahead and store it in a clean airtight jar.

2 COOL SMOKE RUB. Mix all the ingredients in a bowl. You can make this ahead and store it in a clean airtight jar.

3 COOL SMOKE INJECTION. Thoroughly whisk all the ingredients together in a bowl. It will keep in the refrigerator for up to 5 days.

4 COOL SMOKE BARBECUE SAUCE. In a 4-quart saucepan, whisk together all the ingredients. Bring to simmer over medium heat and cook, stirring constantly, until the sauce thickens, about 20 minutes. Set aside to cool completely. Transfer to a very clean airtight container and refrigerate until ready to use, up to 2 weeks. The sauce is best if it sits overnight.

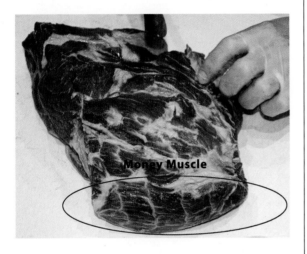

Money Muscle

5 TRIM. Cut most of the surface fat off the butt. You want to put the rub on the meat not on a layer of fat. Cut it so the money muscle, the one with the tiger stripes on the end, is fully exposed. You might have to slice a little above and below it to get it hanging out so it gets browned on almost all sides, but rules don't allow you to remove it. Give this piece special care because it is very tender and juicy.

6 INJECT. Immediately after the inspector has approved your meat, inject as much as possible, repeatedly injecting the meat every 1 inch with about 1 ounce (2 tablespoons). Dust the meat evenly all over with the rub. Refrigerate, uncovered, several hours.

7 FIRE UP. When you are ready to cook, preheat the smoker to 300°F. If you don't have a smoker, heat the grill to 300°F in the indirect zone. I know we usually cook at 225°F, but pork butt is resilient. It can take higher temps and this temp gets you done on time.

8 SMOKE. Tuffy cooks only with logs, but in competition you can use charcoal with wood chunks or with pellets or gas with wood at home. Place the meat in the smoker away from the heat, or on the indirect side of a grill, add 3 wood chunks, and close the lid. Cook for 1 hour, and if the wood burns up, add 3 more chunks. Then spray the meat with apple juice. Cook for 3 hours, spraying the meat every 30 minutes and adding wood when the smoke dies.

9 TEXAS CRUTCH. Cut two 18 × 24-inch pieces of heavy-duty aluminum foil and join them lengthwise by crimping the edges. Lay the resulting piece out flat on your work surface. Remove the butt from the cooker and fold it tightly in the foil. Return it to the smoker or to the indirect side of the grill. At home you can use your indoor oven. Cook for another 2½ to 3½ hours, checking the temperature after 2½ hours. It's done at 195°F.

10 SET THE SAUCE. Remove the meat from the foil and brush with ½ cup of barbecue sauce. Return it, uncovered, to the smoker and cook for another 10 minutes to set the sauce.

Remove from the heat, wrap it in foil, and let it rest for 45 minutes in a beer cooler.

11 BOX. Now you are faced with a conundrum. What do you want to turn in? As judges, Clint and I like to see some variety in the turn-in box, usually some slices of the money muscle and pulled or shredded pork. Many cooks give it a shine and some moisture by painting the cut surface with jus from the foil mixed with a little Kansas City–style BBQ sauce.

12 TURN-IN. Be sure to get it to the turn-in table by 1 p.m.

DORIE'S PORK À LA NORMANDE

Normandy, on the north coast of France, was the site of bloody D-Day, June 6, 1944. No matter what you have heard about the French, they have not forgotten our sacrifice and they love Americans in Normandy. It is also replete with glorious rolling hills, apple orchards, and wondrous food. The region is famous for seafood and its four Cs: Cider, Calvados, Cream, and Camembert.

Dorie Greenspan is the author of numerous brilliant cookbooks with recipes that never fail. Dorie and her husband, Michael, have lived in Paris on and off for decades. Her award-winning book *Around My French Table*

is a favorite around our American table. I was flattered when Dorie and Michael came to one of my book signings, bought a copy of my last book, and asked me to sign it.

So many French recipes start by searing meat in a pan, then building the sauce on top of the fond, the browned bits that stick to the pan when the meat is done searing. But surprisingly Dorie's recipe doesn't depend on fond. It allows us to build the sauce separately from the meat, perfect for a griller. And let me tell you, smoke and cream go together like pork and apples, or cream and mushrooms. And they're all here. Here's my version of her recipe, slightly modified for the grill. You want thick chops for this method.

MAKES *2 servings*

TAKES *1 to 2 hours to dry brine, 30 minutes to prep, 30 minutes to smoke, 30 minutes to finish on the grill*

- 2 pork chops, 2 inches thick
- Morton Coarse Kosher Salt
- 4 mushrooms (the best you can find)
- 1 medium onion
- 2 tablespoons good olive oil
- 8 large fresh sage leaves
- 1 teaspoon cornstarch
- 1 medium apple
- 2 tablespoons (1 ounce) unsalted butter
- ½ teaspoon rubbed (ground) sage
- Fine-grind black pepper to taste
- ½ cup low-sodium chicken broth

3 tablespoons Calvados, apple jack, Cognac, Armagnac, or other straight brandy

½ cup cream or half-and-half

1 cup brown rice

Flaky finishing salt, such as Maldon

SERVE WITH. A sparkling hard cider from Normandy.

1 PREP. Trim the chops of excess fat and dry brine them 1 to 2 hours in advance, ½ teaspoon Morton Coarse Kosher Salt per pound. Clean the mushrooms and peel the onion, then chop them kinda small.

2 FRY THE SAGE LEAVES. In a small skillet, heat the olive oil over medium-high heat until the oil shimmers. Meanwhile, lightly coat the 8 large flawless sage leaves with cornstarch. When the oil is ready, push them down into the oil so they are coated. After about 30 seconds remove with tongs and place on a paper towel to drain. They should be nice and crispy.

3 FIRE UP. Fire up your smoker and aim for 225°F. Or set up a grill for 2-zone cooking, aim for about 225°F on the indirect side, and get some smoke going.

4 SMOKE. Add the pork chops and smoke at 225°F indirect, lid down, until internal temperature reaches 120°F, about 45 minutes.

5 BUILD THE SAUCE. While the meat is smoking, peel the apple and cut it into bite-size chunks. In a 12-inch frying pan, melt the butter over medium heat. Add the rubbed sage, apple, onion, and mushrooms and sprinkle with salt and pepper. Turn up the heat and sauté until the onions and mushrooms are limp and the butter begins to brown, about 15 to 20 minutes. Add the broth, Calvados, and cream and simmer until reduced by half, about 5 minutes. Keep the sauce warm on low.

6 RICE. When the meat hits 120°F it is time to start cooking the rice.

7 GRILL. Move the meat over to the direct heat side. Grill, lid up, until the chops are golden or brown all over and they hit 135° to 140°F.

8 SERVE. Put a bed of rice on each plate, spoon the sauce on top, and nestle a chop on top of that. Scatter with the large-grain salt like Maldon and garnish with the fried sage leaves.

GRIDDLED HAM-N-SHRIMP HASH

Here's a Sunday brunch perfect for your griddle. All the elements cook fast.

MAKES *2 servings*

TAKES *About 45 minutes*

SPECIAL TOOLS *Griddle, large spatula*

1 pound potatoes, such as Yukon Gold or russet

¼ teaspoon baking soda

- 1 jalapeño pepper
- 1 small red bell pepper or ½ medium red bell pepper
- 6 ounces Canadian bacon or ham
- ¼ pound medium shrimp
- 2 garlic cloves
- 4 scallions
- 2 tablespoons good olive oil, plus more as needed
- ¼ teaspoon Morton Coarse Kosher Salt
- ¼ teaspoon medium-grind black pepper
- 3 tablespoons bacon grease, duck fat, or clarified butter
- 1 tablespoon fresh thyme
- 2 large eggs

OPTIONAL. Your favorite hot sauce, for serving

ABOUT THE CANADIAN BACON. Canadian bacon is cured pork loin and is closer to ham than bacon, so feel free to use ham.

1 PARCOOK THE TATERS. We'll use a technique taught by Kenji. Scrub and peel the potatoes and cut them into 1-inch cubes. Add 1 quart of water to a 2-quart pot, add the potatoes, and bring to a boil. Add the baking soda and simmer until the potatoes are tender but still a little firm in the center, 10 minutes or so. Drain in a colander and toss them around a bit in order to make the surfaces fluffy and easier to get crunchy.

2 MORE PREP. Cut the jalapeño in half, remove the stem and seeds. If you want it spicy, leave in the ribs. Mince it. Cut the bell pepper in half and remove the stem and seeds. Cut it into ½-inch chunks. Cut the Canadian bacon or ham into ½-inch chunks. Peel the shrimp and remove the vein that runs down its back. Peel and mince or press the garlic. Clean the scallions, peel off any dead skin, chop off the roots, cut the white parts into ½-inch lengths. Gently toss all these goodies in a mixing bowl with the olive oil, salt, and black pepper. Thinly slice the green parts of the scallions and set aside about 2 tablespoons for garnish.

3 FIRE UP. Fire up your griddle or set up the grill for 2-zone cooking; aim for Warp 7 on the hot side. If you are using a griddle topper, place it over direct heat. Add the bacon grease to the griddle and when it is hot spread it out to cover the entire surface with a spatula or brush. Toss on the potatoes and let them sit until they start to brown, then roll them around.

4 ADD THE REST OF THE HASH. Add the Canadian bacon, shrimp, bell pepper, garlic, scallion whites, and jalapeño on the griddle and cook, lid down, until things start getting golden in places. Every 3 to 4 minutes use a spatula to turn the hash over so it can brown on the other side. The shrimp is ready when it starts to curl and turns opaque.

5 FINISH. As it gets close to being done, sprinkle the thyme and about 2 tablespoons of the scallion greens on top and stir it up. Move the hash aside, add more oil to the griddle, crack on the eggs, close the lid, and fry to your liking. As they approach doneness, scoop out the hash and plate it. Top with the eggs. If desired, serve with hot sauce.

CHICKEN AND TURKEY

"The best comfort food will always be greens, cornbread, and fried chicken."

—MAYA ANGELOU

How many cookouts have you been to where the chicken was black on the outside and raw on the inside? We can fix this.

Sorry to tell you this, but because of the way chickens are grown, processed, and shipped nowadays, there is a high likelihood it has pathogenic bacteria in it. But if you handle it with care, there is nothing to fear. Be careful about drips and splatters. Be careful to clean counters and utensils that touch raw poultry. A digital thermometer is an absolute requirement.

According to USDA, the per capita consumption of chicken is 97 pounds, 46% of all our meat consumption (beef is next with 27% and pork weighs in at 24%). Makes sense. Chicken is much less expensive, it's lean, and its pale meat is a blank canvas, willing to accept a wide range of flavorings. It is also easier on the environment.

BREASTS have little fat or connective tissue. They dry out easily and should not be cooked past 160°F. This is a cut that really benefits from sous-vide-que and pasteurized/finished at 155°F.

THIGHS AND DRUMSTICKS have more myoglobin, fat, and connective tissue, which makes them darker and a little more flavorful. They are best at about 170°.

WINGS are technically white meat, but their charm is the skin and you want that crispy. They cook faster than other parts, but once the skin is done, rest assured the meat is done.

POULTRY TERMS TO KNOW

BROILER OR FRYER. These are interchangeable terms used for the most common all-purpose chicken in the grocery. It is a chicken usually younger than 10 weeks old, of either sex, weighing 3 to 4 pounds.

CORNISH GAME HEN. Also common in US groceries, this is not a game bird or even a different breed. It is just an immature chicken younger than 5 weeks, of either sex, with a ready-to-cook weight of 2 pounds or less.

ROASTER OR ROASTING CHICKEN. Between 8 to 12 weeks old, of either sex, weighing 5 pounds or more.

FRESH. A deceptive term that means the carcass has never been below 26°F. At this temperature ice crystals can form and the carcass is as hard as a bowling ball. I have no idea how the USDA can consider this fresh, which to most consumers means never frozen.

ENHANCED, BASTED, SELF BASTED. USDA allows producers to add up to 8% of the carcass weight with an injection of salt water, flavor enhancers, and tenderizers. They improve water retention, taste, texture, and profits by about 8%. Do not brine these birds.

CAGE-FREE. The birds are not kept in cages, but they are kept in hen houses, safe from predators but crowded and stuffy.

FREE-RANGE, FREE-ROAMING. Another misleading label term. Most poultry is raised in a row of connected "battery" cages in a hen house. They can be humane and safe for the birds if they are not overcrowded, but in reality they are often cramped, leaving birds barely enough room to turn around. "Free range" animals may be roaming loose in crowded buildings, but a door is open and they have been *allowed access to the outside*, often a small fenced-in yard. That doesn't mean they actually walked out the door, and often they don't. It is bright out there. No roof. Scary. But they can go for a short stroll if they want to.

PASTURE-RAISED. Has no legal definition. Might be true. Might not.

NO HORMONES. Hormones are not allowed to be given to hogs or poultry so the claim "no hormones" on a label is like saying "no milk added" on an orange juice bottle. If you see it on a label, somebody is trying to bamboozle you.

NO ANTIBIOTICS. The term may be used on poultry if documentation is provided to the USDA demonstrating that healthy animals were raised without antibiotics. Sick animals may still be treated with antibiotics.

USDA ORGANIC. Antibiotics can only be used on newly hatched birds less than 2 days old or sick birds, although *Consumer Reports* says that "some farmers stretch the boundaries of what is medically necessary." They must be fed only certified organic feed for their entire lives once they reach 2 days old. Organic food cannot contain antibiotics, they cannot be genetically engineered, they cannot be grown with the use of pesticides or chemical fertilizers, and they cannot contain any animal by-products. They must have outdoor access.

NATURAL. There are no standards or regulations for this word. It is meaningless.

BREAST LOBES. Technically, poultry has one breast, made up of two lobes, or two muscles, on either side of the keel bone. Most chefs incorrectly call a lobe a breast, when it is really only half a breast. So to avoid confusion, I specify how many lobes in my recipes.

COOKING CHICKEN

Cooking whole animals is usually a mistake because they don't cook evenly, and this is true for poultry. They are like fleshy tubes, hollow inside with thick breasts, medium-thick legs, and thin wings. They also have all sorts of bones, some thick, like leg bones, and some very thin, like rib bones. Then there is the skin, whose chemistry is significantly different than the flesh, but it is all on one side.

When you roast a whole bird, the inside of the tube remains pale, but when you butterfly or break down the bird into parts, you expose all sides to heat and they can brown, and brown is beautiful. Breaking the bird into parts allows you to cook each piece to optimum temperature and get crispy skin on all sides of the wings.

Start by removing the packet of neck, gizzard, heart, and liver from the cavity. Put the liver in a zipper bag and save up a bunch of livers in the freezer to use for making chopped chicken liver spread (you'll find a nice recipe linked to AmazingRibs.com/mm). Put the rest in another bag in the freezer and when you get a bunch of them and some bones, dump them in a pot of water and simmer for hours to make stock.

The secret to cooking poultry is to reverse sear. Start with a dry brine and then cook with the skin side up on the indirect side of the grill at 325° to 350°F. We need this higher temp to render fat under the skin and crisp the skin. At about 145°F internal temperature, you move it to direct heat, skin side down, lid open, and brown the skin, then flip, and brown the meat side.

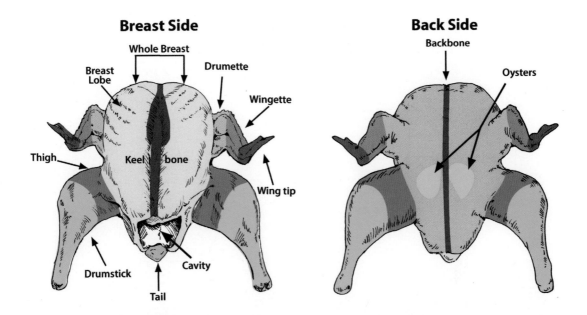

Breast Side

Whole Breast

Breast Lobe

Drumette

Wingette

Thigh

Keel bone

Wing tip

Drumstick

Cavity

Tail

Back Side

Backbone

Oysters

SPATCHCOCKING (BUTTERFLYING) WHOLE BIRDS

Spatchcocked is an olde English term that sounds quite naughty. Americans often prefer the less iffy term: *Butterflied*. That's a spatchcocked chicken in the image on page 258.

In addition to the bird getting browned on both sides, when you spatchcock you can season all sides more evenly, it cooks faster with less loss of moisture, carving is easier, it looks impressive, and it is just plain fun to say "spatchcock." Here's how to do it:

Place the bird on a cutting board breast side down. You can tell because the wing tips and drumsticks are pointing down. Cut alongside the backbone on both sides with kitchen shears or clean tin snips. Remove the backbone and spread open the cavity. Clean the brown goop (kidneys) off the backbone and freeze the backbone for making stock.

Flip the bird over skin side up and press down on the breastbone between the wings. You'll hear it crack. That's the collar bone (wishbone). You now have a spatchcocked bird. I like to slip a blade under the ribs and remove them. I also like to remove the wing tips because they are so thin they easily burn, and I toss them into the bag in the freezer with the backbone, neck, heart, ribs, and gizzard for making stock. Some cooks like to run a metal skewer through the wings and breasts to help the bird stay splayed when flipped.

HALVING THE BIRD

After removing the backbone, you can flip the bird over and cut between the two lobes of the breast along one side of the keel bone. Now you have two halves. Half a bird is a generous portion for a hungry person.

QUARTERING THE BIRD

One of the problems with spatchcocking and halving is that the thighs are very loosely connected and when you want to move the bird around or flip it during cooking, the legs easily tear off. So why not remove them before they fall off? You now have four parts. Quartering is also the best way to make sure the dark meat and white meat eaters get what they want. I think cutting the bird into quarters gives you the best chance of cooking all pieces perfectly. If you cut the bird into four parts, you can use your trusty instant-read thermometer and monitor doneness for each and every part and

move them closer or farther from the heat as needed. You will be surprised at how two identical legs will cook at different rates.

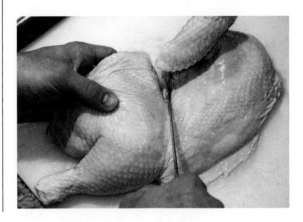

EIGHT OR TEN PIECES

The next logical step, and the method I recommend most often, is to go one step further and cut the bird into 8 or 10 pieces. Do it right and you will have tender, juicy pieces, browned all over, with each piece cooked to perfection, a feat impossible to achieve if the bird is whole or even spatchcocked. Best of all, you don't have to struggle carving up a whole hot bird.

After quartering, remove the wings. The charm of the wings is that they are wrapped in skin that can be made crispy, but because they are so thin, they will easily burn while

attached to the breast. Removing them is a bit tricky because the shoulder joint is buried in the breast. The way to do this is to grab the wing by the drumette near where it joins the body and wiggle it aggressively until the ball and socket pops. You can then easily get a knife in the joint and cut through any meat and skin. You can cook the wings alongside the other pieces. Just get them away from the heat sooner or toss them on after the thighs and breasts have been cooking for about 15 minutes. Or you can freeze them and when you have a lot make a meal of them.

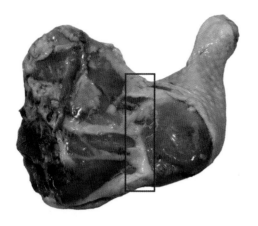

Now separate the drumsticks from the thighs. Turn the leg skin side down and you will see a seam of fat running right where the joint is. Follow that seam with your knife, just a bit toward the drumstick side, and you should hit the knee joint. If you're having trouble, grab both parts and bend until the joint pops and finish the task with a knife. You now have 8 pieces.

If you are deep-frying, it is important that all the pieces be about the same size so they get done about the same time. So when I am doing this, I cut the breast in half chopping right through the rib cage so each half is about the same size as the thigh. That's 10 pieces.

CARVING A WHOLE BIRD

If you must cook the bird whole or spatchcocked, when it is done, place the bird on a cutting board with the wings and thighs down, breasts up. Bend the thighs until you see the hip joint and cut the two bones apart removing the legs. Locate the keel bone down the center of the breast on top. Place a filleting knife on one side of the keel and draw it across the skin and downward a few times until you strike the ribs. Turn the knife blade outward slightly and slide it along the ribs until the whole breast lobe falls off. Repeat on the other side. Cut each lobe into ½-inch slices across the grain so they all have some skin. Roll the carcass on its side and bend the wing back until you see the shoulder joint and cut through it. Repeat on the other side. Flip the bird upside down and feel along the backbone near where the hip joints were. You should find two little hunks of succulent meat, one on either side of the backbone, called the "oysters" because they are about the size of an oyster's meat. Make sure nobody is watching and then eat them. You earned it. Take the remaining carcass, break it into pieces, and set it aside for making stock.

Beer Can Chicken

BUSTED. No doubt about it, a whole golden chicken sitting on its butt like Buddha is delicious looking. And it absolutely is delicious, because it is a roasted chicken and roasted chicken is a culinary wonder. But the beer can has nothing to do with its wondrousness, and in fact, it keeps the chicken from reaching its pinnacle. Here's why.

First, chicken meat is fully saturated with water. There's no room in there for any beer. Take a chicken, pat it dry, weigh it, and soak it overnight in a vat of beer. Pat it dry and weigh it again. It has absorbed next to nothing.

Even if the chicken could absorb beer, vapor escaping the top of the can is all the way up the cavity just below the shoulders, so any vapor would contact only a tiny part of the

bird. Alas, there are no superhighways in the chicken for the beer to travel through to get to the breast or thighs. The vapors hit the cold shoulders, condense, and roll back into the can.

But that's if the beer could get outta the can. The bird comes out of the fridge at 38°F. Wrapped around the beer it becomes a very effective beer coozie. As it cooks it heats up to 160° to 170°F, and so does the beer. They are a single thermal mass. Even if you waaaaay overcook the bird to 190°F, at which point it is cardboard, beer doesn't really evaporate much until it hits 212°F. Then, to make matters worse, chicken juices and fat drip into the can and fat floats on the beer making a very effective lid on the liquid. Meanwhile, with a metal object up its butt, the meat in the cavity stays a boring tan with no tasty Maillard browning.

Besides, who wants chicken that tastes like beer? Whoever thought this up musta had one too many. Ask yourself this: Have you ever seen a fine-dining restaurant serve Beer Can Chicken?

Truss poultry legs

BUSTED. All the cookbooks and TV chefs tell us to tie the tips of the drumsticks together so the thighs hug the body. So what happens when you do? Well, you make the thermal mass larger by adding weight to the body, slowing the cooking and increasing moisture loss. You are also pulling the insides of the thighs close to the crotch, a recipe for a rash in humans, but in poultry it means a lot of tan, flabby skin. But if you let the thighs hang out, hot air can get in there and brown the skin. It can also cook the thighs from all sides, and that's good because we like to get the thighs 10°F warmer than the breasts. The only time I truss is when I am cooking poultry on a rotisserie. If the legs aren't trussed on a rotisserie they will flop around and break off.

CHAMPIONSHIP CHICKEN

If you want to compete, here's the way a lot of the champs do chicken. Not surprisingly, they don't do this at home. But if you want to taste killer chicken, here you go.

So many competition cooks say chicken is their Achilles' heel and it is the category that causes them the most stress. Most cook thighs, because dark meat is more forgiving and juicier. Judges expect "bite through" skin coated in sweet red sauce. Bite through means soft skin so that when you take a bite your teeth cut through and you get skin and meat in every bite. If the skin pulls off, there goes your score. This ain't easy. Personally, I prefer herbs to Kansas City Red sauce on poultry, as in my French Rub (page 167), and I prefer crispy skin.

But sadly, for some reason, that approach will not win in a competition.

Some actually stuff boneless thighs into muffin tins that mold them into a uniform shape and then braise them in butter in their smokers. There are many creative techniques,

Chicken thigh turn-in box by Todd Johns, Plowboys BBQ.

so for this recipe I have stuck close to one by Hall of Famer Darren Warth and his wife Sherry of Smokey D's BBQ restaurant in Des Moines, Iowa. They are the best team I know. Their meats are always simply amazing. At the time I am writing this, they have more than eight hundred category wins and more than seventy-five state championships, including prestigious wins at the American Royal World Series of BBQ, the Jack Daniel's World Championship Invitational BBQ, the Kingsford Invitational, the Sam's Club National Championship, the King of the Smokers, and the Houston Livestock Show and Rodeo World's Championship Bar-B-Que Contest. Here is my interpretation of their procedures.

MAKES *12 thighs (although you need only 6 for the judges, everyone makes more and only the pick of the litter make it into the turn-in box)*
TAKES *20 minutes to prep, 2 hours to cook*

> 12 medium bone-in, skin-on chicken thighs
>
> 3 tablespoons Meathead's Memphis Dust (page 165)
>
> Morton Coarse Kosher Salt
>
> 2 cups low-sodium chicken stock
>
> 1 stick (4 ounces) unsalted butter
>
> 3½ cups Kansas City Red (page 180)
>
> ½ cup apple juice

ABOUT THE CHICKEN STOCK. This is to be used as an injection. Darren and Sherry use ¼ cup Bird Booster Rotisserie Flavor injection by Butcher BBQ mixed with 2 cups water. Butcher BBQ injections are cleverly formulated with compounds that amplify flavor and enhance moisture retention. You can find them online.

ABOUT THE RUBS AND SAUCE. Darren and Sherry use three different commercial rubs, all of which have salt in them. They even have one for the skin side and one for the meat side so the tongue and roof of the mouth get different flavors! Most teams go with one rub and it is a lot like my Meathead's Memphis Dust with a little salt. In fact, that's a recipe that many teams use. The Warths also mix two commercial sauces.

1 PREP. Remove the fat on the edge of thighs, the blood vein nestled beside the bone, any loose tendons and membrane, and excess skin. Chop off the "knuckle," the joint end of the bone. Some cooks use garden shears for this task and some even remove the bone entirely. Dig out the piece of cartilage under the bone. The goal is to even the edges and make all the thighs same size so that they cook equally, weighing about 5 ounces after trimming. Many cooks do the trimming at home before they leave for the event, but the rules say you can't season until the meat is inspected on site.

2 SKIN. There are several ways to approach the skin problem once you are at the event. Here are three:

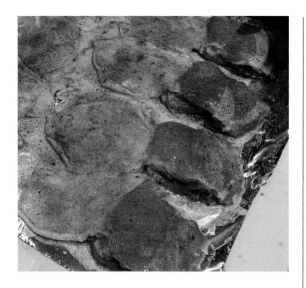

2A As shown in the image above, pull the skin back, trying hard not to detach it from one edge, and then scrape off the subcutaneous fat layer with a knife, being careful not to break the skin. Sprinkle the exposed meat and underside of the skin with the rub and lay the skin back on top like a toupee. Sprinkle the rub on the top side. Sprinkle Morton Coarse Kosher Salt on both sides now. If you are using a commercial rub that has salt, skip the additional salt.

2B Some cooks remove the skin altogether, scrape off the fat, sprinkle rub on the naked meat and the underside of the skin, place the thigh in the center of the skin, and roll the skin up so it covers almost all the flesh. Some competitors will pin the edges of the skin with toothpicks or metal poultry pins to keep it from shrinking away from the meat as it cooks. Be careful to remove toothpicks or pins before placing the thighs in the turn-in box because foreign objects will disqualify your entry.

2C At this time, some cooks even use a jaccard, a device with several dozen needle-like knives, to puncture the skin and the flesh. Jaccarding a steak cooked to 135°F is unsafe, but jaccarding chicken cooked to 160°F and up is safe.

3 INJECT. Place a wire rack in a pan and set the chicken thighs on the rack skin side down. Dissolve 1 teaspoon of Morton Coarse Kosher Salt in the chicken stock and inject about 1 ounce (2 tablespoons) of the chicken stock into each thigh from the underside. Aim to inject each thigh twice on either side of the bone. At this time, some cooks put the thighs in zipper bags with a marinade or the leftover injection. Instead, the Warths place the thighs in the refrigerator uncovered for 2 hours. Leaving them uncovered helps the skin dry, especially if there is salt in the rub.

4 FLAVORIZE. Remove the thighs from the refrigerator. Turn them skin side up and shape them so they all look uniform. Sprinkle the tops of the skins with Meathead's Memphis Dust and then sprinkle salt lightly on the skin. If you are using a commercial rub with salt, don't add more. Return the thighs to the refrigerator uncovered for 1 hour.

5 FIRE UP. Start your smoker and aim for 325°F or set up a grill for 2-zone cooking, 325°F on the indirect side, and get some smoke rolling. Apple and cherry seem to be favorite woods with champions.

6 SMOKE. Place all 12 thighs skin side up in a pan. (Some folks put the chicken on a wire rack, some coat the bottom of the pan with liquid margarine and place the thighs on the margarine. Some even put the meat in muffin pans and poach it in margarine or butter.) Cut the stick of butter into 12 equal pieces and place a piece on top of each thigh and place in the smoker or on the indirect heat side of a grill until the thighs reach an internal temperature of 160°F, about 1 hour.

7 COOK. Cover the pan tightly with foil and cook for an additional 40 minutes or so until the thighs reach an internal temp of 185°F. Yes, I know 170° to 175°F is the temperature I recommend for most dark meat, but if a judge sees any pink juices, you're doomed. Thighs can take the higher temps and there will be no pink.

8 GET READY. As the thighs cook, in a small saucepan, warm the Kansas City Red and apple juice over low heat. Prepare the turn-in box with whatever garnish is allowed. I like curly leaf parsley.

9 GRILL. Remove the pan with the meat from the cooker. Remove the thighs from the pan, put the thighs on a wire rack skin side up and sprinkle lightly on both sides with more rub. Place the rack over direct heat and watch it like a hawk. You are done when the bottoms of the thighs are dark and almost crispy, 10 to 15 minutes total. The goal is to combine the backyard grilled flavor with the juicy tenderness of an oven-cooked thigh.

10 TURN-IN. Remove the thighs from grill and let them sit for 5 minutes. Then submerge each thigh into the warm Kansas City Red. This coats all sides and there are no brushstrokes. Place the sauced thighs back on a rack to drip and then gently place 6 to 8 thighs into the turn-in box on top of the greens. Turn-in is noon. Don't be late.

11 TEST. Go to AmazingRibs.com/mm to find a link to the exact commercial products the Warths use. Test which path to follow with the skin. Test salt quantities. Test finish temperature. Practice many times before competing. And have fun.

EXTRA CRISPY CHICKEN WINGS WITHOUT FRYING

Once thought to be an inferior cut, wings are so popular now that prices have skyrocketed. When it comes to finger foods while watching the game, they are hard to beat.

They have three parts and I separate them. I remove the tips because they are thin and a lot of skin wrapped around some soft bones. I save them for making chicken stock. This leaves the meatier parts of the wings, the drumette (the mini drumstick) and the wingette (the flat with two bones) and their skin.

Contrary to popular belief, chicken skin is not all fat. Yes, there is a lot of fat in it and more under it, but there is also a lot of water and protein. To get crunchy skin, several things need to happen. The connective tissues need to break down, creating a lot of moisture, then that moisture needs to go away and the proteins need to get tough. Baking powder can be used on poultry skins and pork skins to help make them crackling good. When wet, baking powder produces CO_2 gas. It makes thin blisters in the skin, which make more surface area to crisp, especially when baked in an oven like a grill. For this recipe, inspired by J. Kenji López-Alt, I add salt for dry brining.

MAKES *1 pound wings, about 1 serving*
TAKES *24 hours to air-dry, 10 minutes to hair-dry, and about 30 minutes to cook*

1¼ pounds whole chicken wings

1 teaspoon baking powder

1 teaspoon Morton Coarse Kosher Salt

OPTIONAL. Drizzle with a sauce. Kansas City Red (page 180) is a good choice, so are Alabama White Sauce (page 183), Teriyaki (page 175), Tare (page 203), and my fave, Thai Sweet Chile Sauce (page 200).

1 PREP. Break the wing into three parts. Freeze the tips for use in making stock. Stab each remaining segment with a sharp knife about 8 times so the fat will drain when cooking.

2 AIR-DRY. In a small bowl, mix the baking powder and salt. Sprinkle it on the meat. Space the wings on a wire rack over a pan in the fridge. Leave it uncovered in the fridge for 24 hours so the air will start to dehydrate the skin and the salt will penetrate.

3 FIRE UP your grill in 2 zones and get the indirect zone as hot as you can, 400°F or more if possible.

4 HAIR-DRY. Just before cooking, use a hair dryer on the chicken for about 10 minutes to drive off more moisture.

5 COOK. Roast, lid down, in the indirect zone until the meat reaches 145°F.

6 SEAR. Then sear both sides over direct heat with the lid open until the internal temp hits 160°F and the skin is golden and crunchy. Serve right away, no resting or tenting. If you like, drizzle with a sauce.

PARIS CHICKEN

Boneless chicken breasts make great sandwiches or even center-of-plate entrées. You can buy breasts boneless (more expensive than bone-in) or easily cut them off the bones and save the bones for stock. But then you have a pear-shaped piece. Thick at one end, tapering to thin at the other. And you know what happens. Either the thin end is dry and crunchy, or the thick end is stringy. The solution? Make both ends the same thickness by pounding the thick end flatter. You can do this with turkey breasts, too. Best of all, pounding chicken and turkey breasts is a great outlet for your aggressions.

I first tasted this extraordinary prep in Paris and it is an example of French cooking at its best. The creamy mushroom sauce is heavenly and I amp up the umami by adding MSG. Don't be spooked. Read about MSG on page 5. You can do this recipe with chicken or even pork chops.

MAKES *2 servings*

TAKES *1 to 2 hours to dry brine, 10 minutes to prep (if the mushroom sauce is done in advance or during the dry brining), 20 minutes to cook.*

> 2 boneless, skinless chicken breast lobes
>
> 1 teaspoon Morton Coarse Kosher Salt
>
> 2 teaspoons French Rub (page 167)
>
> 1 teaspoon MSG, such as Ac'cent or Aji-No-Moto
>
> Mushroom Cream Sauce (page 192)

1 POUND. Place a breast in a zipper bag to keep the juices from flying all over the place,

squeeze out most of the air, and with a frying pan or your fist, pound on the thick end until the whole breast is uniformly about ¾ inch thick. Repeat with the other breast. Sprinkle lightly with the salt and French Rub and hold in the fridge for an hour or two.

2 FIRE UP. Set up the grill in a 2-zone configuration and shoot for Warp 7 in the direct zone. No smoke needed. Sprinkle with the MSG just before cooking. Cook in the direct zone flipping often until it hits 160°F. While it is cooking, warm the Mushroom Cream Sauce.

3 SERVE. Spoon the sauce in two wide bowls and place the meat on top.

CHICK UNDER A BRICK

A Cornish game hen is a perfect little 2-pound package of flavor just right for two people. This young chick cooks quick, evenly, tender, and juicy, and because it is thin, it is perfect for this great method: We spatchcock it and sear it on both sides simultaneously between two cast-iron skillets.

You can do it indoors, but there will be a lot of sputtering and spattering, so it is better on the grill. Spatchcocking puts more of the surface in contact with hot metal where Maillard flavor is made. This method gets its name, Chicken Under a Brick, because it is often done with a brick wrapped in foil, but I prefer to use the superheated cast-iron skillets, one to hold the bird and one on top of

the bird. There's more surface area in contact with the meat, and the pans hold more energy than the brick. Kinda like a panini press.

MAKES *2 servings*
TAKES *2 to 4 hours to dry brine, about 25 minutes to cook*

SPECIAL TOOLS *Sturdy poultry shears, a sharp filleting knife, and two cast-iron skillets (ideally both 12 inches, but one can be 10 inches) or a griddle for the bottom or a brick wrapped in foil for the top weight*

- 1 Cornish game hen
- ¾ teaspoon Morton Coarse Kosher Salt
- ½ fennel bulb
- 1 cup seedless grapes (about 5 ounces)
- ½ cup pitted brine cured Kalamata-style olives
- ½ lemon
- 1 tablespoon white wine
- 4 garlic cloves
- 2 tablespoons good olive oil
- ½ teaspoon medium-grind black pepper
- 1 tablespoon dried rosemary leaves, ground almost to a powder
- 1½ tablespoons duck fat or inexpensive olive oil
- Small handful of roughly chopped fresh flat-leaf parsley

1 SPATCH. Use sturdy poultry shears to spatchcock the bird by cutting out the backbone (rinse out the brown goop and freeze the backbone for stock). Use a filleting knife to remove the rib cage so more meat is in contact

with hot metal and freeze the bones for stock. Sprinkle the bird with the salt. Refrigerate for 2 to 4 hours.

2 PREP. Trim, quarter, and slice the fennel bulb into ⅛-inch half moons. Cut the grapes and olives in half. Squeeze the juice from the half lemon and set it aside in a cup. Add the wine to the cup. Peel and mince or press the garlic. Combine the garlic, olive oil, black pepper, and rosemary in another cup.

3 FIRE UP. Set up your grill for 2-zone cooking. Preheat two cast-iron skillets over Warp 10 direct heat.

4 COOK. Just before cooking, rub the garlic/rosemary blend all over the bird, getting

a little bit underneath the skin of the breast and thighs. Pour the duck fat in one skillet and spread it around with a brush or spatula. Arrange the hen in the oiled skillet skin side up and place the other hot pan on top of the bird. Sizzle for about 10 minutes, lid down, moving the weight if necessary to get the legs in touch with the metal.

5 FLIP AND COOK. When the meat side is golden brown, turn the bird over and rearrange it so the skin is in good contact with the pan and add the second pan on top. Check the meat temperature in several locations and cook until everything is 150°F, about 4 minutes more. Remove the weight, take the bird out of the pan, and move it to the indirect side. Cut it in half to make two servings. Leave the pan where it is on the direct heat side.

6 MAKE THE PAN SAUCE. Add the fennel, olives, and grapes to the pan. If you need a bit more fat, add it. Cook until the fennel slices soften, about 5 minutes.

7 FINISH AND SERVE. Move the pan to the indirect side. Add the lemon juice/wine mixture and as it boils it should loosen the fond, the browned bits of chicken stuck to the pan. Use a wooden spoon or spatula to scrape them up and stir them into the liquid. Move the bird halves to dinner plates. Throw the parsley in the pan and after 30 seconds pour the mixture over the bird.

PASTALAYA

Pastalaya is a traditional Louisiana jambalaya with chicken, andouille sausage, shrimp, or ham, and of course the trinity of onion, bell peppers, and celery. Instead of rice it has pasta. So we had to kick it up a notch with smoked chicken and chicken skin cracklins on top. Yes, certainly you could just carve and eat this tasty smoked chicken all by itself, but here is a way to stretch it and turn it into an outrageously zaftig dish that will feed a big, happy crowd. If unexpected guests arrive, just increase the amount of pasta and broth to make the same basic ingredients feed several extra mouths.

MAKES *A big bowl with 6 to 8 servings*
TAKES *2 hours to dry brine, 45 minutes to cook the chicken, 65 to 70 minutes to cook the pastalaya*

CHICKEN AND SAUSAGE

   1 whole chicken (3½ to 4 pounds)

1½ teaspoons Morton Coarse Kosher Salt

3 tablespoons French Rub (page 167)

¾ pound andouille sausage

CHICKEN CRACKLINS

2 tablespoons vegetable oil or chicken fat

½ teaspoon Morton Coarse Kosher Salt

PASTALAYA

1 medium yellow onion

1 large green bell pepper

1 large red bell pepper

2 celery stalks

6 garlic cloves

1 tablespoon vegetable oil

2 dried bay leaves

1½ teaspoons Cajun Seasoning (page 166)

1 teaspoon Morton Coarse Kosher Salt

2 cups crushed tomatoes, fresh or canned

6 cups low-sodium chicken broth

12 ounces cavatappi or other similarly-sized pasta tubes

1 PREP THE CHICKEN. Gently peel off the chicken skin, salt it, and refrigerate. Break the bird down into 8 pieces: 2 breast lobes, 2 drumsticks, 2 thighs, and 2 wings, as on page 263. Sprinkle the salt and the French Rub evenly all over the chicken, both sides. Refrigerate on a rack, uncovered, for at least 2 hours.

2 PREP THE REST. Peel and roughly chop the onion. Core, seed, and roughly chop the green and red peppers. Chop the celery. Peel and mince or press the garlic.

3 FIRE UP. Light your smoker and get it up to 325°F or set up your grill for 2-zone cooking and get the indirect side to 225°F. Get some smoke rolling.

4 SMOKE. Put the sausage and the chicken in the indirect zone. Cook the sausage until it is 155°F in the center and remove it, after about 30 minutes. Cook the chicken until the temperature in the thickest part of the breast reaches 145°F and the chicken is a deep, golden brown, perhaps 45 to 60 minutes (we will cook it further in a bit).

5 CUT UP THE CHICKEN AND SAUSAGE. Let the chicken cool a bit so you can pull all the meat off the bones and then cut into bite-sized chunks. Freeze the carcass and wing tips for a future smoky stock. Halve the sausage lengthwise, then slice it crosswise ½ inch thick.

6 MAKE THE CRACKLINS. Pour 2 tablespoons of oil into a frying pan and crank it to medium-high heat. Lower the skins into the oil, spread them out to flatten them, and fry until golden and crisp. Take them out and place them on some paper towels to drip dry.

7 START THE PASTALAYA. In a large saucepan, heat the oil over medium-low heat. Add the onion, green pepper, red pepper, celery, garlic, bay leaves, Cajun Seasoning, and salt and cook, stirring occasionally, until the vegetables are soft, about 20 minutes.

8 FINISH THE PASTALAYA. Add the tomatoes, chicken chunks, sausage chunks, and broth. Cover the pot, increase the heat, and bring the mixture to a simmer but not a boil. Stir in the pasta and push it down under the liquid to be sure it all cooks evenly. Partially cover the pot and reduce the heat to very low. Simmer gently until the pasta is done, perhaps 25 minutes. Frequently check the liquid level and push the pasta down under. Add ½ cup extra broth or water if necessary. The mixture should be juicy but not wet.

9 SERVE. Discard the bay leaves and ladle the dish into big serving bowls, break the cracklins into bite-sized chunks, and scatter them over each serving.

TANDOORI CHICKEN

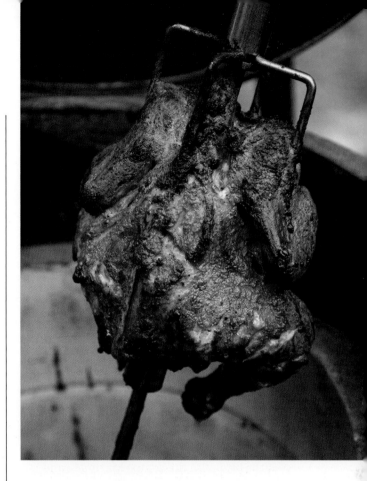

No nation does spices like India. They are master blenders with several masala and curry recipes. Many homes and practically all restaurants cook with a tandoor, usually a charcoal-burning ceramic cylinder, like a kamado similar to the Big Green Egg. In a tandoor, food dangles into the scorching hot chamber on skewers or spits. Heavily seasoned yogurt marinades and meat juices drip on the charcoal and vaporize back up to flavor the meat. It's not hard to make India's most famous dish, tandoori chicken, on any grill or rotisserie, but if you have a kamado, you can make it into a "kamadoor" as described on page 98. I do not smoke this dish.

MAKES *4 servings*

TAKES *1 hour to dry brine, 12 to 24 hours to marinate, 10 minutes to prep, 45 minutes to cook, as an estimate, though cooking time will vary depending on what you cook it on and the temp*

- 1 whole chicken (3½ to 4 pounds)
- 1 tablespoon Morton Coarse Kosher Salt
- 1 cup Yogurt Marinade (page 175)
- **OPTIONAL.** Chopped fresh cilantro leaves for garnish

SERVE WITH. Basmati rice with peas and a mango lassi.

YOUR APPROACH. You can do this several ways:

IF YOU HAVE A TANDOOR OR KAMADO: You can do a whole chicken on a spit easily. I skewer the whole bird with the rotisserie spit from my gas grill and stand it in my kamado with one end in the bed of coals and the other sticking out the top vent. The wet yogurt is a pretty good heat shield, so you can get your kamado pretty hot.

IF YOU HAVE A ROTISSERIE ON A NORMAL GRILL: Just spear the chicken, lock it in place, and turn it on like you would any rotisserie chicken. Just keep the heat down if there is a dedicated rotisserie burner because that is intense IR. If you are cooking on a rotisserie with heat from below, you may need to crank it up.

IF YOU HAVE A GAS OR CHARCOAL GRILL WITHOUT A ROTISSERIE: You can simply spatchcock or break the chicken into parts and reverse sear it.

1 PREP. The chicken skin won't get crispy, so I remove it, toss it on the grill or in a pan, and make cracklins as a garnish as described in Pastalaya (page 274). This also has the benefit of allowing the yogurt to work on the muscle tissue. With a sharp knife, cut ¼-inch-deep gashes in the breast every ¾ inch. Sprinkle the salt all over the chicken, even in the cavity. Give it 1 hour in the fridge to penetrate before applying the marinade. While you're waiting, make the Yogurt Marinade.

2 MARINATE THE CHICKEN. After the hour, coat the bird with the Yogurt Marinade and

put it in or over a pan in the fridge for 12 to 24 hours. If you are going to cook this on the grates of a gas or charcoal grill, you can break it down into parts (see Eight or Ten Pieces, page 263).

3 COOK THE BIRD. Skewer the bird if you are doing it whole. In a kamado, leave the legs and wings loose so heat can get into the armpits and crotch. Keep the heat down around 225°F. On the rotisserie, tie or pin the wings to the breasts and cinch the drumsticks together so they don't flop around and tear off. Again, keep the heat down around 225°F. If you cook it on a grate in the indirect zone of a grill, because you are not exposing it to IR you can get it as hot as 400°F. You're done when it hits 155°F in the thickest part of the breast; it will continue to rise to 160°F+.

4 SERVE. Garnish with chopped cilantro, if desired.

REAL FRIED CHICKEN ON A GAS GRILL (IT'S SAFE!)

So hear me out before you order the straitjacket. The best place to deep-fry is on a gas grill. It solves all the problems of deep-frying. And it is totally safe. Frying on a grill is less risky than frying on a stovetop indoors. If you do it in a large cast-iron Dutch oven, there's no way you can knock it over, and even if there's an earthquake, which would you rather have burn down: Your grill or your house? Doing it outdoors, there are no smoke alarms, no spattered stovetop, no smell.

Different oils begin to smoke at different temperatures. For deep-frying, which is best starting at 375°F for chicken and most other things (donuts and fish like a 350°F start), you need to pick a neutral-tasting high-smoke-point oil. There is a discussion of different oils and their smoke points beginning on page 154.

For most frying, I recommend a vegetable oil blend. They are good choices because they have a high smoke point, a low price point, and neutral flavor. When you're done, if the oil looks dark brown, has a lot of flotsam and jetsam, starts to smoke, and smells a little reminiscent of crayons (a sign of rancidity), it's time to change the oil. If not, filter it and save it.

When the oil is spent, never pour it down the drain. It can coagulate there, mix with hair and other waste, and clog your pipes or even the city sewage system (they're called fatbergs—Google it). I keep a small jar of spent oil that I drizzle on newspaper to start my charcoal chimney, a log-burning smoker, campfires, and fireplaces.

MAKES *10 pieces, 2 servings*
TAKES *1 hour to marinate, 20 minutes to prep, 20 minutes to cook*

SPECIAL TOOLS *A large heavy Dutch oven (4-quart or larger), frying thermometer (although you can make just about any handheld digital*

Coating the Chicken

DRY: DREDGES AND BREADINGS.
I usually do chicken with a dredge, not a batter. The simplest dredge is to pat the chicken with a wet paper towel and then dip it in all-purpose flour. It works surprisingly well. To make sure it doesn't pop off in the fryer, don't let the floured meat sit around too long or the flour will absorb water and gelatinize.

For a thicker crunchier crust, there's the old triple-dip breading: flour, egg, flour. Or flour, egg, bread crumbs (or panko, cornflakes, Cheez-Its, or Cheetos). Or flour, egg, flour, egg, flour. Some chefs mix in rice flour, potato flour, cornstarch, and baking powder. So many options to play with.

WET: BATTERS. Batters are usually a wet flour-based goop, thick and runny like a milk shake. Many are made with beer or another carbonated beverage. Because batters are a liquid, it takes longer for the oil to drive off all the moisture. But the steam bubbling within the batter makes the cooked coating lighter and crunchier. You must dunk and fry rapidly before the carbonation goes away. Onion rings and fish are often battered, and Japanese tempura is always battered.

thermometer work), protective gloves, tongs, a spider or slotted spatula like the one in the image on page 94. I prefer to do this on a gas grill, but it can be done in a charcoal grill.

- 1 cup dill pickle brine (or 1 cup distilled white vinegar with 3 tablespoons Morton Coarse Kosher Salt)
- 1 whole chicken (3½ to 4 pounds)
- 2 teaspoons French Rub (page 167)
- 2 teaspoons baking powder
- ½ cup cornstarch
- ½ cup all-purpose flour
- 6 cups neutral-tasting vegetable oil (or enough to fill the Dutch oven to a depth of about 1½ inches)

ABOUT PICKLE BRINE. You want the brine from classic dill pickles. Nothing sweet.

NO BUTTERMILK OR EGGS. Practically every Southern chef either marinates or dips chicken in buttermilk or eggs before the flour. I explain why on page 175. The problem is that they tend to make the crust turn brown when fried, before the meat is cooked through. It then must go into an oven to finish cooking to a safe temperature. Personally, I think fried chicken should be golden, not brown.

But just about any acid can have the same effect on protein as buttermilk. A quick soak in pickle juice, vinegar, or lemon juice has the same effect and creates better flavor without the browning. With vinegar, the meat gets a nice tang reminiscent of salt and vinegar potato chips. Try pickle juice. You

For French Fries

The best French fries are made in two frying steps, usually from russet Burbanks, because they are low in moisture and high in starch.

1 Soak. Scrub the potatoes and cut them into strips about ⅓ inch wide. Then soak them in cold water with 1 tablespoon of distilled vinegar per quart of water in the fridge for an hour. This removes excess starch.

2 Drain, pat them dry, and deep-fry in a Dutch oven for about 3 minutes in 300°F vegetable oil, just until the surface begins to darken in color and firm up. Take them out, let them drain and cool. You can put them in the fridge for hours, even a day. Some cooks even freeze them.

3 Just before serving, fry again in 375°F oil for 3 to 4 minutes (depending on thickness), until golden brown. Salt generously and serve immediately.

get the tenderizing of vinegar and the flavor and moisture benefits of salt. Trust me. Fried chicken works fine without the buttermilk.

ABOUT EVERCRISP. Many restaurants use a product called EverCrisp for fried chicken. It makes dredges and batters crunchier and keeps it crunchy longer. It's a powder

made from a starch called wheat dextrin, that's all. Just substitute it for 20% of the flour/cornstarch mix.

SERVE WITH. I like to put salt on the table in case someone wants more. I also like to put some things on the table to drizzle on. Honey is traditional in the South. Sweet vs. salty is a natural combo. Why not Hot Honey? You can put out some barbecue sauce, but my personal favorite is Thai Sweet Chile Sauce (page 200).

1 MARINATE. If you are not using pickle juice, make the brine by mixing the vinegar and salt. Cut the chicken into 2 drumsticks, 2 thighs, 2 wings, and 2 breasts, and cut the breasts in half so they are about the same size as the thighs. Instructions for breaking down a bird are on page 263. Pour the brine into a large zipper bag or a bowl large enough to hold the chicken. Put the chicken in there and let it soak for an hour or two.

2 SEASON. Sprinkle the French Rub and baking powder on the bird. I have no idea why so many recipes tell you to season the flour with salt and spices and not the bird. If you season the flour, you can't control how much salt and spices get on the bird. It's also a waste because you never use all the flour. Also, salt in the breading can degrade the oil.

3 DUNK. Pour the cornstarch and flour into a 1-gallon plastic bag and mix. Drop the wet chicken pieces one at a time into the bag and shake to coat thoroughly. Take it out and put it on a sheet pan. Designate a dry hand and a wet hand. Put the dry hand behind your back and with the wet hand, one at a time, dunk the chicken pieces back in the brine, and back into the flour bag, coating thoroughly. Shake off the excess. Place the chicken back on the sheet pan.

OPTIONAL. For super shaggy coating, add another bowl to the process. In it put the same amount of dredge as above and lightly mix it with ¼ cup brine. This makes it lumpy. Tack on the lumps by hand. They stick to the chicken and make crunchy shards.

4 FIRE UP. Set up your grill for 2-zone cooking.

5 Add about 1½ inches oil to a large Dutch oven. You don't want the oil any deeper because you want some of the meat in contact with the hot metal on the bottom of the pot and some of it above the oil. Leaving some

of the chicken above the oil allows steam to escape and that helps prevent the steam from knocking off the crust. Put the pot on the hot side, close the lid on the grill, and bring the oil temperature to about 375°F (stir the oil before taking its temperature). If it goes higher, dial down the gas or move the pot off the flames onto the indirect side. At lower temperatures the coating can absorb too much oil. But don't worry: The oil can't enter the meat. Food is mostly water and the steam coming out does a good job of keeping the food from getting greasy.

6 STAY SAFE. When you add cold food that is 75% water, it immediately creates steam, and that's what all that bubbling is. Use long-handled tongs. Slide the chicken in slowly so it doesn't go nuts and splash you and so the exterior has a few seconds to firm up. This keeps the pieces from sticking together. Handle hot oil with respect. Keep children and pets away. Wear an apron or clothes you

don't mind getting a little grease on. I find it's a good idea to wear my glasses.

7 DON'T CROWD THE OIL. Cold chicken really knocks the dickens out of 375°F oil all the way down to 300°F or so. If you are frying in batches, let the oil come back up to temp before putting in more food. You want it 350°F minimum. Close the lid of the grill, but don't put a lid on the pot. After 4 minutes you can look at the bottoms of each piece. When they are Golden Brown and Delicious (GBD), flip them over. After you flip they should take another 4 to 5 minutes. Let the color of the crust, not the meat temperature, decide when to take it out of the oil.

8 GBD. Remove the chicken pieces from the oil when they are GBD all over, and put them on the indirect side of the grill. Sprinkle with salt. If you probe the meat you may learn that it is not 160°F yet. But if you pull it out when it is the perfect color and place it on the rack in the indirect side, when you close the lid it will continue to bake to perfection without burning the crust. Then temp it before serving it.

9 KEEP THE OIL CLEAN. Skim off bits and chunks that come off the food. When the oil gets cloudy, time to change. Now add more chicken to the oil and close the lid of the grill.

10 Place on a wire rack or paper towels to drain and then serve.

IMPROVED NASHVILLE HOT CHICKEN

This zippy variation on fried chicken was invented in Nashville at Prince's Hot Chicken Shack and is now served nationwide. Even KFC restaurants are serving a version.

The origin story is a doozy. Apparently, Thornton Prince was a ladies' man and when he got home late one evening after consorting with a lady friend, his live-in girlfriend was livid. According to the restaurant's website, "Instead of a lecture the next morning, Prince awoke to the sizzlin' smell of fried chicken. The trap set, Prince's cuckolded lover served up a plate of homemade fried chicken. Without noticing the devilish amount of peppers and spices she had sprinkled on the chicken, Prince dug in. Much to her dismay, Prince didn't fall over weeping in pain. Nope, he asked for seconds, and, at that moment, the legend was born."

Here's my interpretation of the recipe. It works with the recipe for Real Fried Chicken on a Gas Grill (page 278). You can do it with any piece of chicken, but if you use a breast lobe pounded flat or a boned thigh, you have the core of a killer sandwich. Tradition says to serve this with pickle slices on a single slice of white bread, but it's mighty good on a bun topped with pickle slices or even creamy coleslaw.

A cheat: Just make the fried chicken and anoint it with Hot Honey (recipe on AmazingRibs.com/mm).

MAKES *2 to 4 servings*
TAKES *1 hour to marinate, 35 minutes to prep, 20 minutes to cook*

- 1 cup vegetable oil
- ¼ cup brown sugar
- 3 tablespoons ground chipotle
- 3 tablespoons hot pepper flakes
- 2 tablespoons smoked mild paprika
- 2 tablespoons mustard powder
- 1 tablespoon garlic powder
- 1 tablespoon medium-grind black pepper
- ½ teaspoon Morton Coarse Kosher Salt
- 1 recipe Real Fried Chicken on a Gas Grill (page 278)

ABOUT THE OIL. Many restaurants use the oil they fried the chicken with. But I prefer fresh oil or bacon fat.

1 BRING THE HEAT. In a small pot or pan, heat the vegetable oil over medium heat. Add the brown sugar, all the spices, mustard powder, garlic powder, pepper, and salt and stir to mix. Turn the heat to its lowest setting and let the mixture stay warm for 30 minutes or so to extract the flavors. Set it aside. You can refrigerate for a day or three. Pour through a fine-mesh sieve

and push down on the pulp with a spoon to extract as much flavored oil as possible.

2 FRY. Make the Real Fried Chicken on a Gas Grill as directed. Please note that the chicken is fried in fresh oil, not the hot oil from step 1.

3 FINISH. While the chicken is frying, rewarm the hot flavored oil. When the chicken is done, keep it warm on the indirect side of the grill until just before serving to preserve the crunch. Then, at the last minute, paint the hot oil on the chicken with a basting brush.

BUTTERED-UP TURKEY BREAST WITH DRUNKEN CRANBERRIES AND CRUNCHY SKIN

Let's start by defining some terms. Technically a turkey breast is a whole breast section with two pear-shaped lobes separated by the breastbone also known as the keelbone. You can buy it bone-in with both lobes, but you can also find a single boneless lobe. That's what we're cooking here, but you can easily adapt this for a whole bone-in breast section.

The problem with turkey breasts is that they have very little fat and so they get dry when cooked to a safe temperature of 160°F. So we solve the problem four ways:

ONE. We remove the skin, which, although it contains fat, the fat cannot penetrate the meat. To crisp the skin for easy slicing and better texture, we usually cook turkey at 325°F or more, but that can dry out the meat because the skin is so thick. With the skin removed we can smoke low and slow at 225°F. But don't worry, we won't waste that tasty skin. We're going to make cracklins from it and sprinkle them on top when we serve.

TWO. Then we pound the meat flat. We do this because turkey (and chicken) breasts are much thicker at one end and as a result the thin ends way overcook and get dry and hard. Pounding them flat makes them cook more evenly.

THREE. Then we dry brine hours in advance. As you know, the salt helps retain moisture.

FOUR. And finally, we inject with butter! Can you see Julia smiling?

MAKES *6 to 8 servings*

TAKES *15 minutes to prep, 2 to 4 hours to dry brine, 1 hour to cook*

SPECIAL TOOLS *Meat injector*

- 1 boneless turkey breast lobe (about 3 pounds)
- 1 teaspoon Morton Coarse Kosher Salt, plus more for sprinkling
- 1 tablespoon French Rub (page 167)
- 4 ounces sweetened dried cranberries
- ¾ cup inexpensive American port wine
- 1 stick (4 ounces) unsalted butter

1 POUND. 4 to 6 hours before serving, remove the turkey tender if there is one. It is a small loose muscle nestled against the breast on the side with no skin. Sometimes it has been removed already. It is just as tender as its name and a perfect serving size, so freeze it aside for another meal. Remove the skin and set it aside. Trim any excess fat from the meat. Put the meat in a 1-gallon zipper bag and with a frying pan pound that sucka until it is pretty much an even thickness, about 1 inch thick.

2 DRY BRINE. Now sprinkle half the salt all over the meat. You can also sprinkle on the French Rub now, too. Try to give the salt 2 to 4 hours to penetrate.

3 REHYDRATE. Take the cranberries and the port wine and put them in a saucepan over low heat for 10 minutes. This will allow the berries to soak up the wine and rehydrate. Turn the heat off and let them sit.

4 FIRE UP. Set up your grill for 2-zone cooking with Warp 5 on the direct heat side.

5 MAKE THE CRACKLINS. Stretch the turkey skin over the grates on the direct side. Sprinkle them lightly with salt. Keep the lid open and stand right there and watch. As soon as the skin starts to turn golden, flip it. Remove it when it gets stiff and crackly like a potato chip. When it cools, chop it into ½-inch chunks. Set aside.

6 SMOKE AND BUTTER UP. Get your grill or smoker to 225°F on the indirect side and get some smoke rolling. Put the meat on the smoker and start checking its temperature after about 30 minutes. We have waited until now to inject the butter because if we try to inject it into cold meat the butter solidifies in the needle and clogs it. But when the meat has warmed, it will go where you want it. Melt the butter and pour it into a tall narrow glass. This allows you to suck more into the injector than a short fat glass. Fill the injector and start pumping it into the meat through the sides at intervals about ¾ inch apart.

7 SERVE. The meat is done when it hits 155° to 160°F, after about 1 hour. It will go up another 5° to 10°F after you take it off. Put the breast on a cutting board with the skinny edge facing you. Slice it across the grain and plate. Spoon some cranberries on top and sprinkle with some chopped cracklins. Or make a sandwich.

GROUND AND CHOPPED

"Sausage is a great deal like life. You get out of it about what you put into it."

—JIMMY DEAN

I think of hamburgers, hot dogs, sausages, meatloaves, kofta, seekh kebab, and pâté as cousins. They are all ground or chopped meat. As such, they are somewhat interchangeable. There's nothing stopping you from cramming a meatloaf recipe into casings and making links or taking a sausage recipe without the casing and wrapping it around a stick and making kofta with it.

There is something else they all have in common: Because of the risk of contamination from the surface of the meat being ground up and mixed into the center, they all must be cooked to 155°F minimum unless the meat has been pasteurized.

- I use these definitions for burgers:
  - Fast-food burger: 3 to 4 ounces
  - Pub burger: 5 to 7 ounces
  - Steakhouse burger: 8 ounces or more

- 5 to 7 ounces is the ideal weight for grilled burgers.

- 4½ inches wide is the ideal patty for a 4-inch bun. It will shrink.

- Avoid buying meat labeled "hamburger" or "ground beef." They are substandard quality. Freshly ground meat is better than frozen. You can ask your butcher to grind it for you. If you pick a nice piece of chuck, you'll likely be in the 20% fat range. Ask her to add enough fat to make it 30%. If you have a choice, take a hardworking cut like flank steak or short plate, because it tends to have more flavor and have her grind that with added fat. Or do it yourself. There is a handy calculator linked to AmazingRibs.com/mm that will tell you how much fat to add to meat to get to whatever percentage you want.

- You can add butter, but I prefer beef fat. You want fresh white fat, not yellow fat. I've been known to freeze fat trimmed from my briskets and add it to my burger grind. Wrap it tightly in plastic first, then foil, and don't keep it for more than 60 days or it can start to taste funky. You can also use beef fat to coat your griddle or pan if you are making smash burgers.

- Don't bother with wagyu for burgers. Wagyu's great benefit is tenderness and fat. A burger is ground. You can't get much more tender than that. And you can control the fat. If you can get it, however, wagyu fat is nice.

- Cooking burgers to less than 155°F is a health risk. But you can eat a medium-rare burger safely if you buy irradiated meat or buy a steak and some fat and dunk them in boiling water for 20 seconds or so to pasteurize, then grind a 30% fat blend yourself.

- Don't grind it too fine. Many recipes say grind twice. Don't do it. You want to leave sections of muscle intact for moisture retention and texture.

- Don't pack the patty tight. You want air pockets in which juices can hide.

Never squish a burger with your spatula while it is on the grill

BUSTED. When we squish a burger on a grill, melted fat, meat juices, and seasonings that hit flame or glowing coals vaporize and land on the burger and flame jumps up to sear the meat. This is good. The problem is that if you have a burger that is 20% fat, the common blend, a few squishes results in a hard dry patty. The way to combat this is to up the fat content of the raw burger to 30%. Now when you squish you can afford to lose some fat. This, however, causes another problem: 30% burgers might break or crumble when squished. So cook them on a grill topper so the juices can drip through.

- Chop some onion or bacon and mix it with the meat. They bring moisture, flavor, and sweetness. Bacon brings salt. Do mix in MSG: ½ teaspoon of Ac'cent or Aji-no-moto per pound of meat. If you wish, you can mix in seasoning like garlic powder. But don't go crazy with mix-ins. You want a beef burger, not meatloaf.

- Do not mix salt into the patty. It compacts it and makes it tougher. But absolutely do sprinkle salt on the surface just before cooking.

- Keep the meat cold until you are ready to cook.

- If you have a 20% fat blend, throw a chunk of beef fat on the flame or coals. It will vaporize and flavor the meat. If you have 30%, press it with a spatula while cooking.

- Butter and toast your buns over direct heat. I prefer brioche buns.

- Use the 2-zone system. This gives you temperature control. You can move the burgers from side to side. Steakhouse burgers can be reverse seared.

- Give your burger 15 minutes on a smoker or in a cloche with a smoking gun at its lowest possible temperature before searing.

- It will take several practice runs to get your technique down. Be patient and persistent. Get it right and you are a hero.

THE ULTIMATE SMASH BURGER: OKLAHOMA ONION BURGER

During the Depression, meat was too expensive to be consumed casually, so, legend has it, Ross Davis of the Hamburger Inn on Route 66 in El Reno, Oklahoma, came up with a Nobel Prize–worthy innovation. He decided to beef up his 3-ounce burger by griddling it on a mound of onions. This not only saved money, but the caramelized onions make the meat surprisingly tasty. Not to mention the crispy edges on the patty.

Although the original Hamburger Inn in El Reno closed, Sid's Diner, Johnnie's, and Robert's Grill (since 1926) have carried on the tradition. In about 1938, a second location of the Hamburger Inn opened a two-hour drive south in Ardmore and is still going strong.

You can do this the traditional way on a griddle or, if you don't have a griddle, you can do this in a frying pan. It spatters, so put the frying pan on the grill.

MAKES *2 burgers*

TAKES *20 minutes to prep, 20 minutes to cook*

SPECIAL TOOLS *Griddle, wide spatula with a sharp edge to get under the patty without leaving meat on the metal (I bought a heavy spatula and ground it to a sharp edge), and as with all griddling, an infrared gun thermometer and a dome are helpful*

- 1 stick (4 ounces) butter
- ½ pound coarsely ground beef chuck (80% lean)
- ½ pound yellow or white onions
- 2 old-fashioned hamburger buns
- 2 tablespoons tallow (beef fat), clarified butter, or vegetable oil
- ½ teaspoon Morton Coarse Kosher Salt
- ½ teaspoon medium-grind black pepper
- 2 American cheese singles
- 8 dill pickle chips

1 PREP. Take the butter out of the fridge and let it come to room temperature. Divide the beef into 2 equal portions and make them into 4-ounce balls. Peel the onions, slice them in half, cut off the ends, and then slice the onions as thin as possible. If you have a mandoline, use it. Divide the onions into two piles, each about 3 ounces.

2 FIRE UP. Fire up the griddle to about 375°F as measured with an infrared gun. If you don't have one, a few drops of water on the griddle should dance around and vaporize rapidly. You will need to work fast, so make sure you take all the ingredients and tools out with you.

3 TOAST THE BUNS. Put a light coat of softened butter on the cut side of each bun and place them cut side down on the griddle. Press them down with your spatula to make sure they are intimate with the hot metal and brown evenly. After a minute check to see how they're doing. When they are golden to light brown, set them aside.

4 COOK THE BURGERS. Put 1 tablespoon of beef fat for each burger about 6 inches apart on the griddle. When it has melted, place the underside of your spatula in the fat so it is greased on one side. Slap each ball of meat onto the oiled metal, and pile the onions on top so you can't see any meat. Smash down on the onions and meat with the spatula until each patty is about 4 inches in diameter. Tilt the spatula and work the edges so they are as thin as possible. If some onions hang over, that's OK. Sprinkle the salt and pepper on top. Let the patties cook for a few minutes until the edges turn browned and crispy.

5 FLIP THE BURGERS. Work the spatula gently under the edges of the patty all around, pressing hard against the griddle until the burger lifts free. Try not to leave any of that hard-earned Maillard behind. Flip the burgers so the onions end up underneath. A second spatula comes in handy to help with this, or just place your hand on the onions to hold

them in place. They're not too hot. Place the top buns on top of the meat and leave them there for a minute until they steam and warm and absorb onion and beef aromas.

6 ADD THE CHEESE AND SERVE. Lift the buns, place the cheese on the burgers, and if you have a dome or a pan, place that over the burgers to help melt the cheese. Otherwise just close the lid. When the onions are good and caramelized and the cheese is melted, place the bottom buns on top of the cheese, and then flip the burgers onto the bottom buns cheese side down. Add the pickle slices on top of the onions, then place the top buns on, and slide the whole thing onto serving plates. Some folks in Oklahoma put mustard on their burgers, but that masks the meat and onion flavors too much for me.

BEST CHICKEN BURGER EVER

Ground chicken and turkey are too lean, so we amp them up with ground bacon. This burger is great on a bun, or it sits handsomely atop a salad.

MAKES *2 burgers (6 to 8 ounces each)*
TAKES *30 minutes to prep, 20 minutes to cook over direct heat*

SPECIAL TOOLS *Food processor*

2 ounces Smoked Tomato Raisins (page 339) or dry-packed sun-dried tomatoes

1½ tablespoons high-quality olive oil

¼ teaspoon dried thyme

1 medium onion

3 tablespoons mayonnaise

¼ teaspoon toasted (brown) sesame oil

2 ounces (about 2 slices) thin-cut bacon

12 ounces ground chicken

1-inch finger of fresh ginger

½ small Asian pear

2 scallions

10 fresh cilantro leaves

¼ teaspoon dried oregano

½ teaspoon medium-grind black pepper

¼ teaspoon Morton Coarse Kosher Salt

2 buns (I like onion rolls and brioche rolls, don't you?)

ABOUT THE DRIED TOMATOES. If yours are leathery, soak them in warm water for at least 1 hour before making into the jam.

ABOUT THE ASIAN PEAR. Asian pears have the taste of a pear with the texture of an apple. Feel free to substitute a slightly underripe pear or a ripe apple.

1 PREP. In a bowl, combine the Smoked Tomato Raisins, oil, and thyme and make a smooth spreadable "tomato jam," adding a little more olive oil if necessary. Peel and thinly slice the onion and make Quick Pickles (page 342). Mix the mayo and sesame oil. Put it all in the fridge for serving.

2 MAKE THE BURGERS. Mince the bacon into BB-sized bits. Put them into a large bowl. Add the ground chicken. Mince or grind the ginger to get 1½ teaspoons. Cut the Asian pear into quarters and peel and core it and then cut it into ¼-inch dice until you get 3 ounces. Eat the rest. Peel dead skin off the scallions, chop off the roots, mince the whites. Save the greens for another dish. Chop the cilantro into small bits until you have about 2 tablespoons. Everything into the bowl. Add the oregano and black pepper. Don't add any salt. It compacts the meat. We'll salt later. With a light hand, mix everything and form 2 loosely packed patties. Refrigerate for at least 15 minutes and up to 6 hours.

3 FIRE UP. Prepare a grill for 2-zone cooking and aim for Warp 10 in the direct zone. Place a grill topper over direct heat. When you are

ready to cook, toss a wood chunk or a handful of chips or pellets on the fire.

4 COOK. Season the burgers generously with salt and put them on the topper on the direct side. Cook, lid down, flipping gently every 4 to 5 minutes until the temperature reaches 165°F at the thickest point. The surface should be nicely crusty and golden brown. Grill the buns cut sides down over direct heat until golden, 1 to 2 minutes.

5 SERVE. Dollop a nice chunky spoonful of the tomato jam on each bottom bun with some quick pickled onions. Top with the burger, spread some sesame mayo on top of the burger, and then add the top bun.

MARY'S MEATBALLS

I have never understood why more people don't grill meatballs. The grill, with a little wood smoldering below, gives them an added dimension. They can then be used in a wide variety of venues: On pasta, in soup, on pizza, as stuffing for pot stickers or other dumplings, on couscous, on a bed of steamed greens topped with grilled veggies, straight on frilly toothpicks, like candy, or my favorite, meatball subs.

This recipe is based on one that my wife's Italian American mom passed on to her. She died some time ago, but if you are making her meatballs, she lives on. My hope is eternal life for Mom through you.

One secret is the panade, or milk-soaked bread, an old Italian trick that not only saves money by stretching the meat but improves texture and moisture. Another is to make them smallish, about the size of golf balls. The small size allows a higher ratio of crust to interior than a burger, and the crust is the best part. Not to worry about overcooking them, because if you use the right ingredients, they will be as tender as a baby's bottom, as juicy as a gossip column, as complex as a Chopin piano cadenza, and they don't need any more sauce than does a ribeye steak. But if you must, gahead and put marinara sauce on them. Another secret is that I skip the veal. I think beef and pork make a better tasting meatball without the expensive bland veal.

You can prep them a day or two in advance, chill, and cook them when you need them. Or they can be cooked and stored for 3 or 4 days and then reheated. Or cook them and freeze them for 2 to 3 months. They'll be there waiting when you are too busy to cook from scratch. If you cook them in advance, you can reheat them in the microwave.

MAKES *About 2½ pounds, enough for 20 small meatballs, or enough for 10 sandwiches*

TAKES *About 45 minutes to prep, 15 minutes to cook*

- 3 ounces white bread, crusts and all
- ¼ cup milk
- 1 large egg
- 2 ounces Parmigiano-Reggiano cheese
- ½ teaspoon French Rub (page 167)
- 1 teaspoon Morton Coarse Kosher Salt
- ¼ teaspoon medium-grind black pepper
- ¼ teaspoon garlic powder
- 3 tablespoons chopped fresh parsley
- 1 small onion
- 1 pound coarsely ground beef chuck with 20% fat
- ½ pound coarsely ground pork
- ¼ cup vegetable oil

SERVE WITH. If you want marinara sauce with the meatballs, figure on about 2 tablespoons per ball.

1 FIRE UP. Set up your grill for 2-zone cooking and preheat to about 225°F on the indirect side. Smoke is optional. I skip it.

2 PREP. Cut the bread into ½-inch chunks and add to a bowl. Soak it in the milk and squeeze it until it is almost dry. Set aside.

3 MIX. In a large bowl, beat the egg with a fork for about 20 seconds. Grate the cheese and add it. Add the French Rub, salt, pepper, garlic powder, and parsley. Peel and mince the onion fine and add it. Add the meat and bread. Mix everything in the bowl with your clean hands. When all the ingredients are more or less evenly distributed, pack into meatballs about 1½ inches in diameter. I use a ¼-cup measuring cup. Put them on a plate or cookie sheet.

4 GRILL. Pour the vegetable oil into a bowl and one at a time roll the meatballs in it to lightly coat them so they won't stick to the grill. Place the meatballs on the indirect heat side, close the lid, but check frequently. When they hit about 130°F, roll them over to the direct heat side until they are dark on all sides and 155°F in the center. If they stick a bit, and they probably will, just leave them alone and eventually they should let go if your grates are clean (and they are clean, aren't they?). If they get close to burning, move them back to the indirect side.

5 KEEP WARM. Put them in a pan on the indirect side to stay warm until you are ready to serve.

NOT GRANNIE'S MEATLOAF

Meatloaf is the ultimate comfort food, and perhaps the least glamorous. So I wanted to design the ultimate meatloaf. The first question is, how far outside the loaf pan can I go? I dreamed up all sorts of exotic creative approaches, but then I slapped my face and reminded myself I want meatloaf, not sausage. Oh, and as with the meatballs, I skip the expensive, bland veal.

One important concept: Never cook meatloaf in a loaf pan or a pan with high sides. Loaf pans keep the sides and bottoms from browning, and the brown crust is the best part, right?

We'll cook at 325°F, which is hot enough to brown the exterior, but not so high as to cause the proteins to get their undies in a bunch and squeeze out all the moisture. At that temp, it should take about 1 hour for the loaf to cook through, but please do not use a clock to tell when this recipe is done. It is done when the meat hits 155°F in the deepest part of the center.

Like you, I serve meatloaf with mashed potatoes and sweet peas.

MAKES *8 servings, 6 ounces each*
TAKES *1 hour to prep, about 1 hour 30 minutes to cook*

SPECIAL TOOLS *Grill topper*

WET STUFF

- 3 large eggs
- 2 tablespoons soy sauce
- 2 tablespoons Dijon-style mustard
- 2 teaspoons your favorite hot sauce, more or less
- ½ cup whole or 2% milk

DRY STUFF

- 1 cup unseasoned dried bread crumbs
- 1 tablespoon crumbled dried basil
- 1 tablespoon freshly grated orange zest (from about 1 large orange)
- 2 teaspoons dried oregano
- 1 teaspoon Morton Coarse Kosher Salt
- ½ teaspoon garlic powder
- ½ teaspoon dried thyme
- ¼ teaspoon medium-grind black pepper

VEGGIES

- 1 small onion
- ½ small red bell pepper
- 2½ ounces Smoked Tomato Raisins (page 339), sun-dried tomatoes, prunes, raisins, or apricots

MEAT

- ½ pound bacon
- 1 pound coarsely ground beef chuck with 20% fat
- 1 pound ground pork

ABOUT THE DRIED TOMATOES. If yours are leathery, soak them in warm water for at least 1 hour before chopping or else they will pull liquids from the loaf.

GLAZE. If you want to paint with a glaze, try Kansas City Red (page 180), Balsamic Syrup (page 198), or Tare Sauce (page 203).

1 MIX THE WET STUFF. In a big bowl, mix the wet stuff thoroughly.

2 MIX THE DRY STUFF AND COMBINE. In a small bowl, mix the dry stuff thoroughly and then mix them into the big bowl with the wet stuff. Let it soak for a minute or two.

3 ADD THE VEGGIES. Peel and coarsely chop the onion and add it to the big bowl. Remove the stem and seeds from the bell pepper, discard them, and chop the pepper's flesh into ¼-inch bits. Add it to the big bowl. Coarsely chop the Smoked Tomato Raisins and add to the big bowl.

4 MIX THE MEAT. Chop the bacon into chunks about ¼ inch and mix with the beef and pork. Add this to the big bowl and mix. Resist the temptation to overwork the mass and pack it tight. We want meatloaf, not meat bricks. If it doesn't mix thoroughly, that's OK. Bursts of flavor are fun.

5 FORM THE LOAF. Move the mixture to a platter and form it into a cylinder about 3 inches in diameter. Yes, I know this sounds weird, but this shape will ensure that it cooks evenly on all sides and it makes it easier to roll around on the grill to brown on all sides. Place the meatloaf in the refrigerator to firm up while you prepare the grill.

6 FIRE UP. Start your smoker and aim for 325°F or set up a grill for 2-zone cooking, 325°F on the indirect side, and get some smoke rolling.

7 COOK. Place the loaf on a grill topper on the indirect side of the grill or on the smoker. Roast until the temperature in the center of the loaf is 160°F, which is the temperature at which the eggs are safe. It should have a nice crust. If you want a glaze, paint it on now.

SEAFOOD

"Give a man a fish and he'll probably get it stuck to his grill. Teach a man to grill and he'll become a big fish among his family and friends."

From a culinary standpoint, most finfish, bivalves (clams, oysters, mussels, scallops, cockles), cephalopods (octopus, squid, cuttlefish), and crustaceans (lobster, shrimp, crab, crawfish) are delicate in flavor, high in water content, low in fat, and extremely sensitive to heat. So my approach is generally: Get outta the way. Let them speak for themselves. Don't mask their beauty. And for goodness' sake, don't overcook them—in general, that means they're done in the 125°F range. As with other meats, thicker is better. Small and thin pieces give you a shorter window to squeeze through.

Freshness and meticulous handling are far more important for seafood than any other meat. They spoil fast. When you buy them, lobsters and crawfish should be alive, and so should most clams, oysters, and mussels. Some stores even sell live shrimp and scallops. When shopping for dead seafood, the first thing to ask the clerk is when was it caught. Believe it or not, this information is often available, because government regulations are trying to make our food chain traceable in case of problems.

Judge fishmongers by how they store seafood. The best way to pack it is on top of shaved ice as shown at right. When fish is in contact with ice, a thin layer of cold water forms between the fish and the ice that helps preserve moisture. The best stores put it right on shaved ice, not in plastic wrap or on Styrofoam trays on top of the ice. Buying

or making ice is expensive, so when you see seafood on ice you know the management cares about quality. The best fishmongers may even dump shaved ice in your plastic bag. Keep it cold and get it into your fridge and onto the dining table ASAP.

There are thousands of seafood species and subspecies; they are all fairly mild and they soak up flavoring like a sponge.

FINFISH

Fish should smell like the ocean, not fishy. Finfish contain a compound called trimethylamine N-oxide (TMAO). When fish die, an enzyme and bacteria begins to convert it to trimethylamine (TMA), and that is the stinky fishy smell. The process moves faster at higher temperatures, so it is crucial that fish be kept at 33° to 38°F, the lower the better.

When you encounter TMA, rinse the fish with cold water, remove the skin and guts, soak the fish for 30 minutes in milk or an acidic solution, such as lemon juice, vinegar, or tomato juice, or salt, and cook it ASAP. The casein in the milk binds to TMA. If you plan to batter-fry the fish, use buttermilk in the process.

Try to get whole fish, head on. That's the best way to tell if it really is what the label says it is and if it is fresh. There is a great deal of fraud. Without the head it is impossible to tell if a fillet is really red snapper. The head tells you how fresh it is. Its eyes should be clear and gills red. But the most important indicator is smell.

Salmon and tilapia are popular fish to farm, although farmed shrimp is also popular in Asia. Consumers have become wary of fish farming. Fish are confined to huge pens, often acres in size. They are fed specially formulated pellets. Uneaten fish pellets are a pollutant. Fish poop is a pollutant. Concentrated populations are often infested with sea lice. New breeds could escape and mate with wild stocks.

The good news is that, from my chair, it appears that the industry has made strides and continues to improve. New methods reduce pollution, disease, and escapees. Canada is a leader. The problem is that there are plenty of bad actors tainting the game. Attempts are underway to regulate the industry in some countries while others close their eyes. Trade associations have begun to label sustainable systems with certifications. I am optimistic for the future.

The biggest problem most of us have with grilling fish is that it loves to stick to the grates. There are several remedies that help, but nothing is surefire. First, make sure your grates are really clean. Get all grease off, and if you can, get the carbon off. Preheat the grates. Oil the fish immediately before it goes on the grill. An all-over slather of mayo works better than just plain oil and adds no flavor. Another technique is to put citrus or fresh herbs between the fish and the metal.

Unlike other meats, you should not flip fish often. Flipping breaks the fish apart. I love GrillGrate® inserts (see page 121). They have a special spatula that gets below the fish like a forklift. If you don't have them, get the biggest spatula you can find and grind the leading edge to knife sharp. Fish baskets that hold the fish between wire mesh, like the one shown in the image below, work beautifully.

In general, freshwater fish are milder than saltwater. Bottom feeders are among the least

Removing Pin Bones

Many fish fillets contain "pin bones." They aren't actually bones. They are calcified nerves along the side of the fish. These nerves allow a school of fish to turn together as if they were connected. You can leave them in, but just be careful when you are eating. Or you can remove them yourself: Just drape the fillets one at a time over an upside down mixing bowl so the pin bones stick out and then yank them out with tweezers or needle-nose pliers.

What's with the Gray Flesh in My Fish?

If you eat fish, especially salmon and mackerel, you have no doubt encountered a layer of gray meat just beneath the skin. It is not dirt. It is an outer layer of muscle with a lot of fat stored in it. If you taste it straight, it can be stronger than the flesh. A lot of people remove it, but it is perfectly edible. Because fat can rancidify with age and oxygen, it may spoil faster than the rest of the fish.

desirable, firm fish are easier to grill, and belly meat is fattier and often tastier.

Don't let fish sit longer than an hour at room temperature at any time during prep. Follow my salting instructions. Don't reject flash-frozen vacuum-sealed fish. In fact, freezing fish for 10 days can kill most parasites (but not bacteria, cooking does that). Consume smoked fish within 36 hours if possible or freeze it. Frozen fish can keep for as long as a year, but they deteriorate in flavor with time. A word of caution: Raw fish often have parasites like tapeworms. You can sometimes see them, but their eggs are hard to spot. Cooking kills them. I have never understood the fascination with raw fish. The flavors and textures of cooked fish are much more appealing to me. And I really don't want a 5-foot tapeworm growing in my gut or brain.

PLANKING

I'm not a fan of planking. When I'm grilling, I am constantly pursuing Maillard and smoke flavors. But putting a wet salmon fillet on a ¼-inch-thick cedar insulator is not getting me there. The wood is pretty effective at blocking heat, so the underside of the food never browns nor does skin crisp. In fact, the wood rarely gets hot enough to give off any of its aromatics. Some recipes tell you to preheat the planks and let them char on the bottom with the idea that smoke will curl up over the meat, but not a lot gets there and it is cedarwood, a pine tree. Every pitmaster on earth tells you to avoid terpene-laden pine smoke for good reason. Clint Cantwell says it makes his fish smell like his grandmother's cedar closet.

Then, when you're done, the plank is charred on one side and full of fish oil on the other. You must throw it away. A waste of money. And don't think about cheaping out and using cedar shakes designed for roofing. They often have a stain and preservatives. Like beer can chicken, this method seems to be more about appearances than quality.

SMOKING FISH

Fish and smoke go together like Romeo and Juliet. For some reason, many barbecue lovers fail to consider smoked fish as barbecue, yet the etching of Tainos cooking fish on a barbacoa is proof. It's a shame fish is ignored by most barbecue competitions and restaurants, but the good news is that there are still numerous small artisanal seafood smokehouses around the country. There are at least a dozen along the Eastern shore of Lake Michigan and many more on the West Coast, in New England, on the Jersey Shore, on Chesapeake Bay, and Florida. Because I was raised in Florida, early on I developed a taste for smoked mullet. In my last book I have a fine recipe for smoking trout with the mullet method.

Coldwater fish that are fatty absorb smoke better and have a more pleasing texture than low-fat fish. Seafood is higher in water content than other meats, lower in fat, connective tissue, and myoglobin. They are also more likely to harbor parasites and harmful bacteria. They require careful handling, proper sanitation, control of temperature, and smoking time.

There are typically two approaches to adding elegant smokiness: Cold smoke and hot smoke. Cold-smoking is done at a very low temperature so the meat is not cooked but the smoke flavor is on the surface. It is delicious, but many people do not like the slippery texture (think sushi or Nova lox), and there is an element of risk to eating uncooked fish. Cold-smoking properly requires careful inspection and great care from the moment the fish is caught, in the handling, and all through the process. Even commercial smokers occasionally must issue recalls. So, operating under the belief that it is bad business for me to kill my readers, I recommend that you do not attempt to cold-smoke seafood.

Hot-smoked fish is a marvel. Some basic principles apply. You must use high-quality fish. Smoking will not salvage old fishy product. Clean your fish thoroughly.

I avoid strong flavored woods such as hickory and mesquite. As always, too much smoke is worse than too little. On a charcoal grill or smoker or an electric smoker, a handful of wood will probably be enough. On a gas grill, because it is has a lot of ventilation, double that.

By now you know that I prefer dry brining to wet brining in most cases. Fish is an exception. Fish absorbs brines and sugar better than land animals, and in a liquid, the salt is more evenly distributed within the meat.

A word of caution: Fish oils permeate everything. They can get into the carbon coating on the walls of your cooker. It is a good idea to give your smoker a thorough washing after smoking fish. Use a power washer or take your smoker to the car wash. If you do a lot of fish, it might be worthwhile having a separate smoker just for fish. I have an inexpensive electric smoker just for seafood.

Smoked seafood is heavenly on a cracker, toast point, or mixed with egg in a scramble or omelet. It goes well with cream sauces and egg sauces, especially on pasta or rice dishes. It is also an elegant hors d'oeuvre on top of slices of boiled or grilled potato with a dollop of sour cream with a little horseradish in it.

Instead of canned tuna with mayo on a sandwich, put some smoked fish in a bowl and flake it with a fork, add a very tiny splash of toasted (brown) sesame oil, some mayo, and spread it on rye bread. Go easy on the mayo.

BIVALVES

Bivalves are clams, oysters, mussels, and scallops. They all have two shells hinged at one end. Their flavor and size depend on the waters in which they are grown. Some are small, some salty, others large, and some mild.

In my last book I went to some length to discuss how to buy clams, mussels, and oysters, so here's the short version: Get them live in their shell. Scrub the shells well under cold running water. If you have dug them yourself, the best way to purge them of sand is to bring them home in a bucket of seawater. If you bought them in a store, put them in a 4% brine by mixing 1 liter (1 kilogram) of water with 40 grams of any salt for 4 hours, max. Then dump the salt water and store them in the fridge in a bowl covered with a wet kitchen towel.

When it is time to cook, sniff each and every one of them. They should smell like the ocean. Throw out any that smell like barf and any that rattle or are too heavy. You do not have to throw out the ones that are slightly open or cracked if they smell fresh. If the two shells slide horizontally, they are probably dead. When in doubt, throw it out.

When cooking, try to save the liquid within the shells, called the liquor. It is rich, and mopping it up with crusty bread is the essence of hedonism. The juice can also be used for stock, soup, or pasta sauce.

CLAMS. Rarely served raw, clams are usually steamed or grilled or made into a soup/chowder. There are many different types and sizes and they are found digging in sand and mud flats when the tide goes out.

MUSSELS. Mostly farmed on ropes dangling into the ocean, these black-shelled mollusks are usually steamed, and they cook in mere minutes.

OYSTERS. Farmed or harvested wild clinging to rocks and piers and seawalls, they are often eaten raw on the half shell. Lately the risk of foodborne illness from raw oysters has increased as the Gulf of Mexico has warmed. But no need to despair. Oysters are delicious steamed, grilled, or smoked.

SCALLOPS. These succulent and sweet little marshmallows of the sea are a joy to eat and simple to cook. Scallops are the adductor muscle, the spring that closes the shell, and they are usually seared on at least one side in a scorching hot pan. They can also be grilled. You will often see them labeled either "sea scallops" or smaller "bay scallops." They love cream sauces.

CEPHALOPODS

Octopus, squid, nautilus, and cuttlefish are called cephalopods and many cooks are as afraid of them as sharks. But they are worth the effort. They all taste similar, a bit briny, a bit sweet, and a bit nutty, but they are rarely fishy. These boneless animals have a lot of connective tissue made of collagen that needs to be converted to silky succulent gelatin. Properly cooked, they have a firm texture; overcook and they turn to rubber; undercook and they are slimy. They all have something else in common: They love to be grilled and come off with a little char.

CRUSTACEANS

Lobster, crawfish, crab, shrimp, langoustines, and all their variations make for succulent meals. In the summer of my tenth year we moved to Maine and Dad and I would go down to the docks when the lobster boats came in and we ate lobster at least weekly. Then we moved to Merritt Island, Florida, and we lived for many years on a channel that connected to the Atlantic Ocean. We spent hours each week fishing with live shrimp as bait. We boiled the unused bait and I popped shrimp like M&M's. After football practice I would take chicken necks and tie them to a rope and throw them into the brackish river and quickly blue crabs would latch on. I would slowly pull the line in and scoop them up with a net. We had crab boil on those nights. My buddies and I would wrap flashlights in plastic bags and snorkel at night in the Sebastian Inlet just south of Melbourne, and snatch spiny lobsters as they peeked out of their sandy holes. And in college we'd go from Gainesville to the tannin-stained tea-colored waters of the St. Johns River and find crawfish hiding under the mangrove roots.

I learned how to cook crustaceans early on, and even though I developed an allergy to crustaceans in the 1990s, I still remember how to cook them. As always, temperature is your guide. Shrimp, langoustines, and crawfish are best at 120° to 125°F, while crab and lobster are best at 130° to 140°F. Err on the lower number because there will be some carryover cooking.

PONASSING SALMON

Native Americans in the Pacific Northwest have been feasting on salmon this way since long before white people came. Called ponassing, in the image above, you can see the method as still practiced by Tillicum people on Blake Island in Puget Sound, where tourists learn of the Native Americans of the region and get a meal of gorgeous fresh salmon cooked the traditional way. The chefs here season the meat, thread wooden sticks through it, jam the stake into the dirt around glowing hardwood embers, and roast the meaty side for about 30 minutes, turn the stake around, and roast the skin side for about 10 minutes. It comes out perfect. Next trip to Seattle, take the ferry to Blake Island on an empty stomach.

At home, try this: Take a fillet of salmon, skin on, sprinkle the flesh side lightly with salt and Meathead's Memphis Dust (page 165). If you can get your smoker to hold steady at 145°F for an hour, you can pasteurize it. Serve it at 125°F and be prepared for gushing praise.

CHARTER BOAT SMOKED FISH

So, here's a method for superb delicate smoked fish done the way charter boat captains like to do it. I serve this on crackers, toast points, bagels with cream cheese, flaked into an omelet or scrambled eggs, or as an entrée.

MAKES *2 entrée servings*
TAKES *20 minutes to prep (depending on the number of pin bones), 3 hours to brine, 2 hours to form the pellicle (optional), 1 hour to smoke*

- 1 pound skin-on fish fillets, preferably fresh
- 3 ounces (113 grams) any salt
- 1 cup hot water
- 2 tablespoons brown sugar (if smoking salmon)
- 1 tablespoon garlic powder
- 1 tablespoon fine-grind black pepper
- 5 cups cold water

1 PREP THE FISH. If you bought or caught a whole fish, remove the guts, head, fins, tail, and scales, but leave the skin on. Fillet it by drawing a flexible filleting knife along the back on both sides of the dorsal fin from head to tail. Keep sliding the knife toward the belly, and when you hit the ribs, let the knife glide along them. There are videos of how to do this online. Then remove the pin bones (see page 301) or not. When you're done you should have 2 boneless fillets and a skeleton. (Freeze the skeleton, and when you have 6 or 8, simmer them to make fish stock.)

2 WET BRINE IT. Add the salt to the hot water and stir until it is dissolved. If you are doing salmon, add the brown sugar. Stir in the garlic powder and black pepper. They will not dissolve much at first. Pour the slurry into a clean nonreactive 9 × 13-inch baking pan. It cannot be made of aluminum, copper, or cast-iron, all of which can react with the salt. Do not use a Styrofoam cooler. Now add the 5 cups of cold water. Submerge the fish skin side up in the brine, cover the pan with plastic wrap, and refrigerate. Gently stir the container occasionally to make sure all parts of the fish come into contact with the brine. Brining time will vary depending on how thick the fillets are. Rule of thumb: 1 hour per inch of thickness. Do not leave the fish in brine longer than 3 hours. If the fillets are thin, brine for less time. Discard the brine. Rinse the fish to remove excess surface salt. Pat dry with paper towels.

OPTIONAL. Form a pellicle: Some folks like to put the fillets on a grate in a pan in the fridge for 3 to 8 hours so that a glossy film or "pellicle" of linked proteins will form on the surface. It makes a pretty sheen and is said to help retain moisture and smoke. It can also suppress "boogers" (see Albumin, page 15). Airflow helps the film form, so some folks put a battery-operated fan in the fridge with the fish. I have tried it with and without pellicle and see no significant taste difference.

3 FIRE UP. Set up your grill in a 2-zone configuration or get the smoker started and get the indirect zone up to about 225°F. Get some smoke rolling. Cut up a brown paper bag or plain white printer paper about the same size as each piece of fish and place the fish on the paper, skin side down. Place the fish on the paper on a grate on your grill or smoker so the pieces are not touching each other. The skin will stick to the paper and when the fillets are done smoking, the paper and skin will peel off easy-peasy.

4 FINISH. Start spot-checking the meat temps after about 30 minutes. Remove the meat when it is at about 125°F internal. Don't overcook. Total cooking time will be about an hour depending on the actual temperature of your grill and the thickness of the meat. Remove the fillets, peel off the paper, and the skins should come right off with it.

SALMON CANDY

Prep the Charter Boat Smoked Fish (page 306) using salmon. You can use other fish, but it is best with salmon. Just before the fish goes into the smoke, sprinkle a layer of Meathead's Memphis Dust (page 165) over it. The sugar and the flavors work superbly with salmon. Another option is to sprinkle brown sugar on or paint it with maple syrup before the cook and once or twice during the cook. This method is a winner for all types of salmon, but is especially terrific with rich, fatty king, coho, and sockeye.

CLOSE PROXIMITY SMOKED FISH WITH POBLANO-BASIL CREAM SAUCE

The recipe calls for expensive Chilean sea bass, but before you reach for less expensive salmon, let me propagate an opinion that I hold so dearly I shall try to pass it off as fact: Chilean sea bass and sablefish are the best-tasting creatures with fins. Chilean sea bass is tender and juicy, somewhat reminiscent of lobster in texture, but much more delicate and buttery in flavor. Fillets are snow white and thick, and the meat, when cooked, has wide flakes, and it remains moist even if overcooked. Yes, I know the photo is not snowy white; the amber comes from the "close proximity smoking," a method I created and that was named by my friend Greg Rempe of *The BBQ Central Show*. I think it makes this incredible fish over-the-top spectacular.

The concept of close proximity smoking is that, because fish cooks fast, if we want to get some smoke on it, we need to place it in close proximity to smoldering wood. So I use GrillGrate® (see page 121). It can hold wood within ½ inch of the food. In addition, their special tongs are perfectly designed for lifting delicate fish.

Chilean sea bass once was overfished, but in recent years regulation has brought the fishery back, so we can now buy it with a clear conscience and a full wallet. Originally called Patagonian toothfish, it can grow to up to two hundred pounds, and is most often found in the deepest cold waters of the Southern Hemisphere. It is not to be confused with other varieties of fish named sea bass. They are not

even kissing cousins. Alas, they are among the most expensive fish, so feel free to substitute other types of fish. This method works well on all of them.

MAKES *2 servings*

TAKES *10 minutes to prep (depending on how many pin bones there are), 1 hour to dry brine, about 15 minutes to cook*

SPECIAL TOOLS *GrillGrate® brand grill grates and their tongs. About 4 ounces of small pieces of wood, pellets, chips, or sawdust.*

> ½ teaspoon Morton Coarse Kosher Salt
>
> 2 Chilean sea bass or other fillets (6 to 8 ounces each)
>
> 3 tablespoons mayonnaise
>
> ½ cup Poblano-Basil Cream Sauce (page 194)
>
> **OPTIONAL.** I have also done this with the Black Garlic Butter (page 186) instead of the poblano-basil cream sauce.

SERVE WITH. Basmati rice to soak up the sauce.

1 PREP. Salt the fish an hour or two in advance if possible. Then completely coat the fish with mayonnaise.

2 FIRE UP. Preheat the grill for 2-zone cooking and bring up to Warp 7 on the direct heat side with a section of GrillGrate® directly over the heat.

3 COOK AND SMOKE. Toss the wood into the valleys directly below where the fish will sit. It should start smoldering quickly. As soon as it does, place the fish directly over the smoke and close the lid. After no more than 4 minutes, flip the fish. If you use the groovy (pun intended) spatula supplied with your GrillGrate® to release the fish, be careful not to scoop up any pellets with it. The underside should have dark grill marks and be a golden color from the smoke. Flip and cook another 4 minutes and test the internal temp. Take the fish off when the internal temperature is between 125° and 130°F. If the wood catches fire, take the fish off the grates and place it in the indirect zone to finish cooking.

4 SERVE. Pool the Poblano-Basil Cream Sauce on each plate and serve the fish atop it with rice alongside.

TEA-SMOKED WHITEFISH

Smoke, butter sauces, and white-fleshed fish are made for each other. Throw some flakes of this fish on a bagel with cream cheese and you'll never spend a paycheck on lox again.

Sablefish, aka black cod, is my all-time favorite fish and it takes to smoke like deep icy Alaskan seawater. It is not related to other fish named cod. It is so much better. Because I live landlocked in the Midwest where fresh fish is hard to find and is often not very fresh or very good, I subscribed to Sitka Seafood Market a few years ago. It is a boat-to-doorstep operation that makes all the right claims about

their sources and delivers pristine frozen seafood to me, including sable.

A word of caution: It seems about 1% of them have a genetic defect called jellybelly and when cooked they disintegrate into a yucky mess. There is no way to tell in advance if the piece you have has jellybelly. Also, their pin bones are almost impossible to remove. You just must carve or eat around them. If this sounds too risky for you, select another white flesh fish.

For this recipe I lightly smoke the skin-on sable on one side only. The skin will separate from the meat when it is served. It is a method I learned from Brooke Orrison Lewis, from one of my all-time favorite BBQ joints, The Shed in Ocean Springs, Mississippi. She lives right near the Gulf of Mexico and loves fishing. Cooking fish, skin on, without flipping is a method she calls "On the Half Shell." It remains delicate, moist, and flaky, and then I anoint it with a gentle nutty lemon brown butter that doesn't hide its natural beauty. But I've gotta tell you, sable can handle overcooking and remain spectacular.

MAKES *2 servings*
TAKES *30 minutes*

  2 sablefish/black cod fillets (6 ounces each)

  ½ teaspoon French Rub (page 167)

  ¼ cup Herbed Lemon Brown Butter (page 190)

  1 orange or tangerine

  ½ cup tea leaves, any type

½ cup dark brown sugar

6 whole star anise pods

2 tablespoons ground ginger

Flaky finishing salt, such as Maldon

ABOUT THE TEA. I don't think the type of tea will make much difference since you are burning it.

1 PREP. Lightly sprinkle the fillets with the French Rub. Make the Herbed Lemon Brown Butter and keep it warm.

2 FIRE UP. Fire up the grill for 2-zone cooking and get the direct zone up to Warp 5.

3 SMOKE. Scrub and zest the orange. Take some aluminum foil and fashion it into a small open-top boat about 6 inches long and fill it with the zest, tea, brown sugar, star anise, and ginger. Place the boat on top of the flames or coals and under the grates on the direct side and place the fish on a well-oiled grill topper or fish basket directly over the boat. Place a metal pan upside down over the fish to trap the smoke. Close the lid and let the smoke roll.

4 FINISH. When the fish hits 130°F just below the top surface, remove it, plate it, and surround it with the brown butter. Pass Maldon salt at the table.

MISO MAPLE SABLE À LA NOBU

With sablefish (black cod) from Alaska, and white miso paste from Japan or China, and dark maple syrup from New England or Canada, be prepared for something cross-culturally brilliant. You can use other fish, especially salmon because it really loves miso and maple syrup. Nobu restaurants are famous for their Black Cod with Miso. I kicked it up a notch by adding smoke, which also loves miso, and maple.

MAKES *2 servings*

TAKES *30 minutes to prep (depending on how many pin bones), 2 hours to marinate, 30 minutes to cook*

SPECIAL TOOLS *About ½ cup of dried herbs, wood pellets, wood chips, or hardwood sawdust. Something that will start to smoke rapidly.*

2 tablespoons white miso

2 tablespoons real maple syrup

1 teaspoon of your favorite hot sauce

2 skin-on sablefish/black cod fillets (8 to 12 ounces each)

Pickled ginger, for garnish

ABOUT THE MISO. White miso is milder and better than red miso in this recipe.

ABOUT THE PICKLED GINGER. This is the stuff you get in sushi restaurants, thin pink slivers. You can find it in most Asian stores and online.

1 MAKE THE BRINERADE. In a coffee cup, mix the miso, syrup, and hot sauce with a fork.

2 PREP THE FISH. Pluck out the pin bones with tweezers or needle-nose pliers. Cut the fillets to lengths that will fit on your largest spatula. This is important so that you can lift it off the grill without the fish breaking apart. I've said it before, but GrillGrate® (see page 121) and their special spatula are perfect for fish.

3 GASH AND PAINT. With a very sharp knife, gash the meat side by drawing the blade across it every ¾ to 1 inch making crosshatches about halfway through the meat. This will help the brinerade penetrate. With a brush or spoon, spread the brinerade on the fish making sure to get it into the gashes. Place the fish on a plate in the refrigerator for 2 hours and every 30 minutes or so paint it again.

4 FIRE UP. Set up the grill for 2-zone cooking, but keep the energy down on the direct heat side, Warp 5. Do not oil the grates (or the fish). Lift the grate over the heat source and toss some dried herbs, pellets, wood chips, sawdust on the fire. Something that will start to smoke rapidly.

5 SMOKE AND SERVE. Lay the fish above the smoke source and place a metal pan upside down over the fish to trap the smoke. Close the lid. After about 5 minutes, take the temperature of the meat just below the top surface. The bottom near the skin will be hotter than the top. When the top layer hits 125° to 130°F, remove it from the heat. The skin may stick to the grate. Great! If not, you should be able to peel it off easily or, what the heck, just serve it and let your guests eat off the half shell. Garnish with the pickled ginger.

SMOKED YOOPER FISH PÂTÉ

Smoked fish spread is a most hospitable way to start a dinner party. Serve on toasted slices of baguette, bagel, bagel chips, crackers,

The smokehouse at Calumet Fisheries in Chicago.

matzoh, potato chips, carrot sticks, or celery, or make the best fish salad sandwich ever.

There are still a smattering of smokehouses around the country like the Calumet Fisheries south of Chicago shown in the image above. Calumet is not fancy, and their smokehouse is among the most primitive I've ever seen, but they have been doing it right since 1948, right enough to win a James Beard American Classic Award.

In Northern Michigan and the Upper Peninsula of Michigan (called the Yooper), there are at least a dozen small family-run smokehouses that make Whitefish Pâté from local whitefish and trout. In south Florida it is called "smoked fish salad" or "smoked fish spread," and is often made with mullet, amberjack, kingfish, mackerel, and even mahimahi. In the Pacific Northwest, it is called "salmon dip," because you can dip crackers and chips in it.

MAKES *About 1½ cups*
TAKES *20 minutes prep (depending on how many pin bones), 1 hour to dry brine, 45 minutes to smoke, 30 minutes to assemble*

- ¼ teaspoon Morton Coarse Kosher Salt
- 8 ounces white-fleshed fish
- 1 teaspoon cooking oil
- 4 ounces cream cheese
- 1 tablespoon full-fat mayonnaise
- 1 tablespoon sour cream
- 1 teaspoon prepared horseradish
- 1½ tablespoons real maple syrup or honey
- ½ teaspoon red pepper flakes
- 1 lime
- 1 large shallot or small onion (size of a golf ball)
- 2 tablespoons capers, drained
- ½ teaspoon mild paprika or smoked mild paprika, for garnish
- Chopped scallions or chives for garnish

FIDDLING. There's lots of room for fiddling with this. To be authentic, you should use a freshwater whitefish, but you can use just about any fish. I've even made it with sunfish and chubs. Add black pepper, fresh dill, or tarragon. Swap lemon for the lime. Swap sour cream, crème fraîche, or yogurt for some of the cream cheese.

1 SMOKE. Salt the fish, oil the skin side, and smoke it in a smoker or on the indirect side of a grill, oiled side down, at 225°F until it hits 135°F, then let it come to room temperature. Skin and bone the fish if necessary and crumble it with your fingers, searching meticulously for rogue bones. Set it in the fridge.

2 MAKE THE PÂTÉ. Pull the cream cheese out of the fridge, cut it into 4 chunks and let it come to room temperature in a bowl. Add the mayo, sour cream, horseradish, maple syrup, and pepper flakes. Scrub the lime thoroughly, scrape the zest off, and add it. Cut the lime in half and squeeze 1 tablespoon into the bowl. With a fork, mix everything together, busting up the cream cheese chunks. Peel and mince the shallot and add 1 to 2 tablespoons to the bowl. Add the capers (if the capers are large, you can chop them). Add the fish. Mix everything.

3 TASTE AND CHILL. Adjust the ingredients to your taste. Scrape it into a serving bowl. You can use it right away, but it is better if you fridge it a few hours so the flavors can be extracted and married, and when chilled it is firmer.

4 SERVE. Just before serving, sprinkle on some paprika and chopped scallion or chives for show.

SMOKED TUNA WITH SESAME-LIME DRESSING

This is the absolute perfect light lunch, and it is darn tasty served hot, room temp, or chilled. The better the grade of tuna, the better the dish. Shoot for "sushi grade" ahi. You must be vigilant and take care to not overcook tuna, which can go from succulent to cardboard in minutes.

MAKES *4 appetizer servings*
TAKES *2 hours to wet brine, about 5 minutes to smoke (with a smoking gun), 5 to 7 minutes to sear*

SPECIAL TOOLS *A smoking gun is my recommendation, but not necessary*

> Sesame-Lime Dressing (page 174)
>
> ½ cup macadamia nuts
>
> 5 tablespoons Morton Coarse Kosher Salt
>
> ¾-pound tuna steak, about 1¼ inches thick
>
> 1½ teaspoons mustard seeds
>
> 1½ teaspoons celery seeds
>
> 1½ teaspoons black sesame seeds
>
> ¾ teaspoon medium-grind black pepper
>
> ⅓ cup full-fat mayonnaise
>
> 1½ cups baby arugula
>
> 1 lime

1 PREP. Make the Sesame-Lime Dressing. Coarsely chop the nuts. In a small, dry skillet, toast the nuts until golden and aromatic. Don't let them burn—1 to 2 minutes is all you need. Make a wet brine by dissolving the salt in 2 quarts water. Immerse the tuna in the brine and refrigerate for about 2 hours. Lightly

Raw ahi tuna is a red meat.

crush the mustard seeds and celery seeds. Whisk them together with the sesame seeds, black pepper, and mayo. Wash and pat dry the arugula. Scrub and zest the lime.

2 GUN IT. Set up a smoking gun according to the manufacturer's directions. Place the tuna in a pan and cover it with cling wrap. Insert the tube of the smoking gun under the cling wrap, light the gun, and run it 4 to 5 cycles, replacing the sawdust as soon as it is spent. (If you wish, you can throw the tuna in a smoker for 30 minutes, but this will partially cook the center.)

3 FREEZE IT. About 30 minutes before you plan to cook, transfer the tuna to the freezer for 30 minutes. To prevent ice crystals, do not leave it in the freezer any longer. We do this to make sure the center is cold so it does not cook too much when you sear.

4 FIRE UP. Set up your grill or griddle for Warp 10. If you are using a grill, preheat a cast-iron skillet or portable griddle until it is as hot as Hades.

5 COOK THE TUNA. Remove the tuna from the freezer and coat it on all sides with the mayonnaise mixture. Place the tuna in the hot pan. Press gently so as much fish as possible is in contact with the hot pan. When the fish has seized, after about 1½ minutes, use a metal spatula to carefully flip to the other side and sear for 1½ minutes more. Immediately transfer to a cutting board.

6 SERVE. Toss the arugula thoroughly with about half of the Sesame-Lime Dressing. Slice the tuna across the grain about ⅓ inch thick and fan them out on individual small plates. Place some dressed arugula on the side and drizzle everything with the remaining dressing. Scatter the macadamia nuts and lime zest over the top.

PETER MAYLE'S FLAMING FISH: A WONDER

In his delightful whodunit *A Vintage Caper*, Peter Mayle (author of the memoir *A Year in Provence*) takes us on a digression that has nothing to do with the plot, inserted no doubt as an excuse for the renowned food and drink lover to share a recipe. He has his hero, Sam, take a walk down by the docks in Marseilles, where he comes face to face with a fishmonger who practically forces him to buy a whole fresh sea bass even though he is staying in a hotel.

She seduces him with this: "Make two deep cuts in your fish, one on each side, and stick two or three short pieces of fennel in each cut. Paint the fish with olive oil. Grill on each side for six or seven minutes. Using a fireproof serving dish, place the fish on a bed of dried fennel stalks. Warm a soup ladle filled with Armagnac, set light to it, and pour it over the serving dish. The fennel catches fire, scents the air, and flavors the fish. *Une merveille*." A wonder.

Fennel fronds are easy to find attached to fennel bulbs in stores. Fronds are the delicate frilly leaves at the ends of the stalks. Better still, sprinkle a few fennel seeds in your garden and before long you have a jungle of fragrant fronds. I also used this method for Lobster with Beurre Blanc (page 328). Save the fennel bulbs for a salad.

MAKES *2 servings*
TAKES *45 minutes*

SPECIAL TOOLS *Two-sided flipping grill basket*

4 bunches of fennel

1 whole fish (2 pounds)

Good olive oil

½ cup Armagnac or other brandy

ABOUT THE ARMAGNAC. Armagnac is a grape brandy as is Cognac. You can even use an American straight brandy or rum or bourbon. I haven't tried it with Scotch yet, but the smokiness might just be what the fishmonger ordered.

1 PREP. Trim the fronds from the fennel. You need about 36 inches of them. Scrape off all the scales and gut the fish. Rinse the cavity and remove all blood. I remove the fins, especially the dorsal fins because they can stab you if you're not careful. You can leave on the head and tail if you wish to enhance the presentation. I think it looks cool but beware many people get queasy when served fish with the head on. Make deep cuts in each side of the fish about 1 inch apart and stuff fennel fronds in the cuts.

2 FIRE UP. Get a grill medium hot, about Warp 5. I use charcoal in a single layer across the bottom of most of the charcoal grate.

3 GRILL. Paint the fish generously with olive oil inside and out and nestle it into a flipping grill basket. Place all the remaining fennel fronds on the grill grates above the coals or flame. Grill the fish on top of the fennel, lid down, on each side for 6 or 7 minutes until the interior temperature is at least 125°F. You will smell the fennel.

4 FLAMBÉ. Pour the brandy over the fish and it should burst into flames as it drips onto the coals. "The fennel catches fire, scents the air, and flavors the fish. *Une merveille.*" Indeed.

SCALLOPS WITH BLACK GARLIC BUTTER SAUCE

Scallops and butter. No-brainer. Add garlic. Duh! But black garlic? Garlic that has fermented until it is black, sweet, and sticky? Oh my.

You will probably buy scallops already shucked and cleaned, white pillows typically about 1 inch across and ½ inch thick, but they can be larger and smaller. When you buy them shucked, the orange roe sack, sometimes called coral, has usually been removed because it spoils rapidly. But if you buy them in the shell, there is a good chance it will still be there. Don't discard it!

BAY SCALLOPS are small, about 1 inch in diameter or less, and are hand harvested from shallow waters.

SEA SCALLOPS are larger, perhaps 2 inches in diameter, and they grow in deeper waters. Both bay and sea scallops are excellent tasting.

DIVER SCALLOPS are usually sea scallops hand-harvested by scuba divers.

DAY-BOAT SCALLOPS are caught on smaller vessels that drag the bottom with chain nets.

WET SCALLOPS are brined in a solution of trisodium phosphate which extends shelf life and adds weight and dilutes the flavor slightly.

DRY SCALLOPS are untreated and the better choice.

For this recipe you will need a smoking gun.

MAKES *4 appetizer servings or 2 entrée servings*
TAKES *15 minutes to prep, 5 minutes to sear*

SPECIAL TOOL *Smoking gun*

- 1 stick (4 ounces) unsalted butter
- 6 black garlic cloves
- 8 kumquats
- 16 large scallops (about 1½ ounces each)
- 1 tablespoon clarified butter (page 188)

ABOUT THE KUMQUATS. Please, don't skip the kumquats. Tangerines or clementines are good, but no substitute. If you can't find them, try harder.

SERVE WITH. Rice, pasta, or something that can be used to absorb extra butter.

1 MAKE THE SAUCE. In a small frying pan, melt the butter and hold it on low. Peel and mince the black garlic. It is sticky so spray oil on the knife. Add it to the butter. Slice the kumquats and discard the seeds. Add them to the butter. Let everyone get to know each other for about 30 minutes.

2 PREP THE SCALLOPS. If they are still in the shells, it is easy to pry them open. One shell is flat, the other curved. Place the scallop on a cutting board flat shell down. Open the shell just a crack with a table knife inserted between the shells opposite the hinge, hold it open with your thumb, insert a sharp flexible filleting knife, and press it along the flat shell, cutting the adductor muscle away from the shell. Now place it curved side down, pry open the shells, and with a spoon cut the muscle away from the curved shell. Pull off the inedible frill and guts with your fingers and with the fillet knife cut them away from the adductor muscle, the edible white marshmallow that we want to eat. Rinse the scallops vigorously and for longer than you think necessary, even if you bought them shucked. Make sure any black specks and film are gone.

3 SMOKE THE SCALLOPS. Place the scallops in a pan and cover tightly with cling wrap. Insert the tube from a smoking gun under the cling wrap and smoke for about 10 minutes, replenishing the sawdust as needed.

4 FIRE UP. Get a grill screaming hot over direct radiant heat. Warm the serving plates and the sauce. In a cast-iron skillet or on a griddle, melt the clarified butter. Pat the scallops very dry and sear them on one side until golden, 3 to 4 minutes. Flip and sear for 1 minute only on the other side. The internal temperature should reach 125° to 130°F.

5 SERVE. Divide the scallops among the warm plates, seared side up. Spoon the chunky buttery black garlic sauce over the top.

MUSSELS WITH SMOKED FETTUCCINE

This recipe starts with a French classic, *moules marinière*, that I have amped up with cream, and then goes the extra mile by serving it on pasta cooked with smoked water. If this sounds weird, remember that smoke and cream go together like Bonnie and Clyde. And if you can't find mussels, this works just fine with clams or oysters.

MAKES 4 servings
TAKES 2 hours to smoke the water, 1 hour for the rest

- 2 to 3 pounds ice
- 2 pounds mussels in their shells
- 1 medium onion
- 3 garlic cloves
- 1 lemon
- 2 tablespoons high-quality olive oil
- ½ teaspoon coarse-grind black pepper
- ¼ cup dry white wine
- 2 tablespoons (1 ounce) unsalted butter
- 2 tablespoons mascarpone or cream cheese
- 2 tablespoons heavy cream or half-and-half
- 1 teaspoon Morton Coarse Kosher Salt
- 16 ounces dried fettuccine
- 2 tablespoons chopped celery leaves
- **OPTIONAL.** Hot pepper flakes on the side

1 **FIRE UP.** Set up your smoker and aim for 225°F.

2 **SMOKE THE WATER.** Put the ice in a 9 × 13-inch baking pan and put it in the smoker for 2 hours at 225°F. The smoke will be attracted to the ice and when it melts you will have smoked water. Pour it into a 2-quart pot and set it aside.

3 **PREP.** Scrub the mussels thoroughly and discard any that are stinky or too heavy (likely from sand). Sniff every darn one. Remove any of the black stringy "beard" that may still be attached to them with pliers or scissors. Peel and chop the onion and peel and press or mince the garlic. Scrub and zest the lemon.

4 **MAKE MOULES MARINIÈRE.** In a large pot, heat the olive oil and onion over medium heat and gently cook until the onion is translucent, about 3 minutes. Add the garlic, zest, and black pepper and stir. After about 2 more minutes,

add the white wine, drop in the mussels, and cover with a loose-fitting lid, leaving a crack so water vapor can escape. Turn the flame to medium-high and continue cooking for 4 to 5 minutes, stirring at least once, until all the mussels have popped open. Turn off heat and lift the lid.

5 FREE THE MOULES. After about 5 minutes, once the mussels are cool enough to handle, throw away any mussels that have not popped open (these were dead before cooking, taste bad, and are possibly unsafe to consume). Now move all the mussels to a separate bowl, leaving behind the flavorful liquor as a base for the sauce. Pop open the shells and with a paring knife scrape out the meat, being sure to get the white adductor muscles attached to the shells. Put all this in a bowl. Discard the shells.

6 MAKE THE SAUCE. Bring the pot back to a simmer and slowly whisk in the butter, mascarpone, and cream until fully incorporated and the sauce is smooth and creamy.

7 BOIL THE PASTA. Add the salt to the pot of smoked water and bring the water to a boil. Add the fettuccine and cook to al dente according to the package directions. Drain the fettucine, reserving the pasta water.

8 FINISH. Add the mussels to the sauce and then add the drained pasta. Stir to coat the pasta and add about a few tablespoons of the pasta water to make the sauce a little looser if you want. Taste and adjust the salt if necessary.

9 SERVE. Scatter some chopped celery leaves on top and serve with hot pepper flakes at the table if your guests want them.

SMOKE CATCHER CLAMBAKE

On a trip along the coast of Oregon, my wife and I stayed on Siletz Bay, whose bottom was almost completely exposed when the tide went out. Strolling the flats on the east side at low tide, it is easy to dig clams. On the north shore people catch blue crabs with chicken necks.

The "smoke catcher" method (page 98), as I call it, is handy for fast-cooking thin fish fillets and it is practically essential for bivalves. Oysters, clams, and mussels contain precious juices, so you need to place the shells on the grill curved side down, making sure they are level to prevent the liquor from running off. Depending on how your grill grates are designed, you may be able to place the shells right on the grates or you may need to use a grill topper.

You will want to use small chips, pellets, or dried herbs directly on the flames so they will catch, burn, and smoke quickly. Once it is belching smoke, invert a large baking pan over the clams. This cover catches smoke and traps it in contact with the meat much better than the grill lid. If the pan is shiny, it also reflects heat. I keep a large disposable aluminum pan just for this task so I don't have to scrub the smoke off pans we actually use for baking.

Gosh these clams are good. Licking your fingers is better than after eating Cheetos.

MAKES *2 serving*

TAKES *2 hours to purge, 30 minutes to smoke*

SPECIAL TOOLS *Large pan (the "smoke catcher") and 2 handfuls of wood chips or pellets*

  2 dozen clams in shells (preferably live)
  2 medium shallots or 1 medium onion
  1 small red jalapeño

1 small handful of celery leaves
1 stick (4 ounces) unsalted butter
1 loaf crusty bread

1 SHUCK. Scrub the clams thoroughly, purge them of sand by leaving them in a bucket of approximately 4% salt water (4 tablespoons of Morton Coarse Kosher Salt per quart of water or 40 grams of salt for every 1 liter of water) for a couple of hours and discard any that are dead. Slip a table knife between the two shells opposite the hinge and pry them apart, popping the hinge and discarding one of the shells.

2 FIRE UP. Set up your grill for 2-zone cooking.

3 PREP. Peel and mince the shallots. Mince the jalapeño and celery leaves. In a small saucepan on the grill, melt the butter. Add the shallots, jalapeño, and 3 tablespoons of celery leaves. If you wish, grill the bread or make Smoke-Roasted Garlic Bread (page 367). You need bread to sop up the juices.

4 SMOKE. Toss a handful of wood chips, dried herbs, or pellets on the fire. Place the clams on the half shell on the grates on the direct heat side. Spoon the butter over them. Let some butter hit the fire and flare up and smoke. Put the "smoke catcher" over the clams.

5 SERVE. Take a peak and as soon as the butter starts to burble, serve them in a bowl. Slurp them down. Dunk the bread in the juice. Lick your fingers. Be transported to a bay in Oregon.

SMOKED NEW ENGLAND CLAM CHOWDER

Like a clam, the world can be divided in halves: Are you Manhattan or New England clam chowder? Manhattan is tomato-based, New England is cream-based. You can make a good case for either, but if you are going to smoke the clams, there is no argument. Cream and smoke work together like Romeo and Juliet.

MAKES *1 quart chowder, enough for 4 cups or 2 big bowls*

TAKES *2 hours to purge, 1½ hours to prep and serve*

SPECIAL TOOLS *Grill topper (even if you have a smoker, do this on the grill because you want lots of thick white smoke in a hurry). About 1 cup of hardwood sawdust, clean dry hay, tea, or dried herbs. Large aluminum pan (the "smoke catcher").*

- 1 dozen large clams or 2 dozen small clams
- 1 small baguette
- 1 large celery stalk
- 1 large onion
- 1 medium Yukon Gold potato
- 4 slices thick-cut bacon
- 2 garlic cloves

1½ tablespoons all-purpose flour

1 cup bottled or canned clam juice

1 cup 2% or whole milk

¼ teaspoon dried thyme

¼ teaspoon medium-grind black pepper

1 small bay leaf

¼ teaspoon Morton Coarse Kosher Salt

1 lemon

OPTIONAL. 1 tablespoon chopped fresh parsley or celery leaves, for garnish

ABOUT THE CLAM JUICE. I amp this up with bottled clam juice, which is simply the broth that is made by steaming fresh clams with a little salt added. You can buy this at the supermarket or online.

1 PURGE AND SHUCK. Scrub the clams thoroughly, purge them of sand by leaving them in a bucket of approximately 4% salt water (4 tablespoons of Morton Coarse Kosher Salt per quart of water or 40 grams of salt for every 1 liter water) for a couple of hours and discard any that are dead. Slip a table knife between the two shells opposite the hinge and pry them apart. Pop the hinge and discard one of the shells.

2 PREP. Slice and grill the bread or make Smoke-Roasted Garlic Bread (page 367). Dice the celery until you have about ⅓ cup, and peel and dice the onions until you have about ½ cup. Scrub and cut the potato into 1-inch chunks (large enough so they won't fall through the grates), put them in a bowl, and cover them with cold water to keep them from turning brown.

3 FIRE UP. Set up your grill for 2-zone cooking.

4 SMOKE. Place the clams and the potatoes on a grill topper. Toss the sawdust, dried herbs, tea, small wood chips, or even clean hay on the fire, something that will smoke effusively and quickly. Place the "smoke catcher" over the clams and taters. After about 10 minutes, remove the potatoes, and when the liquid in the clams is boiling, remove them from the grill and place them in a bowl to catch any juice. When they cool enough to handle, use a sharp knife to cut them free of the shells and discard the shells.

5 BUILD THE CHOWDER. Slice the bacon into 1-inch-square chunks and throw them into a medium soup pot and cook over medium heat on the stovetop or on the direct heat side of a grill until all the bacon has softened and rendered a bit of fat but not yet crisped. Add the celery and onion and stir until limp. Peel and mince or press the garlic, add it, and cook 1 more minute. Stir in the flour and cook for 3 to 4 minutes to form a roux. Whisk gently until the flour is incorporated, turns slightly blonde, and there are no lumps. Stir in the potatoes, clam juice from the clams and the bottle, milk, thyme, black pepper, bay leaf, and salt. Cover with a tight-fitting lid and simmer until the potatoes are soft enough a toothpick slides through. Add the clams and simmer for another 3 minutes. Reduce the heat to low. Taste the soup and add salt if it is needed.

6 SERVE. Spoon the chowder into bowls and garnish with chopped fresh parsley or celery leaves, if using. Cut the lemon into wedges and serve them and the grilled bread on the side.

BRINE-SIMMERED AND CHARRED OCTOPUS

"I'd like to be under the sea in an octopus's garden with you."

—RINGO STARR

In English translations of Jules Verne's 1870 science fiction classic *Twenty Thousand Leagues Under the Seas*, Captain Nemo's submarine, the *Nautilus*, is attacked by a giant squid that devours a member of his crew. Actually, Verne called it a *poulp*, the French word for octopus, but for some unknown reason most English translations call it a squid.

BTW, *octopus* is a Greek word and the plural for *octopus* is *octopodes*, not the Latin *octopi*, or the James Bond version, *Octopussy*. But according to my editor, who quotes *Merriam-Webster*, in English *octopuses* is fine.

Octopuses have a balloon-like head and eight arms with suckers on the underside, all of which are delicious. The eyes and beak are not.

Baby octopuses are not much larger than squid, and they are a more tender than adult octopuses, but they are expensive. Adults are typically 2 to 4 pounds with tentacles up to 18 inches long, and that is what I usually buy. Avoid the giants that can run up to 100 pounds. **THIS IS IMPORTANT:** Buy twice as much octopus as you think you need. They shrink as much as 50%.

You might find fresh uncleaned adult octopuses at the docks or at a specialty fishmonger, but most stores sell cleaned and frozen. I prefer frozen, because freezing helps tenderize it. If your octopus arrives fresh, and has never been frozen, you can freeze and thaw it to tenderize it a bit. After thawing, it should not smell at all fishy.

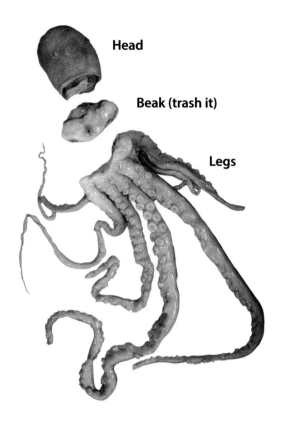

Head

Beak (trash it)

Legs

If you buy fresh, you can ask the fishmonger to clean it for you, but you can easily do it yourself. First, remove the legs by cutting where they meet the body below the eyes. This is easiest done with kitchen shears. Now decapitate just above the eyes. You will now have a section that contains the eyes and the beak. Discard this section. Slit open the head and remove anything within. If you find an ink sac, don't try to save it. The best ink is harvested from fresh specimens. Rinse and scrub everything aggressively to remove any sand or grit.

The challenge now is tenderizing. Chefs have devised numerous schemes: In Greece and Italy (and on YouTube) you can still see fishermen beating them on rocks. If you order from The Octopus Garden in Brooklyn, they tenderize in a large stainless steel tumbler. Avra Madison restaurant in New York City gives their octopuses an hour in a washing machine on the rinse cycle! Some chefs poach them gently in olive oil. At estiatorio Milos in Las Vegas they braise it in wine in a Dutch oven at 350°F for 2 hours and the results are superb. A popular traditional recipe calls for simmering it with a wine cork, a superstition whose science escapes me.

I've had good luck with sous-vide-que. First dry brine, then in the sous vide bag with 2 tablespoons of butter, and into the bath at 170°F for 4 hours. The oil helps keep the skin on. Next, pat dry, toss with olive oil, and sear for 3 minutes or so on either side until you get a little char on the thin ends of the legs.

Simmering in a brine, on the other hand, allows salt to penetrate thoroughly. In a side-by-side test, my tasters and I preferred the taste of this method to the sous-vide-que even though the sous vide test had all manner of garlic and other flavorings in the bag. Simmering in brine produces a slightly, only slightly, firmer chew than sous-vide-que, but it is simpler, quicker, and still wonderful. "Less is more" wins again.

MAKES *2 servings.*

TAKES *15 minutes to prep, 4 hours to simmer, 15 minutes to cook*

SPECIAL TOOLS *Grill topper*

2 tablespoons Morton Coarse Kosher Salt

1 whole octopus (2 to 3 pounds)

1 tablespoon smoked mild paprika

3 tablespoons inexpensive olive oil

Chopped fresh parsley, for serving

Best-quality olive oil, for serving

1 SIMMER. In a large pot on your stovetop or grill, bring 3 quarts water to a low simmer (at 170°F) and add the salt. Stir until the salt dissolves. Slide the octopus into the brine. The cold meat will cool the water. Wait for it to come back up to temperature and hold it at 170°F for 4 hours.

2 FIRE UP. Set up your grill for 2-zone cooking and get the direct heat side up to Warp 10. Put a metal grill topper over the heat.

3 PREP THE OCTOPUS. Pull the meat out of the pot and pat it dry. If the skin starts to come off, go ahead and rub it with a paper towel to remove it. Cut the legs into thick ends and skinny tips so you can get the skinny tips off the grill before they overcook. If your octopus has thick legs, you can butterfly them by slitting down the length. This creates more surface area for more flavor. Sprinkle on the smoked paprika. Toss the meat with the inexpensive olive oil in a bowl until coated.

4 GRILL. Spread the pieces of meat on the grill topper on the direct heat side. With the lid open, stand there and watch carefully. Roll them around until the edges begin to crisp and you get a little char (not too much). If they flare up out of control, move them to the indirect

side until the flames go down. The head should lie flat. Cook both sides. The thin legs will cook fast, 5 minutes per side or less, and if you walk away, they will almost certainly burn. Take them off when they start to char.

5 SERVE. You can serve the octopus hot from the grill, garnished with minced parsley and a drizzle of best-quality olive oil. It is also delicious at room temperature or chilled. Here are some more ideas:

5A CENTER OF PLATE. Serve it warm, garnished with fresh parsley or chives, and squeeze on some fresh lemon juice. Try with my Ladolemono Sauce (from my last book) or drizzled with melted butter and black garlic (see Scallops with Black Garlic Butter Sauce, page 317).

5B WITH A DIP. Serve nekkid and provide a flavored mayo as a dip. Try a few drops of toasted (brown) sesame oil in mayo, Sriracha in mayo, or chipotle in adobo in mayo.

5C OCTOPUS GARDEN. Serve it chilled on a salad of your favorite greens with some fennel slices, pickled onion, sun-dried tomato, boiled potatoes, oil-packed olives, minced red peppers, green jalapeño, and Lemongrette (page 172).

5D OR . . . Serve on a jicama slaw with a citrus dressing, on a bed of polenta, on pasta, on risotto, on boiled potatoes with pesto, on Smoke-Roasted Garlic Bread (page 367), drizzled with melted Black Garlic Butter (page 186), or on a sandwich.

LOBSTER WITH BEURRE BLANC

When I was 10, I lived a summer on the coast of Maine and I participated in an ancient primitive feast. It begins by digging a hole in the beach, throwing in some large rocks, then some logs, and setting the wood on fire. It will burn down to embers and the rocks will retain the heat. Then you toss on a pile of wet seaweed and on top of that, the lobstahs (and clams and corn) and more wet seaweed to hold in the heat and for steam. Primitive becomes decadent.

Although I am now allergic to lobster, the taste has lingered. So, for my Midwestern wife, I try to re-create the experience on my grill by laying down a bed of fresh herbs from her garden, splitting a lobster in half, and grilling it on the aromatics. The herbs add a seductive whisp of complexity. I like to do it on Valentine's Day to show how I will sacrifice for her like a good little martyr.

In the US, there are two kinds of lobster: Maine or cold-water lobster, and Florida or Caribbean spiny lobster. Spiny lobsters lack the big meaty claws of their northern kin, although some maintain they are not really related. The good news is that they taste similarly delicious.

Many frozen meats can be as good as fresh meats. Not so for lobster. When lobsters die, enzymes begin the process of degrading their unique proteins and the meat becomes mushy in a hurry. So plan your lobster meal in advance. Always buy live Maine lobster and keep them alive in a pot in the refrigerator with wet newspapers or towels until prep time. Do not submerge them in water unless it is the right salinity and is aerated.

Because spiny lobsters don't have claws, when you buy lobster tails, that is what you are probably getting. They should be sold frozen.

MAKES *2 decadent servings*

TAKES *45 minutes to prep, 6 to 8 minutes to cook*

SPECIAL TOOLS *Heavy-duty kitchen shears, nutcrackers for serving, ½ pound fennel fronds or tarragon or other herbs*

- 6 tablespoons (3 ounces) unsalted butter
- 2 live lobsters
- 2 teaspoons Morton Coarse Kosher Salt
- Medium-grind black pepper
- ½ cup Beurre Blanc (page 190)
- 2 teaspoons minced fresh tarragon, flat-leaf parsley, or finely snipped fresh chives
- 2 lemon wedges
- OPTIONAL. 3 tablespoons breadcrumbs if there is roe

1 FIRE UP. Set up your grill for direct heat cooking and aim for about Warp 7.

2 PREP. Melt the butter. Kill the lobsters by stabbing them in the brain right behind the eyes with a sharp knife. This is the fastest and most humane method. The legs and tail might continue to move a bit from contracting nerves but don't worry, the lobster is quite dead and feeling no pain. Cut down to the cutting board and through the rest of the head and then flip it onto its back and cut through the tail (use kitchen shears if the shell will not yield to your knife). You will now have two halves ready to be prepped. Take the rubber bands off the claws and crack them with the back of the knife blade so there are cracks that butter can enter.

3 ROE. In the head you may see roe and tomalley. The roe is bright orange or red and the tomalley is tan or green. Both are delicious. The roe is a mass of eggs only in females and it is perfectly safe. Scoop out any roe plop it in a frying pan with 1 tablespoon of the melted butter and the bread crumbs. Fry it and set it aside. Tomalley is part of the digestive system and because of pollution, doctors recommend you discard it. My cheffy friends tell me that because they eat lobster so

infrequently, and because the waters are less polluted, they mix the tomalley with butter and bread crumbs and fry it, too. They will both get used later in this recipe.

4 GRILL. Brush the exposed lobster meat and the cracks in the claws with the melted butter (not the Beurre Blanc). Season with salt and pepper. Spread the fennel on the grill, wait for it to start smoking, and then grill the lobster halves meat side down on top of the fennel, lid down, for about 5 minutes. Flip and grill until the meat is 135°F in the tail.

5 SERVE. Immediately dollop a spoonful or two of Beurre Blanc atop the meat in each lobster half, fill the cavity near the head with fried roe if you have it and tomalley if you want to try it, and serve with the remaining Beurre Blanc on the side. Scatter with minced tarragon, flat-leaf parsley, or finely snipped fresh chives. Place a lemon wedge on each plate. Oh, and two nutcrackers with lots of napkins.

SHRIMP WITH GRAND MARNIER BUTTER

Shrimp is tricky because they are small and thin. When properly cooked to 131°F most shrimp meat is opaque milky white with some pink or orange on the surface, and if the shell is on, it should be orange to pink. If the flesh goes to bright white, it is overcooked. It should be curved like an apostrophe, but not tight like a fist.

Shrimp come in a wide range of sizes. The standard measure is how many per pound, so 16/20 means there are 16 to 20 in a pound. U/10 means under 10 in a pound. They often have fanciful descriptors such as "Extra Colossal" and "Extra Small," but the numbers are more meaningful. For example, 16/20 are called "Extra Jumbo," but think about it. At 16 to 20 per pound, each one weighs an ounce or less—not exactly jumbo in my world. Sorta like the fanciful names for condoms. Nobody asks for small. Keep in mind that heads and shells have weight, so if you remove the head, a 16/20 will go up 2 counts to 18/22 and if you remove shells and legs they go up another count to 19/23.

I usually go for larger shrimp because they take longer to cook and that means they are less likely to overcook, they will retain more moisture, and a slower moving target is easier to hit. I usually buy them whole and remove the heads as soon as I get them home because

one doesn't eat the heads and the contents of the head spoil more quickly. I leave the shell on because they remain moist longer and cooking them in the shell is sometimes the best way to go, especially on the grill, because the shell contributes flavor, protects the delicate insides from heat, and retains moisture.

Shrimp have a long digestive tube, incorrectly called the "vein," running from the head to the tail, and it is sometimes filled with grit or waste. With the shell gone it is easy to remove the vein. It's trickier with the shell on. Start by snipping the shell open along the back from the end of the head to the end of the tail. Then run a sharp knife down the back meat to the level of the vein. Now it is easy to pull it out with the tip of the knife or tweezers.

Don't skip the Grand Marnier Garlic Bread. It's special. The tarragon, Grand Marnier, and zest give it a real zing.

MAKES *2 entrée servings*

TAKES *Under 1 hour if the shrimp have shells on*

SPECIAL TOOLS *Metal grill topper, large pan or metal bowl (the "smoke catcher")*

> 2 batches of Grand Marnier Butter (page 368)
>
> 2 pounds 16/20 or larger heads-on shrimp
>
> 1 French-style crusty baguette at least 8 inches long

1 BUTTER UP. Make two batches of the Grand Marnier Butter. Take the pan off the heat and let it cool a bit so it won't cook the shrimp when you dunk them. Divide it into 2 bowls.

2 PREP THE SHRIMP. Snap the heads off the shrimp and discard them. With scissors, cut the shell down the back. With a sharp knife cut through the meat down to the vein. Slip the tip of the knife under the end of the vein and lift it out.

3 FIRE UP. Preheat a grill for 2-zone cooking and get the direct heat side up to Warp 7.

4 MAKE THE GARLIC BREAD. Cut the baguette into ½-inch slices on the bias and paint them on one side with the Grand Marnier mix, all the way to the edges so they don't burn. Place the slices, butter side down, on the direct heat side and stand right there. Check them every 30 seconds and turn them 90 degrees. When the edges are on the edge of black, they're done.

5 SMOKE. Throw some wood on and get some smoke rolling. Place a metal grill topper on the direct heat side. Dunk the shrimp in the Grand Marnier Butter and roll them around so they all get a good coating and put them in a bowl. When they are all dunked, dump them onto the grill topper. There will be flareups, so quickly put the "smoke catcher" upside down over the top to trap some smoke and stifle the flames. Leave the grill's lid up and stand right there. Every 2 to 3 minutes check the shrimp. When the undersides of the shells start to turn pink, the meat turns opaque, and the tail starts to curl, all at about 110°F internal, paint them with more Grand Marnier Butter, flip them, and paint the cooked side with the butter.

6 SERVE. When they are opaque, curling, and hit 131°F in the thick end, scoop the shrimp and some Grand Marnier Butter that hasn't touched shrimp into serving bowls and serve the bread alongside. Encourage your guests to peel and plop a shrimp on a slice of bread with plenty of Grand Marnier Butter and chomp down.

CRAWFISH ON DIRTY RICE

Dirty rice is a Louisiana Bayou classic made with white rice that gets a dirty color from being cooked with chicken livers, gizzards, necks, wings, and backs. Add andouille sausage, seasonings, the Holy Trinity (peppers, celery, onion), and whatever else is in the fridge for full effect. There's a lot of leeway here, so I went my own way. My rice gets dirty by being smoked beneath sausages and crawfish and catching the drippings. It gets a little heat and plenty of flavor from my Cajun seasoning. Special thanks to Chef Ryan Udvett who was fresh out of culinary school and my righthand man back in 2015 when I was writing my first barbecue book. We created this recipe for that book, but it hit the cutting room floor for space reasons. He is now a Product Developer for US Foods. So proud of him.

MAKES *2 servings*

TAKES *2 to 2½ hours*

SPECIAL TOOLS *9 × 13-inch baking pan and a wire rack that will sit on top of it*

- 2 tablespoons (1 ounce) unsalted butter
- 2 stalks celery
- 1 medium onion
- 1 small green bell pepper
- 1 small red bell pepper
- 1 tablespoon Cajun Seasoning (page 166)
- 1 teaspoon garlic powder
- ½ teaspoon onion powder
- 1 teaspoon Morton Coarse Kosher Salt
- ½ teaspoon medium-grind black pepper
- ½ cup canned black beans
- 1 cup white rice (preferably basmati)
- 1 bay leaf
- 3 pounds whole crawfish
- 1 pound andouille sausage
- **OPTIONAL.** ¼ cup chopped fresh parsley, chives, or scallions, for garnish

ABOUT THE RICE. I prefer basmati from India.

ABOUT THE CRAWFISH. They are best live in season from December through July, but frozen are almost as good. When you buy frozen, they usually come headless.

1 PREP. Cut the butter into pea-size chunks. Dice the celery until you have about ½ cup. Peel and coarsely chop the onion. Chop the two peppers. In a small bowl, mix together the Cajun Seasoning, garlic powder, onion powder, salt, and pepper. Drain and rinse the beans.

2 FIRE UP. Set up the grill for 2-zone cooking and preheat the indirect zone to 225°F or fire up your smoker to 225°F. Get some smoke rolling.

3 START THE RICE. Pour 2½ cups water into a 9 × 13-inch baking pan and bring it to a boil. Add the rice. On top of the rice add the butter, celery, onion, both bell peppers, the beans, and the bay leaf and sprinkle on the spice blend. Give the rice a good stir. Place the pan on the indirect side of the grill or in the smoker. Place a wire rack on top of the pan of rice and put the crawfish and sausage on the rack. Don't put the crawfish directly on the rice because you'll need to stir the rice later and you don't want to spoil the presentation by getting rice all over the crawfish.

4 TASTE. After the sausages hit 155°F, give the rice a good stir and taste it. It should be soft and there should be no more liquid in the pan. If it is still hard, add more boiling water and let it go for another 10 minutes or so.

5 SERVE. Spoon the rice onto a large platter and top it with the crawfish. Slice the sausage into coins ⅛ to ¼ inch thick and scatter them around the top of the rice. Sprinkle the plate with the optional garnish.

6 EATING. Eating crawfish is a messy affair and many folks just spread newspaper on the table and have at it. Begin by twisting the head off and peeling the tail. The good stuff is in the tail, but some folks like to suck the head. Try it. And when you eat, make sure you try the sausage and crawfish meat on the same fork.

VEGETABLES

"There is no real need for decorations when throwing a barbecue party—let the summer garden, in all its vibrant and luscious splendor, speak for itself."

—PIPPA MIDDLETON

My wife is a Master Gardener with a certificate from the University of Illinois, and in August, when her garden is pumping out tomatoes, eggplants, squash, beans, and more, we eat meat-free many nights. Meathead becomes Veghead. Grilled veggie sandwiches, ratatouille, eggplant parm, stuffed zucchini, stuffed tomatoes, oh my!

One thing to keep in mind about storing fruits and vegetables: Even after harvest, most are still taking in oxygen and giving off carbon dioxide. Many can continue to ripen, grow higher in sugar and lower in acid, after harvest.

Vegetables love fire and smoke, and although I have admonished you against charring, doggone it, some, because of their high sugar content, love a little char. Can we talk about lightly charred corn on the cob? Onions? Bell peppers? Asparagus?

In general, and there are exceptions, plants have a lot of cellulose and carbs, less protein, and the fats are very different. Some vegetables and fruits can contain as much as 95% water.

Some veggies benefit from being tossed in oil before grilling. The oil brings flavor and helps brown and crisp the surface. Play with ghee and clarified butter.

Foods with high water content like tomatoes or zucchini can take high heat. More fibrous, less watery things like carrots, winter squash, and cauliflower are best started over indirect heat or even simmered, steamed, sous vided, or microwaved before grilling and smoking. Potatoes should be parboiled if for no other reason than they take forever on the grill if you don't.

It's a good idea to cut broccoli and cauliflower into florets, carrots and potatoes into coins. Most grilled veggies need salt, and most take well to herbs and spices. Many grilled veggies like sauces, especially cream sauces, egg-based sauces, such as hollandaise, Béarnaise, and Baconnaise, cheese sauces, aioli, and romesco (recipes for many of these are on my website). They also enjoy the company of other grilled vegetables such as onions and sweet bell peppers as well as bacon bits, toasted bread crumbs, pickles, and pine nuts. In a hurry? Make maque choux, a simple combo of grilled corn kernels, red bell peppers, and onions.

Many grilled vegetables make excellent soups. Fire-roasted tomatoes make a killer gazpacho and can really amp up a chili.

If space is limited on your grill, cook plant matter before the meats. They are usually just fine when served warm rather than hot. Grilled asparagus, maybe with some shaved parm, a scattering of toasted bread crumbs, and a drizzle of balsamic syrup, is dynamite at room temperature. There's a recipe for this in my last book and linked to this page: AmazingRibs.com/mm.

Make sure your grill grates are very clean. Hamburger grease will not enhance corn. Grill toppers are often necessary to add color and keep the vegetables from diving into the flames. Many vegetables are best cooked over direct heat, but not Warp 10. Dial back to Warp 5 to 7.

SMOKE-ROASTED GARLIC

Raw garlic is harsh and sulfurous, strong enough to ward off vampires. But it gets mellow, nutty, and sweet when cooked. But not *too* mellow. It still retains its unique character. One of the best ways to mellow garlic is to roast it with some olive oil. It becomes soft and spreadable. Smear it on bread, toast, or crackers, in mashed potatoes, salad dressings, soups, and sauces. Make extra. You can pop the cloves out of the skins, put them in a plastic bag in the fridge for a week, or even freeze them. Keep some on hand. You can even puree and freeze them. You'll be glad you did. Use this in the Smoke-Roasted Garlic Bread (page 367).

MAKES *1 head garlic, about 12 cloves*
TAKES *1 hour*

  1 head garlic
  1 tablespoon inexpensive olive oil

1 FIRE UP. Start your smoker and aim for 225°F or set up a grill for 2-zone cooking with 225°F on the indirect side. Get some smoke rolling.

2 PREP THE GARLIC. With a sharp knife, cut off the pointy end of the garlic head about ½ inch below the top of the head. This should be deep enough to expose the flesh of most of the cloves within. If not, cut a little farther down. Squeeze to separate the cloves slightly so there is a bit of space between them. Drizzle some olive oil over the bare garlic meats and let it run down between the cloves.

3 ROAST. Place the whole shootin' match on the grill on the indirect heat side. After about 30 minutes stick a thermometer probe into one of the center cloves. If it meets resistance, cook another 15 minutes. If it slides in like buttah, it is done.

SMOKED TOMATO RAISINS

When a guy named Meathead says one of his favorite things to smoke is a vegetable, it is time to pay attention. And nothing is simpler to make on a smoker than smoked tomato raisins. In essence they are just like sun-dried tomatoes, only with a sexy smoky flavor. The dehydration makes them as sweet as raisins. These sweet baubles are superb for salads, on a pizza or focaccia, in pasta, baked into breads, on baked potatoes, in stews, in pot pies, in roulades like porchetta, on a BLT, stuffed into chicken breasts or pork chops, in omelets or scrambled eggs, anything with a cream sauce, or anything you would do with raisins or sun-dried tomatoes. I use them in my Best Chicken Burger Ever (page 292) and Not Grannie's Meatloaf (page 296).

It is breathtaking how many cherry tomatoes one plant can produce, and if you have two or three plants, you are swimming in them. This is what to do with the ones you don't pop in your mouth while picking them. This process also works on full-size tomatoes, especially meaty ones like Romas or San Marzanos.

MAKES *As many as you can fit*
TAKES *About 6 hours*

SPECIAL TOOLS *Grill topper (or something to keep them from falling through the grates when they shrink)*

Tomatoes. That's the only ingredient.

1 PREP. Pop the stems off the tomatoes and stab them 3 or 4 times with the tip of a sharp knife so moisture can escape.

2 FIRE UP. Set the temperature for a smoker or the indirect zone of a grill as low as 170 to 200°F. Don't get any hotter than this. Get some smoke rolling and keep it rolling for at least an hour or two.

3 SMOKE. Spread the tomatoes around on a grill topper trying to leave a little room between them. After 3 hours or so, roll them around a bit. If you want, you can bring them inside and finish them in the oven or dehydrator. When they have shrunk to about 25% of their original size yet remain pliable like raisins, they are done. Don't let them get hard. They'll keep in the fridge for weeks depending on the moisture content, but I've kept them in the freezer in zipper bags for a year until the next crop arrived.

FIRE-ROASTED RED BELL PEPPERS (PIMIENTOS)

Many recipes call for roasted peppers or pimientos (and yes, that is the proper spelling): Pimento cheese (that's how they

and the skins blacken, about 10 minutes, turning once or twice.

3 PEEL. When limp, place them in a bowl, cover it with a plate, and let them steam for 15 minutes or so and the skin will be easy to scrape off with a serrated knife. If a little charred skin remains, it's OK. Freeze them in zipper bags with a splash of olive oil.

RED PEPPER RISOTTO

Risotto is a luxurious treat typically made from Arborio rice, a variety of short-grain rice that has a lot of starch. When cooked, the starch produces a creamy sauce. The cooking method involves starting it submerged in liquid, stirring it often to let the grains rub against each other and give up some starch, and when the liquid is absorbed, more is added, more stirring follows, more liquid added until the rice is tender. The fun part is selecting the liquid. You can use water, stock, and in this case, pureed fire-roasted sweet red bell peppers with chicken stock. I have made risotto with the drippings from smoked turkey and other roasts, grilled mushrooms, carrots, and chorizo, but my favorite is with grilled sweet red pepper.

MAKES *2 or 3 servings*
TAKES *About 45 minutes after the pepper is roasted*

spell it down South), potato salad, harissa, Italian sausage sandwiches, Italian beef sandwiches, omelets, risotto (see Red Pepper Risotto, at right), and in several other recipes this book. Here's how to make your own fire-roasted peppers. Saves money, tastes better. Freeze them and have them on hand all winter long. This method works fine on most peppers, even hot chiles. If you use hot chiles, wear rubber gloves or a baggie on your hands.

MAKES *As many as you can fit*
TAKES *About 15 minutes*

> **Bell peppers or hot chiles of whatever color you want**
> **Olive oil**

1 FIRE UP. Set up your grill for 2-zone cooking and aim for Warp 7 on the direct heat side.

2 CHAR. Cut the peppers in half, pop out the seedpod and stem. Grill the peppers skin side down, on the direct side, lid down, until tender

1 red bell pepper

1 garlic clove

2 cups low-sodium chicken broth or stock

¼ cup dry white wine or rosé

1 teaspoon red wine vinegar

½ teaspoon Morton Coarse Kosher Salt

2 tablespoons (1 ounce) unsalted butter

¾ cup Arborio rice

6 fresh basil leaves, for garnish

1 FIRE UP. Grill the bell pepper as directed in Fire-Roasted Red Bell Peppers (page 340).

2 MAKE THE SAUCE. When the pepper is done, peel and chop it into about 8 chunks. Toss in a blender or food processor. Coarsely chop the garlic and add that too. Add the chicken broth, wine, vinegar, and salt. Take it for a spin until it is pureed. It can remain a little grainy. Pour it into a saucepan and warm it over medium-low heat.

3 MAKE THE RISOTTO. In a 12-inch skillet, melt the butter over medium heat. (I know many risotto recipes are made in saucepans, but a larger skillet helps the grains get more liquid contact, which speeds absorption so it cooks faster, which means less stirring.) Add the rice and stir until it is coated with butter and let it cook for a minute or two to get some toastiness. Add 1 cup of the bell pepper blend and stir almost constantly until most of the liquid has been absorbed. Add ½ cup more bell pepper blend and stir often until it is almost all absorbed. Repeat until almost all the liquid is absorbed. Taste it. It is done when the rice is no longer crunchy. It should be slightly firm. If it is still hard and all the liquid has been used, add hot water and keep stirring.

4 SERVE. Roll the basil into a cigarette and chop across the diameter making thin chiffonade strips. Garnish the risotto with the basil chiffonade.

QUICK PICKLES

Need zing? In a hurry? Quick pickles can be ready in less than an hour. Perk up tuna salad, burgers, tacos, and salads with a topping of something bright and cleansing that can be ready to serve in a hurry. Quick pickles made from fruits and vegetables do not require the complex sterile canning process, but they must be refrigerated. I usually have a jar of thinly sliced pickled onions in the fridge, and another

of jalapeños. But the method works just dandy on cucumbers, radishes, string beans, ginger, summer squash, carrots, asparagus, beets, and just about any fresh vegetable or fruit. I've even done it with crab apples purloined from my neighbor's tree. You can flavor them with seeds or herbs. Use pickled onions on my Best Chicken Burger Ever (page 292). Do not try to quick pickle meats.

MAKES *As much as you want*

TAKES *25 minutes to assemble, overnight to age (but you can start using it in as little as an hour)*

- 1 part vegetables or fruits
- 1 part distilled white vinegar
- 1 part water
- ½ part granulated white sugar
- ½ teaspoon Morton Coarse Kosher Salt per ½ cup sugar

OPTIONAL ADD-INS. Garlic, dill, dill seeds, cardamom seeds (not pods), cloves, mustard seeds, black pepper, coriander seeds, hot pepper flakes, bay leaf

ABOUT THE VEGETABLES. Onion is the standard starting place, and just plain onion is all you need. But after that, try cucumber slices, jalapeño, carrot coins, radish coins.

ABOUT THE VINEGAR. I prefer distilled white vinegar. Rice vinegar or white wine vinegar works nicely. Rice vinegar is lower acidity, so use about 20% more. Wine vinegar has a more distinctive flavor that you might love or hate. You can try cider vinegar if you wish, but it has a strong distinctive flavor that sometimes dominates. Ditto for red wine vinegar, sherry vinegar, or balsamic. You can even use lemon or lime juice. There may be a recipe that will benefit from these strong flavors, but as a rule of thumb, plain old distilled white vinegar is the best bet.

ABOUT THE SUGAR. When you taste it, if the vinegar is too strong, you can cut it with more water, or balance it with more sugar.

1 PREP. Thoroughly wash the jars and lids, preferably in a dishwasher. Scrub the veggies or fruits well and slice them to about ⅛ inch thick, although some things can be left whole.

2 PACK. Pack the veggies or fruits into jar(s) leaving at least 1 inch headspace at the top.

3 MAKE THE BRINE. In a saucepan or pot, combine the vinegar, water, sugar, salt, and any optional add-ins. Bring it to a boil and boil for at least 3 minutes, stirring to dissolve the sugar and salt.

4 POUR. Add the hot brine to the jar(s), making sure to cover the ingredients yet leaving about ½ inch of headspace. Put on the lid, but don't tighten it yet. Bang the jar on the counter to get the air bubbles to the top, then tighten the lid. Let it cool for about 1 hour.

5 REFRIGERATE. Thin items like onions can be used within an hour but I usually move them into the fridge for at least 24 hours to allow the pickling liquid to penetrate. Then serve on a salad or sandwich.

CABBAGE DRESSED IN HONEY MUSTARD

There's something transformative about roasting vegetables that concentrates and transcends their humble flavors. Boiling veggies dilutes their flavors and, in the case of cabbage, turns its signature crunch to mush. When you serve this, your guests will slap their foreheads and say, "Why didn't *I* think of this?" Now is the time to serve grilled corned beef.

MAKES *4 large servings*

TAKES *15 minutes to prep, about 2 hours to cook (depending on the size of the cabbage)*

- 6 garlic cloves
- 2 sticks (8 ounces) unsalted butter
- ¼ cup honey
- 2 limes or lemons
- 2 tablespoons Dijon-style mustard
- ½ teaspoon Morton Coarse Kosher Salt
- 1 small head cabbage

1 MAKE THE BASTE. Peel and press or mince the garlic. In a saucepan, melt the butter over medium-low heat. Add the garlic and cook for about 2 minutes. Stir in the honey, squeeze in the citrus juice, mustard, and salt. Turn off the heat.

2 PREP THE CABBAGE. Peel off the wilted outer layer or two of the cabbage and remove about ¼ inch of the dirty edge of the stem. Slice the cabbage in half top to bottom through the core and in half again through the core into 4 wedges. If you have a large head, cut it into 6 or 8 wedges. Leave the core intact. It will hold the leaves together (and count me among those who love the core).

3 FIRE UP. Set up your grill for 2-zone cooking and shoot for about 225°F on the indirect side.

4 COOK. Brush the cut surfaces of the cabbage wedges with the baste and place one of the cut faces down directly over the high heat to get some color. Lid down. When the first cut side has browned but not blackened, paint the other cut side and flip it so it is facing down. When it is browned, flip it onto the curved side, move it to the indirect side, and paint the cut sides again. Close the lid and monitor the temperature of your grill and try and keep it in the 225°F range on the indirect side. Keep painting. When the center of the cabbage is tender, a thermometer probe should slide in easily and read about 160°F in the center.

5 SERVE. Give the cabbage one last coat of paint, lift the lid, and move them back over direct high heat to give it a little char. Serve.

BBQ BEANS

Boston-style baked beans laced with molasses and barbecue sauce pair with low and slow smoked meats like bacon pairs with eggs. Toss in some leftover smoked meat and you've taken it to the next level. The pinnacle is what happens when you put it under

cooking meats so drippings laden with smoky rub dive in.

You can do this with either dried beans or canned. Dried beans take many hours to cook, so I use them with pork butt and beef brisket, which can take 8 to 12 hours. They also require more flavorings. They are cheaper than canned and some folks like the taste better. You can find dried beans in any grocery, a nice selection in Mexican groceries, and a massive selection on Rancho Gordo's website.

Ribs take fewer than 6 hours, so for them I use canned beans like Bush's Original Baked Beans. They are navy beans baked with bacon and brown sugar. You can use them on butt and brisket, too, if you wish. Just take them off after about 4 hours and reheat them at dinner time.

MAKES *6 (1-cup) servings*
TAKES *3 hours*

6 slices bacon

2 medium onions

3 medium red, yellow, or orange bell peppers

2 garlic cloves

1 tablespoon (½ ounce) unsalted butter

½ cup Kansas City Red (page 180)

¼ cup molasses

¼ cup apple cider vinegar

2 (28-ounce) cans Bush's Original Baked Beans

OPTIONAL. 2 chipotle peppers in adobo sauce or your favorite hot sauce

1 PREP. In a 12-inch skillet, cook the bacon over medium heat until the meat is firm and the fat is still soft, but it has rendered a lot of fat. Remove them from the pan, and when they cool, chop them into ½-inch chunks. Peel and chop the onions. Chop the peppers and chipotles (if using), and peel and mince or press the garlic.

2 COOK. Keeping the bacon fat hot, add the butter, onions, and bell peppers to the pan and cook until they are soft. Add the garlic and cook another 2 minutes. Add the Kansas City Red, molasses, vinegar, beans, and chipotles (if using). Stir and pour them into a large baking pan.

3 FIRE UP AND BAKE. Fire up the smoker or indirect side of the grill to 225°F. Place the pan of baked beans under the grate with the meat on it and bake for 2 to 4 hours. Keep an eye on them, stir occasionally, and don't let them dry out, stick to the bottom, or burn. Add water if needed.

4 TASTE. After 2 hours, taste them and add sugar, salt, vinegar, or hot sauce if you want. When they are the tenderness and moisture you like, take them out and hold them in a very low oven until the meat is ready, or chill and reheat before serving.

RATATOUILLE

Yep, Meathead makes vegan dishes occasionally. You may never make ratatouille the Disney way again. I like to serve it over pasta, but you can serve it on couscous or rice. It is even good cold, so bring it to a party. Try the Eggplant Parm on the website and in the last book.

MAKES *6 servings*

TAKES *15 minutes to prep, about 20 minutes to cook*

1 pound eggplant

1 pound zucchini or yellow squash

1 pound meaty tomatoes, such as Romas or San Marzanos

1 large red, yellow, or orange bell pepper, or a mix

1 large onion

Morton Coarse Kosher Salt

Inexpensive olive oil

2 garlic cloves

2 teaspoons fresh thyme

¼ teaspoon red pepper flakes

Medium-grind black pepper

16 ounces dried pasta (I like farfalle/little bow ties)

Fresh basil, oregano, chives, and/or thyme, for garnish

Grated Parmigiano-Reggiano cheese, for garnish

OPTIONAL. High-quality olive oil

1 PREP THE VEGGIES. With a potato peeler, peel 3 or 4 stripes of skin off the eggplant, top to bottom. You want to get some of the skin off but leave at least half on. The skin has flavor and nutrients, but if you don't take some off, it is more difficult to cut and chew. Cut the eggplant crosswise into discs ½ inch thick. Cut the zucchini lengthwise into ½-inch-thick planks or crosswise into discs. Cut the tomatoes in half from top to bottom. Cut the bell pepper in half lengthwise and remove the seeds and stems. Peel the onion, cut off the stem and top ends, and cut it in half across the

equator. Insert two toothpicks into each half from the sides to hold them together.

2 **FIRE UP.** Get the grill running in 2 zones and get the hot side up to about Warp 7.

3 **GRILL THE VEGGIES.** Dump the eggplant, zucchini, tomatoes, and onion onto a sheet pan, sprinkle one side lightly with salt, and paint all sides with olive oil. Add the veggies to the grill grates over the hot side and grill until they get a few brown stripes, but don't cook until they turn to mush. Leave a little crunch. As they finish, you can snatch the skins off the tomatoes and discard them; they should come off easily with a pinch. Chop the rest of the veggies (not the tomatoes) into bite-sized chunks and slide them into a bowl.

4 **MAKE THE TOMATO SAUCE.** Peel and mince or press the garlic. In a cast-iron skillet set over direct heat on the grill (or on a burner), heat 1 tablespoon inexpensive olive oil until hot. Add the garlic, thyme, pepper flakes, and black pepper to taste, and cook only for about a minute, enough to kill the garlic's rawness. Add the tomatoes, smush them, and cook until the sauce thickens a bit, perhaps 5 minutes. Add the rest of the veggies and liquid from the bowl, stir, and warm over medium heat.

5 **PASTA.** On the stovetop, bring 1 gallon of water to a boil for the pasta and add about 1 teaspoon of salt. Add the pasta and cook to al dente according to the package directions.

6 **SERVE.** Drain the pasta, move it to a big bowl, pour on the tomato sauce, then pile on the rest of the veggies. Garnish and gently mix the herbs and Parmigiano in so they warm and exude their perfume. If you like, do as they do in Italy, and drizzle a high-quality olive oil over the top of each individual serving (especially if you plan to serve it cold).

MARIA'S CAPONATA

Caponata is a traditional eggplant-based recipe especially popular in southern Italy. It is served in a sandwich, as an appetizer on rounds of a crusty fresh bread, on toast, on crackers, and as a sauce on grilled seafood, grilled pork chops, chicken, pasta, and even pizza (well, maybe not in Italy).

Eggplant caponata is usually made in a pan or pot, but grilling the ingredients adds a layer of complexity. Once you have the concept, you can riff on it. Serve it warm or chilled. This recipe was inspired by a recipe from my wife's niece Maria Schueler.

MAKES *A bit more than 4 cups*
TAKES *1 hour*

¼ cup pine nuts

½ cup oil-cured black olives

1½ pounds eggplant

¾ pound zucchini

1 medium onion

2 celery stalks

1 medium red bell pepper

6 tablespoons good olive oil, or more as needed

2 garlic cloves

2 cups unseasoned tomato sauce

2 tablespoons salad-grade balsamic vinegar

½ teaspoon your favorite hot sauce

4 teaspoons granulated white sugar

¼ cup raisins

2 teaspoons capers

¼ teaspoon Morton Coarse Kosher Salt

¼ teaspoon medium-grind black pepper

8 fresh basil leaves or 3 tablespoons dried basil

1 PREP. In a dry frying pan, toast the pine nuts over medium heat until they get some color. Be careful not to burn them. Remove the pits from the olives and chop into ¼-inch chunks. Cut the eggplant and zucchini into 1-inch-thick rounds, discarding the ends. Peel the onion, leave the root end on, and cut off the other end. Cut it in half from pole to pole. Remove the leaves from the celery and clean the stalks well. Save the leaves for another dish. Cut the bell pepper in half and remove the stem and seeds.

2 FIRE UP. Prepare a grill for Warp 5 direct heat cooking.

3 GRILL. Paint the eggplant, zucchini, onion, celery, and bell pepper with oil and place them in a single layer on the grill. Cook until the eggplant and zucchini are starting to brown on both sides, until the onions get some grill marks and soften, and until the celery begins to get flexible. When they cool, chop everything into ¼- to ½-inch bits.

4 SIMMER. In a large frying pan or Dutch oven, warm as much of the olive oil as needed to cover the base of the pot with a thin layer. Peel and mince or press the garlic, stir it in, and let it cook until it is soft but not browned, about 1 minute. Add the tomato sauce and bring to a boil. Add the balsamic, hot sauce, sugar, raisins, and capers (if you are using dried basil, add it now). Add the grilled vegetables, salt, and pepper. Reduce the heat to low and simmer, uncovered, until the sauce is thick, about 30 minutes.

5 SERVE. Chop the fresh basil, if using, and scatter it on top. Top with the pine nuts and olives. Stir and taste, adjusting the salt if needed. You can serve it warm or chilled, from the pot or pour it into a bowl.

SQUASH BISQUE

This is a lush and sensuous make-ahead dish perfect for the cool weather. How about Thanksgiving? Buy two pumpkins for Halloween, carve one, and draw a face on the other with Magic Marker. That's the one we will eat. Or you can use kabocha, acorn, or butternut squash. Frankly, they're better. If you have homemade Smoked Bone Broth in the freezer (of course you do, don't you?), use it instead of the chicken broth.

MAKES *4 servings*
TAKES *35 minutes to prep, 40 minutes to smoke, 30 minutes for the soup*

SPECIAL TOOLS *Grill topper, blender or food processor*

- 1 large shallot or small onion
- 4 slices bacon
- 2 pounds of meat from a fresh orange squash such as pumpkin, kabocha, acorn, or butternut
- Morton Coarse Kosher Salt
- 3 cups Smoked Bone Broth (page 217), chicken stock, or beef stock
- ¾ cup dry white wine
- 1½ tablespoons honey
- ¼ teaspoon ground chipotle
- 1 cup cream or half-and-half
- 12 large fresh sage leaves or 6 dried leaves
- ¼ cup plain thick yogurt

IF YOU DON'T DO WINE. Just add more broth.

ABOUT THE CHIPOTLE. I've made this with other chile powders and even gochujang.

1 PREP. Peel and mince the shallot. In a skillet, cook the bacon until it is crisp. Once cool, crumble it. Reserve the drippings. Cut the squash in half. Scoop out the seeds and clean them, removing as much of the stringy slimy stuff as possible. If some remains, not to worry as it will disappear in the smoking process. With a vegetable peeler or paring knife, peel off the skin from the squash. It is tough, so don't worry if you cut off some of the meat. Cut the pumpkin meat into 1-inch cubes. Keep 2 pounds for the recipe and freeze the rest, or use it all and scale up the rest of the ingredients.

2 SOAK THE SEEDS. Dissolve 2 tablespoons Morton Coarse Kosher Salt in 1 cup hot water and add the seeds. Let stand for 30 minutes, then drain and leave wet.

3 FIRE UP. Light your smoker and get it up to 225°F or set up your grill for 2-zone cooking and get the indirect side to 225°F. Get some smoke rolling.

4 SMOKE. Distribute the cubes of squash and wet seeds on a grill topper. Smoke, lid down, for 40 minutes. Reserve the cubes and seeds separately.

5 START THE SOUP. In a large saucepan, heat 2 tablespoons of the reserved bacon fat over medium heat until hot. Add the shallot and cook until tender, about 4 minutes. Add the smoked squash, Smoked Bone Broth, wine, honey, chipotle, and 1 teaspoon kosher salt. Simmer until the squash is tender, 30 to 40 minutes.

6 BLEND. Add the cream and carefully transfer everything to a blender or food processor and puree until very smooth. Don't try this with a stick blender because you need more horsepower to make this stuff smooth. Pour it through a sieve into a pot to remove any chunks. Taste for seasoning and adjust with salt, chipotle, and cream as desired.

7 SAGE ADVICE. Take the sage leaves, cut them in half across their width, and add them to the soup. Heat it to a simmer.

8 SERVE. Ladle the soup into bowls and dollop a spoonful of yogurt in the center. Scatter with the bacon and pumpkin seeds.

CORN SALAD

This is a quintessential summer salad with copious fresh herbs. Serve immediately or refrigerate for up to 24 hours before serving, but be sure to return to room temperature before serving to awaken the flavors.

MAKES *6 servings*

TAKES *About 25 minutes*

- 4 ears sweet corn
- 1 (15.5-ounce) can chickpeas
- 2 ripe tomatoes
- 2 tablespoons (1 ounce) unsalted butter
- 1 teaspoon Morton Coarse Kosher Salt
- 1 cup Smoked Tomato Vinaigrette (page 173)
- ¼ cup chopped fresh dill

1 teaspoon granulated white sugar

¼ cup chopped fresh chives

3 tablespoons chopped fresh parsley

OPTIONAL. ½ teaspoon Tabasco sauce, or to taste. Top with 4 tablespoons of softened goat cheese.

1 PREP. Shuck the corn and remove all the silk. Drain the chickpeas and rinse them in a strainer or colander. Chop the tomatoes and save 1 cup for this recipe. In a saucepan, melt the butter over low heat and mix in the salt.

2 FIRE UP. Prepare a grill for 2-zone cooking and aim for about Warp 7 on the direct heat side.

3 CARAMELIZE. Brush the corn with the butter and grill over direct heat, lid down, for about 2 minutes or until it starts to get some brown spots. Roll it a quarter-turn and keep going until all sides have been grilled and have some browning. This caramelizes the sugars a bit and develops rich flavors.

4 GET THE KERNELS. When the corn is cool enough to handle, cut off the kernels, keeping the blade angled so you get the whole kernel but none of the tough cob. You can do this on a cutting board, but I like to do it by sticking the pointy end of the cob into the hole of a Bundt pan and slicing downward so all the kernels collect in the pan.

5 SERVE. In a large bowl, combine the corn, Smoked Tomato Vinaigrette, chickpeas, tomatoes, dill, sugar, and chives and toss to blend. Chop the parsley and sprinkle on top.

I really like hitting it with some hot sauce and crowning it with a scoop of fresh goat cheese.

BROCCOLI WITH PANKO PEBBLES AND YUZU BUTTER

Grilling or roasting broccoli and cauliflower brings out a level of richness you never taste when they're boiled or steamed. There's a secret to cutting up broccoli and cauliflower to grill them. We don't start hacking away at the top end of the bundle of florets. You'll just end up with bits. Read on.

MAKES *2 side-dish servings*

TAKES *20 minutes to prep, 8 to 10 minutes to grill*

SPECIAL TOOLS *Metal grill topper*

> 4 tablespoons Yuzu Butter (page 187)
>
> 4 tablespoons Panko Pebbles (page 374)
>
> 1 pound broccoli
>
> ½ teaspoon Morton Coarse Kosher Salt
>
> 3 tablespoons vegetable oil
>
> **OPTIONAL.** Large-grain salt like Maldon if it needs it

1 PREP. Mix the Yuzu Butter and Panko Pebbles together. Pour into a frying pan and brown the panko slightly over low heat.

2 BROCCOLI. Slice off about ¼ inch from the bottom of the broccoli stem where it is woody and probably brown. Cut the stem just above where all the florets branch off so they fall loose. The goal is to break the head down into individual florets that are about the same size so they will cook evenly. Once they are all liberated you can cut some of the bigger ones in half. With a vegetable peeler, scrape the woody bark off the stem. Cut it into ⅛-inch-thick coins.

3 STEAM. Place a steamer basket over simmering water and steam the broccoli pieces, covered, for 5 minutes. If you don't have a steamer basket, dunk the broccoli in a pot of boiling water for about 2 minutes to blanch them. This fixes the color to a bright green and tenderizes the stems. Drain and transfer to a bowl. Sprinkle with the salt and vegetable oil.

4 FIRE UP. Set up your grill for 2-zone cooking and aim for Warp 5 on the direct heat side. Set a metal grill topper on the direct heat side to preheat.

5 GRILL. Lightly paint the topper with oil. Spread the broccoli pieces on the grill topper and cook, lid down, until browned in places, perhaps a tiny bit of char, and crisp-tender, about 10 minutes. Turn with tongs several times.

6 SERVE. Transfer to a platter and scatter the panko over the top. Taste and if it needs more salt, scatter with a large-grain salt such as Maldon.

ASPARAGUS WITH MISO VINAIGRETTE

When the first stalks of asparagus appear in the grocery store you know spring has arrived. A harbinger equal to robins and the Kentucky Derby.

Asparagus is a fern and it has roots that must be buried in a trench. Then, in spring, it pushes its helmet through the soil and rapidly grows. In sunlight it turns green or purple. Purple is a recent version from Italy that is slightly nuttier. It can be ready for harvest within a week or two of breaking the surface. But some farmers hide the stalk with a mound of dirt to keep it white. White asparagus is less grassy (because there is less chlorophyll) and slightly sweeter; it's also more expensive.

It is widely believed that thick asparagus isn't as good as thin. But once you get past the woody bark, the interior is just as delectable. So if the store only has thick spears, don't pass them up. Just peel them up.

It's always a great idea to serve asparagus with hollandaise, but this presentation steps outside the box for an Asian-inspired jolt of umami. Serve the asparagus warm, room temp, or even chilled with the warm sauce.

MAKES *3 servings of 6 spears each*
TAKES *15 minutes to prep, 4 to 8 minutes to grill*

SPECIAL TOOLS *Flat metal skewers or a grill topper*

> 18 asparagus spears
> Vegetable oil
> ⅛ teaspoon Morton Coarse Kosher Salt
> ⅛ teaspoon medium-grind black pepper
> Miso Vinaigrette (page 174)
> **OPTIONAL.** 1 tablespoon black sesame seeds

1 PREP. Snap off and discard the tough bottom of each asparagus spear by bending it gently until it snaps at its natural point of tenderness, an inch or two from the bottom. If you wish, you can peel the woody stringy lower 2 inches of each spear with a potato

heat. A little char and crunch is OK. You might need to do them one raft at a time. Just keep the finished ones warm on the indirect side.

5 SERVE. Transfer the spears to a platter, remove the skewers, and spoon the creamy miso vinaigrette over the top. If desired, scatter with black sesame seeds.

ZUCCHINI SALAD

While on vacation in Spain, my friend and collaborator Brigit Binns was served a simply elegant, light lunch. The zucchini was served raw and it was called carpaccio of zucchini, but doncha know we had to throw it on the grill, just for a moment. It makes it sweeter and deeper. To make this, try to get a fresh tender zucchini that is less than 10 inches long, because, as they age, the seeds get tougher and the meat gets mealy. You can use either green zucchini or yellow squash (sometimes called straightneck squash). Both are long skinny summer squashes. Don't use any of the hard gourd-like squashes, such as acorn, butternut, Hubbard, kabocha, or pumpkin.

MAKES *2 servings*
TAKES *30 minutes*

SPECIAL TOOLS *Grill topper (nice but not required), tongs with flat metal tips*

peeler. This will help them cook more evenly and make them more tender.

2 STICK 'EM UP. Line up 6 spears close together and parallel. Slide two skewers through them to form "rafts." This makes it easier to flip them all at once on the grill and get even browning. It also stops any spears from committing suicide by jumping into the fire. (Repeat with the rest of the asparagus.) Lightly coat both sides of each raft with oil. Season all over with the salt and pepper.

3 FIRE UP. Set up your grill for 2-zone cooking and aim for Warp 7 on the direct heat side. I skip the smoke.

4 GRILL. Grill the asparagus over direct heat until they get some dark splotches, perhaps 4 minutes on each side, then flip the raft to the other side. Move them to the indirect side if they are browning too fast. Take them off when they still have some crunch. Watch out that the tips don't burn badly. If they start to burn, I slide them so the tips are not over direct

4 **SERVE.** Lay the zucchini on dinner plates. Don't worry if they cool off a bit, they are fine at room temp. Drizzle with the Lemongrette, drain the capers and scatter them on top, and crown with curls of cheese shaved from a block with a veggie peeler. Sprinkle on a little salt like the rain in Spain. If you like, you can make a sandwich with this, too.

Lemongrette (page 172)

1 medium zucchini or yellow squash (less than 10 inches long)

2 tablespoons the best olive oil you can find

2 tablespoons capers

2-ounce chunk Manchego cheese

Large-grain finishing salt, such as Maldon

ABOUT THE CHEESE. We are partial to Manchego, but we have made it successfully with provolone, cheddar, and kefalotyri.

1 **PREP.** Make the Lemongrette, taste, and adjust the salt and vinegar to your preference. Slice the zucchini on a steep diagonal into thin ovals less than ¼ inch thick. Use a mandoline if you have one. Put them in a bowl and toss with the oil.

2 **FIRE UP.** Preheat your grill to about Warp 7. Clean the grates thoroughly. Preheat a grill topper if you are using one.

3 **GRILL.** Lay the zucchini on the hot grates or on the hot metal grill topper. Leave the lid up and stand right there, because all we want to do is warm and wilt and get some grill marks. Two minutes on one side only should be plenty.

MUSHROOM SANDWICHES WITH TERIYAKI SAUCE

Step away from your email app. I know mushrooms are not a vegetable, but I had no other section to put this recipe. They are a fungus and fungi are a whole "kingdom" in biological parlance, just like Animalia and Plantae. Apologies to the biologists.

Most mushrooms are more than 90% water and they are very porous, so they drink up marinades like a freshman at a fraternity party. These babies are meaty crowd-pleasers on a sandwich all by themselves, or for garnishing a steak or burger, chicken, on a salad, in a soup, in pasta, or on baked potato, or you can make cream of mushroom sauce for steak. You can cook them on a grill or a griddle. Oh, the imagination soars!

MAKES *2 sandwiches*
TAKES *10 minutes to prep, 1 hour to marinate, 20 minutes to cook*

SPECIAL TOOLS *Metal grill topper*

- 8 garlic cloves
- 2 tablespoons fresh thyme
- 1 large red bell pepper
- 1 pound mushrooms
- ½ cup salad-grade balsamic vinegar
- ½ cup Teriyaki Brinerade and Sauce (page 175)
- 2 of your favorite sandwich buns

ABOUT THE MUSHROOMS. My favorite is maitake, aka hen of the woods, as shown in the photo above. They have frilly jagged edges that get crisp. You can use inexpensive buttons, cremini, porcini, even huge portobello caps, whatever, but they just won't have the meatiness.

ABOUT THE HERBS. Take a flyer here. You pick. You can't go wrong with thyme, but I can't think of an herb that won't work here. If you can't get fresh, dried will work, just use half as much.

1 PREP. Peel and mince or press the garlic. Strip the thyme leaves off the stems. Cut the bell pepper in half and remove the stem and seeds. Cut ⅛ inch off the bottoms of the mushroom stems. Clean the mushrooms with cold running water. Yes, you can and should wash mushrooms (see page 147). Leave small mushrooms whole; cut larger ones in half.

2 MARINATE. In a bowl, stir together the garlic, thyme, balsamic, and Teriyaki Sauce. Add the 'shrooms and roll them around until they are coated. Leave them at room temperature for 1 hour and roll them around whenever you walk by.

3 FIRE UP. Fire up the grill for 2-zone cooking. Warp 10 on the direct heat side. Preheat a metal grill topper.

4 GRILL. Tumble the mushrooms onto the hot metal. After they get some color on side A, flip them. If you are grilling large portobello mushrooms, start with the gills down. If you cook gills up, the cup will fill with liquid and it will not brown. As the 'shrooms are cooking, cook the pepper halves until limp.

5 SERVE. Toast the buns and pile on the mushrooms and roasted pepper.

POTATOES

"Every single diet I ever fell off of was because of potatoes and gravy."

—DOLLY PARTON

They are fat-free, cholesterol-free, and only about 100 calories per 5-ounce serving on average. They grow underground, bulbous parts of a plant's roots called tubers. The skin can be tan, rusty, and brown, but also out there are red, purple, blue, pink with yellow spots, and yellow with pink spots. The flesh is usually white, but it can also be yellowish, purple, or even red. Loaded with starch and sugars, they get sweeter with age. The skin is usually edible if washed thoroughly, and it should be eaten, because many of the nutrients are in the skins or just under them. For culinary purposes, potatoes can be divided into three categories:

STARCHY POTATOES. These are low in moisture and high in starch, and they are especially good for baking, mashing, frying, and roasting, but they are not ideal for boiling because they can disintegrate if boiled too long. Most common is the russet Burbank.

WAXY POTATOES. These are good, all-purpose spuds because they have moderate moisture and starch. When boiled they get soft around the edges but hold their shape, making them especially good for potato salads and casseroles. Most red-skinned potatoes are waxy. Yukon Gold is a popular example. They are medium-sized with a tan skin and a pale-yellow cream-colored interior. They have a winning mild buttery flavor.

NEW POTATOES. These are young spuds harvested in late winter and early spring. They are usually small, about the size of a golf ball or the size of your thumb, thin skinned, and low in moisture and starch. They are best for boiling, skin and all. Use in stews, soups, or boiled, buttered, and sprinkled with parsley, your favorite herb, or pestos. Many fingerling

potatoes are new potatoes, but some are special breeds.

Try to buy potatoes individually rather than by the bag so you can inspect them. When selecting potatoes, pick those that are firm, without sprouts, bruises, scars, or green patches. As they age, they can develop green spots, the eyes can sprout, and they can get wrinkly and soft. When this happens, they can develop compounds called glycoalkaloids (GAs), which can be mildly toxic. Most people will be unharmed by small quantities of GAs, but some may react poorly. Do not eat potato sprouts. Removing them and the meat around them will remove GAs. Green spots are harmless chlorophyl, but they can be an indicator of the presence of GAs. After you wash them or peel them, sniff them. If they smell moldy or musty, throw them out. As always, when in doubt, throw it out.

According to the University of Idaho, ideal storage is 42° to 50°F, dark, and high humidity. Root cellars and crawl spaces are good, the refrigerator is bad. In the fridge the starches turn to sugar quicker, and the cold can also darken the flesh. Put them in paper bags, not plastic bags. They need to breathe or they can rot. Do not freeze. Potatoes and onions go together well on the dinner plate, but not in storage. They each emit a gas that can spoil the other.

Potatoes come out of the ground covered in dirt. They are cleaned before being sold, but there can still be a bit of dirt and microbial contamination, so they need to be scrubbed vigorously. Most cookbooks recommend a brush, but I prefer a scrubby sponge, no soap.

Potatoes can be baked over indirect heat on a grill or smoker, but they take forever to reach the ideal temperature of 208° to 210°F. I usually kick-start the process by simmering them first or heating them for a few minutes in the microwave. To do chunks or slices, simmer them first in salted water. Potatoes can absorb salt deep and evenly.

Potatoes have a lot of pectin, which holds them together. Pectin is made of complex polysaccharides and is found in the cell walls and green parts of plants. There is an enzyme that breaks down pectin and it is heat sensitive and is inactivated at about 140°F. If you plop spuds into boiling water, boom, the enzyme is done. If they go into cold water, the enzyme has time to go to work before it's inactivated. Starting in cold water allows the potatoes to warm slowly and gelatinize more evenly. Add a pinch or two of baking soda to the simmer and that will raise the pH, which helps them brown. Toss them around a bit when you drain them and their exteriors get soft and fuzzy and that helps make them crunchy when oiled and baked, fried, or grilled. Ideal serving temperature is 210°F.

GREEKISH POTATO SALAD

Every Greek restaurant I've ever eaten in has a lemony, oregano-laced potato dish that goes exceedingly well with lamb.

MAKES *2 servings*

TAKES *1 hour*

SPECIAL TOOLS *Metal grill topper*

- 1 pound new or fingerling potatoes
- Morton Coarse Kosher Salt
- Coarse-grind black pepper
- Inexpensive olive oil
- 1 small bulb fennel
- 2 celery stalks
- 1 large shallot or small onion (size of a golf ball)
- 4 jarred oil-marinated artichoke halves
- 2 tablespoons capers, drained
- 2 tablespoons loosely packed small fresh dill leaves
- 1 tablespoon dried oregano
- ¼ cup Lemongrette (page 172)

1 SIMMER. Scrub the potatoes and halve them lengthwise. Season lightly with salt and pepper and coat lightly with olive oil. Cut the fennel bulb in half and set one half aside for later use. Thinly slice the remaining fennel and the celery. Peel and thinly slice the shallot. Cut the artichoke halves lengthwise into quarters.

2 FIRE UP. Prepare a grill for 2-zone cooking and aim for Warp 10 on the direct heat side. Place a grill topper over the direct heat.

3 GRILL. Grill the potatoes, cut sides down first, then flipping with tongs, until golden brown, crispy on the edges, and easily punctured by a thermometer probe. If they are still hard, move them to the indirect side and let them roast until tender. Take them off and let them cool slightly.

4 MAKE THE SALAD. In a large bowl, combine the potatoes, fennel, shallot, celery, capers, dill, and oregano. Toss thoroughly, drizzle with ¼ cup of Lemongrette, toss again, and taste. Season with salt and pepper.

GRIDDLED POTATOES WITH GOODIES

This dish is perfect for a large stand-alone griddle, but it can be done with a portable griddle on a gas grill or large frying pan. Just remember, potatoes take a while to cook, even when sliced thin. Short prep, long relaxed cooking time, copious quantities of savory starch, melty aromatic cheese. Add a reverse-seared steak, and you've achieved Valhalla. You in?

MAKES *2 side-dish servings*

TAKES *15 minutes to prep, 30+ minutes to grill*

SPECIAL TOOLS *Griddle or large cast-iron skillet*

- 1 large shallot (size of a golf ball)
- 2 garlic cloves
- 1 ounce thinly sliced prosciutto
- 1 tablespoon fresh rosemary leaves
- ¾ pound Yukon Gold potatoes

3 tablespoons tallow (beef fat), bacon fat, duck fat, or unsalted butter

¼ teaspoon medium-grind black pepper

1½ ounces smoked Gouda cheese

1 tablespoon fresh tarragon leaves

Large-grain salt, such as Maldon, for serving

OPTIONAL. Near the end, drizzle on ¼ teaspoon truffle oil.

ABOUT THE POTATOES. This dish peaks with Yukon Golds, but works well with others.

ABOUT THE SHALLOT. You can use an onion if you wish.

ABOUT THE PROSCIUTTO. Because it is usually sliced thin, it tends to get crispy during the cook. I like this. If you wish, you can switch to capocollo, guanciale, pancetta, pepperoni, or even American bacon.

ABOUT THE ROSEMARY. If you don't like rosemary, use thyme, chives, or oregano.

1 PREP. Peel and mince the shallot and toss it into a largish bowl. Peel and slice the garlic as thin as you can and add it to the bowl. Chop the prosciutto into ¼-inch chunks and in it goes. Mince the rosemary and add it. Scrub the potatoes, cut out the eyes, and slice them about ⅛ inch thick. Add them to the bowl. Melt the fat and combine it with the potatoes, shallot, garlic, prosciutto, rosemary, and pepper. Toss to make sure all the slices are evenly coated with the aromatic mixture. Grate the cheese and chop the tarragon leaves and set them aside.

2 FIRE UP. Preheat your stand-alone griddle on high. Or get your grill to Warp 10 and preheat a large cast-iron skillet or a griddle that can go on top of the grates.

3 COOK. Scoop the potato mixture onto the griddle and spread into an even layer. Try to get each slice of potato in contact with hot metal. Close the lid and sizzle for about 30 minutes. Check the undersides and when they get golden, if you sliced them thin, they are done. Taste one to see. If you wish, drizzle with truffle oil and mix it in. Scatter the cheese and tarragon evenly over the top and close the lid until the cheese melts. (For a nicer presentation, you can scoop the potatoes into a serving dish, scatter the cheese and tarragon on top, and microwave for 30 seconds.)

4 SERVE. Serve with a large spatula to keep the lovely top intact and scatter with large-grain salt.

SMOKED VICHYSSOISE

Here's a refreshing chilled soup for summer. Vichyssoise (vih-SHEE-swaz) has its origin in the ubiquitous French potato leek soup and gets its name from the town of Vichy in central France. Historians believe that the more elaborate beloved chilled recipe originated at the Ritz in New York City in the early 1900s.

In the 1963 Elvis Presley movie *Fun in Acapulco*, our hero is eating vichyssoise in the kitchen of a resort hotel he works at when the chef sticks his thermometer in his soup. Dude was ahead of his time. "Hey, Doc, is it catching?" Elvis asks cheekily. Chef takes the thermometer out, shakes his head, grumbles, "Outrageous, vichyssoise must be served at 40 degrees, not warmer not colder," and removes the bowl. Noted.

MAKES *3 to 4 servings*

TAKES *1 hour 30 minutes*

SPECIAL TOOLS *Grill topper, blender*

- **2 pounds starchy potatoes, such as russet Burbank**
- **6 large leeks**
- **4 tablespoons (2 ounces) unsalted butter**
- **1½ teaspoons Morton Coarse Kosher Salt**
- **½ teaspoon finely ground white pepper**
- **6 cups chicken stock**
- **1 cup heavy cream or half-and-half**
- **1 cup sour cream**
- **2 scallions, green tops only, cut on the bias, for garnish**
- **OPTIONAL. Add a pinch of nutmeg when you add the salt and pepper. Just before serving you can streak it with a good fresh high-quality olive oil or, better still, basil oil.**

1 PREP. Scrub, peel, and cut the potatoes into ½-inch cubes. Slice the roots off the leeks and discard them. Then slice the whites of the leeks into ⅛-inch discs until you have 5 cups. Compost the greens.

2 FIRE UP. Set up your grill for 2-zone cooking at 225°F on the indirect side or fire up your smoker to 225°F.

3 SMOKE. Use a grill topper and smoke the potatoes at 225°F until tender enough to crush between thumb and forefinger, about 1 hour.

4 SAUTÉ THE LEEKS. When the potatoes are done, in a 4-quart pot, melt the butter over medium heat. Add the leeks, salt, white pepper, and nutmeg (if using). Cook until the leeks are limp, about 5 minutes.

5 SIMMER. Add the stock and potatoes. Bring to a boil and immediately back the heat down to a simmer. Simmer uncovered for 30 minutes to reduce.

6 PUREE. If you have a stick blender, you can puree the soup right in the pot or you can pour it into a blender or food processor. But be careful of the steam. Blend it into a uniform smooth consistency. If you really want it creamy smooth, press it through a sieve.

7 CHILL. Let the soup come to room temperature and then stir in the cream and sour cream. Taste and add more salt and white pepper if needed. Refrigerate until well chilled ("40 degrees, not warmer not colder!").

8 SERVE. Garnish with scallion greens. If desired, streak it with some high-quality olive oil or basil oil.

CLINT'S HASSELBACK SWEET POTATOES WITH MAPLE PECAN BUTTER

Sweet potatoes have never tasted better thanks to AmazingRibs.com president and champion cook Clint Cantwell, who shares his recipe with us. For the Hasselback method, you simply make a series of vertical cuts across the potatoes nearly but not all the way to the bottom. As they cook, the slices begin to open like an accordion, allowing the pecans and maple butter to baste the sweet potato. BTW, the name Hasselback comes from the Hasselbacken restaurant in Stockholm, Sweden, where it is believed the recipe was invented in the 1940s.

MAKES *2 servings*
TAKES *1 hour 30 minutes*

- 2 plump sweet potatoes
- 4 tablespoons (2 ounces) unsalted butter
- 2 tablespoons chopped pecans
- 2 tablespoons real maple syrup
- ¼ teaspoon Morton Coarse Kosher Salt
- ¼ teaspoon finely chopped fresh rosemary or your favorite herb

1 FIRE UP. Start your smoker and aim for 325°F or set up a grill for 2-zone cooking with 325°F on the indirect side. Get some smoke rolling.

Photo by Clint Cantwell.

2 PREP. Scrub a sweet potato and place it between two pencils on a cutting board to create an even stopping point about ¼ inch from the board. Make vertical cuts ¼ inch apart crosswise across the top of the potato, stopping at the pencils. Repeat with the other potato.

3 COOK. In a small saucepan, melt the butter then remove it from the heat. Place the potatoes on a small sheet pan. Brush each potato generously with the butter. Use a knife to gently separate the layers so that the butter can get in there. Place the pan of potatoes over indirect heat. Close the lid and allow them to bake until 210°F, about 90 minutes.

4 MEANWHILE, MAKE THE SAUCE. Add the pecans, maple syrup, salt, and rosemary to the saucepan with the remaining butter and keep it warm.

5 FINISH COOKING. Spoon half the butter/pecan mixture over the tops of the potatoes. Use a table knife if needed to gently separate the layers so that the pecans can fall between them. Close the lid and allow the sweet potatoes to cook for an additional 10 minutes.

6 SERVE. Move the potatoes to individual plates to contain the butter and spoon on any butter/pecan mixture remaining in the pan. Serve immediately.

BREADS

"*The art of bread making can become a consuming hobby, and no matter how often and how many kinds of bread one has made, there always seems to be something new to learn.*"

—JULIA CHILD

Grilled breads, from baguettes to brioche, elevate so many dishes. Grilled bread is so much better than toast from the toaster. Get those grates really clean. Preheat the grill to about Warp 5 on the direct heat side so the metal is hot. Lid up and stand right there. A slight bit of dark brown bordering on black has great flavor. Just don't burn it. If it carbonizes, scrape it off with a serrated knife.

SMOKE-ROASTED GARLIC BREAD

There are a million ways to make garlic bread. The simplest is to simply spread butter or margarine on some bread, sprinkle on garlic powder, and grill face down. Another is to dump powdered garlic into a tub of margarine, melted butter, or olive oil, mix it up, spread it on the cut face of bread, and grill.

Here's my favorite prep, just a bit more effort. It's great as a side dish, for bruschetta, or as a carrier for Maria's Caponata (page 348), fish spread (see Smoked Yooper Fish Pâté on page 313), avocado, or chèvre.

MAKES *as much as you want*
TAKES *30 minutes*

> A good bread such as a baguette
>
> High-quality olive oil
>
> Smoke-Roasted Garlic (page 339)
>
> Ripe tomatoes
>
> Smoked mild paprika, homemade (see page 164) or store-bought
>
> Coarse-grind black pepper

> Flaky finishing salt, such as Maldon
>
> **OPTIONAL.** Hot pepper flakes or ground chipotle
>
> **OPTIONAL.** When it is done, smear it with fresh chèvre (creamy goat cheese).

1 FIRE UP. Set up a grill for 2-zone cooking and get your grill to Warp 5 on the direct side.

2 PREP. Slice the bread crosswise into ½-inch discs. Paint the cut surfaces with some olive oil and make sure you get all the way to the edges to keep them from burning. Take a clove of Smoke-Roasted Garlic and smush it around on the surface. Halve a tomato, squeeze out the seeds and jelly over the trash, then smush it around on the surface of the bread. Discard

the skins. Sprinkle on some smoked paprika and black pepper, and if you wish, sprinkle on some hot pepper flakes or ground chipotle.

3 GRILL. With the lid of the grill up, toast the flavored sides of the bread. Stay right by the grill and constantly check to make sure they don't burn. When you take them off, sprinkle with Maldon salt. Serve warm or room temp.

GRAND MARNIER GARLIC BREAD

This is the bread that accompanies my Shrimp with Grand Marnier Butter (page 331), but it is totally awesome and can be used anywhere anytime anyhow. People love this.

MAKES *12 baguette slices*
TAKES *45 minutes*

- 1½ sticks (6 ounces) unsalted butter
- 1 clementine or tangerine
- 2 garlic cloves
- 1 tablespoon minced fresh tarragon
- 2 tablespoons Grand Marnier (orange liqueur)
- ¼ teaspoon Morton Coarse Kosher Salt
- 1 French-style crusty baguette at least 8 inches long

1 MAKE THE BUTTER. In a small saucepan over low heat, melt the butter. Scrub and zest the clementine and add the zest to the butter. Peel and press and add the garlic. Mince and add the tarragon. Add the Grand Marnier and salt. Keep the heat low so it doesn't bubble and let it steep together for 30 minutes.

2 FIRE UP. Set up a grill in 2 zones and get it up to Warp 5 on the direct side.

3 PREP. Cut the baguette into ½-inch slices. Generously brush the butter on all the way to the edges so they don't burn.

4 GRILL. Place the bread butter side down and stand right there. Every 30 seconds, check them. When the edges are on dark brown, they're done.

EASY NO-FAIL FLAKY BISCUITS

once saw this sign in a restaurant in Alabama: "Mind your biscuits and your life will be gravy."Biscuits are as traditional with Southern barbecue as is sauce. Serve these biscuits nekkid, buttered, with jam, marmalade, or with butter and molasses, as is traditional in some parts of the South. Serve them on Thanksgiving with turkey and all the fixings or on any old Tuesday. Split them and make mini ham sandwiches. For breakfast, smother them in redeye gravy or sausage sawmill gravy (recipes on my website), or topped with a sunny-side egg. Biscuits also make pillowy toppings for fruit cobblers, pot pies, or chili baked in a Dutch oven over a campfire.

MAKES *9 biscuits*

TAKES *1 hour to freeze the butter (see first step), 30 minutes to prep, 45 minutes to chill, 25 minutes to bake*

- 1½ sticks (6 ounces) unsalted butter
- 283 grams (10 ounces) all-purpose flour, plus more for the work surface
- 1 tablespoon baking powder
- ½ teaspoon baking soda
- 2 tablespoons sugar
- ½ teaspoon Morton Coarse Kosher Salt
- 1 cup buttermilk
- OPTIONAL. 4 ounces sharp cheddar cheese

ABOUT THE FLOUR. It is far more accurate to weigh flour than to measure by volume. If you don't have a scale, then 10 ounces of flour is about 2⅓ cups. But if you don't have a digital scale, this is a good reason to get one.

ABOUT THE BUTTERMILK AND HOW TO SUBSTITUTE. Most biscuits are made with buttermilk because it is acidic and the acid reacts with the baking soda and baking powder to create carbon dioxide, which gives the dough "lift," helping it rise and adding lightness. Most of us don't keep buttermilk on hand but pints are in many groceries. To make 1 cup of buttermilk substitute, pour 1 tablespoon distilled white vinegar or lemon juice into a measuring cup and top it off with milk until it measures 1 cup. Stir them together and let them sit for 5 minutes.

ABOUT THE CHEESE. Get a brick of cheddar, not grated cheddar. Some commercially grated cheeses have coatings to prevent them from clumping.

1 FREEZE. Cut the sticks of butter into 3 chunks: 5 tablespoons, 3 tablespoons, and 4 tablespoons. Keep the wrapping paper on the 5-tablespoon chunk so you can hold it without it melting all over your hands. Place the

5-tablespoon chunk on a dinner plate and put it into the freezer 1 hour or so before cooking. The other two chunks of butter will be used later to grease the baking pan and paint the tops of the biscuits before baking, so you can leave them on a plate at room temperature. If you are going to serve your biscuits with butter, set another stick of high-quality butter like Kerrygold on the counter when you start making the biscuits so it can come to room temp and will spread more easily.

2 MAKE THE FLOUR MIXTURE. In a medium bowl, use a fork to mix together the flour, baking powder, baking soda, sugar, and salt.

3 GRATE THE BUTTER. Roll the paper back on the frozen 5 tablespoons of butter and grate it on the large holes of a box grater right onto the cold dinner plate. Then add the grated butter to the flour and mix with a fork until all the butter is coated with flour.

4 MAKE THE DOUGH. Grate the cheddar on the large holes of the grater and mix it into the butter-flour mixture with a fork. Pour in the buttermilk and stir it with a fork until the liquid and all the dry ingredients are absorbed. You should have a wet sticky craggy dough like in the image on the left. Put it in the fridge for about 30 minutes.

5 FOLD. Coat your hands well with flour and toss 3 tablespoons of flour on your work surface. Slide the dough onto the work surface and gently press it into a 6-inch square. It will be about 1 inch thick. Fold the square of dough in half, flatten it out to about 6 × 6 again, and fold it again and again, at least five times. This folding creates layers that will bake into a flaky biscuit. You will probably have to reflour the surface and your hands several times. Finally form the dough into a 6-inch square.

6 CUT BISCUITS. Cut the dough into 9 squares like a tic-tac-dough board. At this point, you can bake them or freeze them between layers of wax paper in a zipper bag. Then you can bake them right out of the freezer, adding an extra 5 minutes or so to the baking time.

7 BUTTER THE PAN. If you have an ovenproof pan that is 6 × 6 inches, packing the biscuits in so they contact the sides is ideal, but if you don't, any baking pan or sheet pan will do. When you are ready to bake, melt the 3-tablespoon chunk of butter and lightly paint the bottom and sides of the pan with it. If you are using a larger pan, just butter a square on the bottom a little larger than 6 × 6. When you are done, pick up the pan and look at it to make sure it is thoroughly coated to prevent

Why Use a Toothpick and Not a Thermometer?

Baked goods are done when they achieve the proper moistness. Sometimes a thermometer will tell us the temperature at which it is likely to be optimum, but thermometers are not the best tool for most baked goods. Undercooked and it is wet and sticky. Overcooked, it is dry. For many recipes a simple toothpick is the answer.

Professor Blonder explains how it works. "The toothpick's surface is slightly rough, and the wood itself is dry. As dough cooks, it goes from wet and slippery to tacky to dry. Wet dough will coat but not stick firmly to the wood, because liquid gets into the rough texture of the surface and it becomes slippery. Dry dough will not stick to dry wood, so the toothpick emerges clean. This often means the cake is overdone. But when it emerges with just a few small clumps clinging to the surface, it is perfect."

Why Is Butter Temperature So Crucial in Many Baking Recipes?

My biscuit recipe and many pie dough recipes call for very cold dough and butter, even frozen. That's because we want solid chips of butter dispersed in the dough to make it flakier. On the other hand, many baking recipes, especially cookie recipes, call for butter at room temp. It's because room temperature butter holds more air when you whip it. Once the dough is made, many recipes ask you to put it in the fridge. That's because the butter melts quickly in the oven and if the dough is chilled, the butter melts more slowly and the biscuits or cookies will hold their shape better.

sticking. Do not be tempted to use a sheet of parchment paper or a silicone baking mat. Butter is crucial to getting a tasty crispy bottom.

8 CHILL THE BISCUITS. Put the biscuits one at a time in the pan or on the buttered baking sheet and push them up against each other (and the sides, if you are using a pan). They like to touch. It makes each biscuit puff up with pride and keeps them from falling over. Pop the pan into the fridge for about 15 minutes.

9 FIRE UP. Preheat the oven to 425°F or the grill for 2-zone cooking with the indirect side at 425°F. You might need to add more coals or turn on an extra burner on a gas grill to get the indirect side that hot. I don't think smoke enhances them so don't use a smoker.

10 BAKE. Just before cooking, melt the remaining 4-tablespoon chunk of butter.

Paint the tops of the biscuits with the butter. Buttering the tops is crucial for the crispiness. You probably won't use it all up, so save the rest for serving. Bake for 10 minutes and then rotate the pan for even browning. If you wish, you can paint them with more butter. Nobody will complain. Continue baking until golden brown on top, a bit darker on the bottom, perhaps another 10 to 15 minutes, depending on how close your cooker is to 425°F. The sides should be getting tan. Insert a toothpick, and if wet dough comes out with it, let it go longer until only a few small dry crumbs come out with it. BTW, if you're cooking indoors, use your fancy thermometers to check your oven. Top, center, bottom. It is probably way off and you may be able to adjust it.

11 SERVE. Let cool for a few minutes and serve them while still warm.

OLD-FASHIONED SKILLET CORNBREAD AND HUSH PUPPIES

Cornbread is a classic sidekick for barbecue with good reason. The flavor and texture are a perfect foil for sweet barbecue sauce and Southern sweet tea. You can eat it straight, or you can butter it. A honey butter or hot honey will generate smiles (recipes on AmazingRibs.com/mm). It also makes a fabulous breakfast substitute for pancakes or waffles, warm, with a dab of butter, and a glug of maple syrup.

Classic cornbread is baked in a cast-iron skillet greased with bacon fat, lard, tallow, duck fat, schmaltz, or other meat grease, although butter works fine. The hot black metal creates a brown crunchy crust that really amps up the flavor and texture. This recipe is designed for a 12-inch cast-iron skillet, but if you don't have one, you can use a 10-inch cast-iron skillet, or any other skillet, or an 8 × 8 × 2-inch baking dish, but it may not brown as well as a black pan. In a narrower pan, the mass of batter is thicker and will take longer to cook, so lower the oven temperature to 325°F.

MAKES *8 nice-sized wedges*
TAKES *20 minutes to prep, 25 minutes to cook*

- 1 cup yellow cornmeal
- 1 cup all-purpose flour
- 1 tablespoon baking powder
- ½ teaspoon baking soda
- 1 teaspoon Morton Coarse Kosher Salt
- 3 large eggs
- 3 tablespoons honey
- ¾ cup sour cream
- 4 tablespoons (2 ounces) unsalted butter
- 2 tablespoons bacon grease or butter for the pan

OPTIONAL: ¼ cup grilled sweet corn kernels and ¼ cup grilled sweet red bell pepper

ABOUT THE SOUR CREAM. Many cornbread recipes call for buttermilk, but this recipe has been formulated for sour cream.

ABOUT THE CORN KERNELS AND PEPPERS. These ingredients are optional, but they really should be required. If you wish, you can amp raw frozen corn up a bit by rolling it around in a hot pan for 2 to 3 minutes until it browns slightly. Don't use canned corn.

ABOUT OTHER MIX-INS. If you wish, you can add 4 slices crumbled cooked bacon or some chunks of cooked sausage. Cracklins are a Southern tradition, too. Or add 5 ounces grated cheddar cheese. A minced jalapeño gives the mix a spice of life. Don't go crazy with the mix-ins. Use just 2 or 3 max.

OPTIONAL. Make cornbread muffins. More crunch! Or make balls and deep-fry at 375°F into hush puppies. Still more crunch.

1 MIX THE DRY INGREDIENTS. In a bowl, mix together the cornmeal, flour, baking powder, baking soda, and salt. If you are using the grilled corn and/or bell pepper (or any other mix-ins), now's the time to mix in.

2 MIX THE WET INGREDIENTS. In another bowl add the eggs, honey, and sour cream and whisk until smooth. Melt the butter (or brown it, as on page 189) and whisk it in.

3 MAKE THE BATTER. Pour the wet ingredients into the dry ingredients. Gently stir until everything is mixed, only about 30 seconds. The batter will be lumpy. That's OK. Do not overmix.

4 FIRE UP. Preheat the oven to 400°F or the grill's indirect side to 400°F and put a 12-inch cast-iron skillet in to preheat it.

5 GREASE THE PAN. Take the skillet out of the heat and add the 2 tablespoons bacon grease or butter. Roll the fat around as it melts, coating the inside of the pan, including the sides. Work quickly so the pan doesn't cool.

6 BAKE. Pour in the batter, level it more or less. Return the pan to the oven/grill. Keep an eye on it to make sure the edges don't burn. Cook until the top is golden and a wooden toothpick inserted in the center comes out dry, about 20 minutes.

PANKO PEBBLES

We use these fun little guys in my Broccoli with Panko Pebbles and Yuzu Butter (page 352). They are great as a breading for fried foods or to sprinkle on chicken, pork chops, salads, and more.

MAKES *About ½ cup*
TAKES *About 15 minutes*

- 1 garlic clove
- 1 teaspoon unsalted butter
- 2 teaspoons high-quality olive oil
- ½ cup panko bread crumbs
- ¼ teaspoon Morton Coarse Kosher Salt
- 2 teaspoons dried thyme
- ½ teaspoon ground chipotle

Peel and mince or press the garlic. In a 12-inch skillet, melt the butter. Add the oil, panko, garlic, salt, thyme, and chipotle and toast the mixture over medium-high heat stirring frequently, until golden brown, 3 to 4 minutes. Store in the refrigerator.

DESSERTS

*"Desserts are like mistresses. They're bad for you.
If you're having one, you might as well have two."*

—CHEF ALAIN DUCASSE

There is no reason why you can't cook your whole meal on the grill, and that means dessert, too. You can bake pies, cakes, and cookies in the indirect zone, but one of the best ways to cook dessert outdoors is to simply grill fruits.

Apples, peaches, apricots, citrus, and bananas transform wonderfully over direct heat. There is a hysterical viral cartoon video on YouTube of me confessing in an interview that my favorite thing from the grill is pineapple. Just search for "Meathead loves pineapple." Most fruits don't need anything because the sugars caramelize, but if you feel adventurous, sprinkle some cinnamon sugar on, or paint them with a liqueur.

To grill fruit, you absolutely must have clean grates. Grease on top will ruin the purity of the fruit flavors, and grease underneath the grates will burn and dress the fruit in acrid smoke.

If you need to amp up the sugar, paint with a liqueur, honey, agave syrup, maple syrup, or another syrup. It is often best to cook slightly underripe fruit. It is firmer and, not to worry, many fruits get richer and sweeter when caramelized with high-temperature dry heat. The heat also makes skins softer and breaks down cell walls so the juices run. Occasionally they get dark and create pops of burnt sugar flavors like burnt marshmallows.

While fruits are best grilled rather than smoked, nuts love smoke. Can you say smoked almonds? The secret is to soak them in water for 15 minutes or so first. This helps the smoke adhere. After soaking you can anoint them with herbs and spices and even make that hard crust like the candied cashews at the state fair. Then there are chestnuts roasting over an open fire. Go to AmazingRibs.com/mm for a link to nut recipes.

WARM FRUIT SALAD

Nothing can be simpler than chopping up some fruit and throwing it on the grill. Your choice. Try fresh figs, tangerine sections, apricots, strawberries, mangoes, peaches, apples, pears. Whatever you do, don't skip the pineapple. It is the best fruit to grill. Just cut your fruits into slices no more than 1 inch thick. Any thicker and you risk burning the sugars before they warm through.

MAKES *8 servings*

TAKES *15 minutes to prep, 30 minutes to cook*

SPECIAL TOOLS *Grill topper*

SYRUP

1 (750 ml) bottle inexpensive ruby port

¼ cup granulated white sugar

FRUIT SALAD

¼ small pineapple

1 apple

1 pear

1 peach

1 tangerine

ANOTHER SYRUP TO TRY. Buy a condimento balsamic or spend a week's pay on some balsamico tradizionale. Or make a Balsamic Syrup (page 198).

1 MAKE THE SYRUP. In a saucepan, boil the port over high heat until it is reduced to syrup,

about 1 cup, about 30 minutes. Let it cool enough so you can taste it and start adding sugar until it is the right sweetness for your tastes. Heat the wine again and stir until the sugar dissolves.

2 PREP THE FRUIT. Peel the pineapple, cut it into quarters, and core it. Quarter and core the apple and the pear. Quarter and pit the peach. Peel and separate the tangerine into sections.

3 FIRE UP. Set up the grill for Warp 5 direct heat. Squirt the hot grates with water to loosen the grease and then brush to get the grill grates really clean. Depending on the spaces in your grates and the size of the fruit, you may need a grill topper.

4 GRILL. Add the fruits and grill on all sides starting skin side down. A few grill marks are a good goal.

5 SERVE. Remove the fruit from the grill, cut into bite-sized chunks, divide among serving plates, drizzle with the syrup, and serve warm.

DRUNKEN PEACHES AND CREAM

Peaches and cream, a classic summer treat and a combo as natural as peanut butter and jelly. The best peaches and cream is made with crème anglaise, which is made with cream, milk, egg yolks, sugar, and vanilla, exactly the same ingredients in vanilla ice cream. So I no longer bother making crème anglaise, I just melt vanilla ice cream, an idea I got from Jacques Pépin, who once told me that my last book reminded him of his magnum opus *La Technique*. Wow. Then again, we both had been drinking.

MAKES *4 servings*

TAKES *45 minutes*

- 1 lemon or lime
- ⅓ cup dark rum, brandy, bourbon, or rye
- 3 tablespoons real maple syrup
- 3 tablespoons dark brown sugar
- ¼ teaspoon Morton Coarse Kosher Salt
- 2 peaches
- 2 tablespoons (1 ounce) unsalted butter
- 4 scoops best-quality vanilla or butter pecan ice cream
- **OPTIONAL GARNISHES.** Candied pecans or plain pecans bring a nice crunch.

ABOUT THE BOOZE. Stick to brown goods, no peach or flavored brandy.

ABOUT THE MAPLE SYRUP. Real maple syrup, please. I like the darker grades, but I have made this with honey, molasses, and even Lyle's Golden Syrup.

ABOUT THE PEACHES. There are many varieties of peaches, but they can be divided into two broad categories: Cling and freestone. Cling peaches cling to the pit and it is hard to separate the meat from the pit. Freestones are loosely attached, and when you cut them in half, the pit comes out freely. For grilling, I prefer freestone. If they are a little underripe,

they will be easier to grill. Really ripe, soft, juicy peaches get mushy.

ABOUT THE ICE CREAM. This is better with vanilla or butter pecan than chocolate or anything else I've tried.

1 FIRE UP. Set up your grill for direct heat cooking about Warp 5. When it is hot, make sure to clean the grates thoroughly so there is no grease top or bottom. If necessary, squirt the hot grates with water to loosen the grease and then brush.

2 MAKE THE SAUCE. While the grill is preheating, extract 1 tablespoon of juice from the lemon or lime and pour it along with the rum, maple syrup, brown sugar, and salt into a saucepan and melt them together over high heat, stirring occasionally, for about 10 minutes to drive off the alcohol and thoroughly dissolve the sugar and salt. Watch

that it doesn't foam and overflow. I know you think burning off the alcohol is a waste, but we want the oak flavor not the alcohol. Alcohol can really detract from the taste.

3 PREP THE PEACHES. Quarter and pit the peaches. Melt the butter and paint the cut sides of the peaches.

4 GRILL. Grill the peaches on the skin sides for about 5 minutes. Flip them over and cook a bit longer on the flesh side until they get some grill marks. Move them to serving bowls or a goblet or a martini glass.

5 SERVE. Scoop the ice cream into the bowl with the peaches, and while the sauce is still warm, pour it over the ice cream so some of it melts into luxurious crème anglaise.

TORCHED FIGS

The judicious use of a torch on a sprinkle of sugar on ripe figs caramelizes it into a crackly surface reminiscent of crème brûlée. You can use a propane torch like the ones used for soldering, but many cooks use an inexpensive, lightweight, handheld butane burner. A butane pipe lighter might work if it puts out a long blue flame.

MAKES *2 servings*
TAKES *3 minutes*

SPECIAL TOOLS *Kitchen torch, pipe lighter, or propane torch*

2 ripe fresh figs

2 teaspoons granulated white sugar

2 tablespoons vanilla ice cream

4 teaspoons balsamico tradizionale, condimento balsamic, or Balsamic Syrup (page 198)

2 teaspoons minced fresh mint leaves

1 PREP. Cut off the stem of each fig and halve the fruit lengthwise. Place on a heatproof surface, such as a baking dish. Sprinkle the cut side of each fig half evenly with ½ teaspoon sugar.

2 FIRE UP. Carefully light the torch according to the manufacturer's directions. Aim the flame directly at the sugar on the cut side of the fig and, holding the torch about 4 inches away, glide the flame back and forth and move in until the sugar turns amber. **Butane torch.**

3 SERVE. Arrange the figs and ice cream artfully on individual plates. Drizzle some balsamic on the plate and scatter with chopped mint leaves.

GRILLED POUND CAKE WITH LEAH'S CHOCOLATE BUTTERCREAM

This recipe is dedicated to my pen pal Leah Eskin, baker extraordinaire. I call her the Poet Laureate of Recipes because she writes the most beautiful lyrical recipe headnotes. Google her and cook anything she writes. Here's an excerpt from her poetry: "Cake preens from its pedestal at the wedding, birthday, and coronation. It marks the moment. Sometimes it *is* the moment. You're not officially six until you've downed the ceremonial bite. Cake can go casual: Coffee cake, tea cake, cupcake. Even layer cake, when forked from the fridge by the enthusiast in pajamas, defines downtime."

So I've grilled up some pound cake and anointed it with ice cream and a slight modification to Eskin's chocolate buttercream sauce. Now this is not your grocery store Bosco chocolate sauce. It is rich, sweet, thick, a touch salty, a touch bitter, and the texture changes as it gets cold. I love it.

MAKES *2 servings*

TAKES *About 45 minutes*

- 4 tablespoons (2 ounces) unsalted butter
- 2 slices pound cake, about ¾ inch thick

CHOCOLATE BUTTERCREAM SAUCE

- 6 ounces high-quality semisweet chocolate
- ¼ cup half-and-half or heavy cream
- 1 tablespoon (½ ounce) unsalted butter
- 1 tablespoon plus 1 teaspoon granulated white sugar
- Pinch of Morton Coarse Kosher Salt

FOR SERVING

- 2 scoops vanilla ice cream

ABOUT THE CHOCOLATE. Using semisweet allows you room to add sugar if you want. If you use sweet chocolate, leave out the sugar. I like 50% to 60% cacao, Eskin recommends 70%. The higher the percentage cacao, the more bitter the chocolate. As the cacao percentage increases, there is less sugar and more cocoa solids producing a more intense and bitter flavor. Other factors such as the quality of the beans and the processing methods also contribute to the overall taste profile of chocolate.

1 PREP THE POUND CAKE. Melt the butter and paint both sides of each pound cake slice with it. Let it soak in.

2 FIRE UP. Set up a grill in 2 zones and get it fired up to Warp 5 on the direct heat side.

3 **GRILL THE POUND CAKE.** Grill over Warp 5 direct heat until the cake is warm, starting to toast golden and developing some nice brown grill marks.

4 **MAKE THE CHOCOLATE BUTTERCREAM SAUCE.** Chop the chocolate to bits. In a small saucepan, combine the chocolate, half-and-half, butter, sugar, and salt. Stir over very low heat constantly until it is smooth, but don't let it simmer or boil. Taste and add more sugar or half-and-half if you wish. You can store the sauce for weeks in the fridge and reheat it in the microwave when it is time to serve.

5 **SERVE.** In wide bowls or plates, lay down the pound cake and top with a scoop of ice cream. Spoon on a generous amount of the warm chocolate sauce.

BAKED APPLES

These apples are delightful hot, warm, or room temp. It is important you get the right apples. My wife is a baker and an apple aficionado, so I asked her and she recommends Honeycrisp, Braeburn, Fuji, Gala, or Granny Smith because they are firm, juicy, and crunchy, and they are also available throughout much the year.

MAKES *2 servings*
TAKES *Less than 1 hour*

- 6 pecan or walnut halves
- 2 tablespoons orange marmalade
- 2 tablespoons raisins
- 2 tablespoons fresh chèvre
- 4 tablespoons mascarpone or cream cheese
- 2 large firm baking apples
- 2 tablespoons dark brown sugar

1 MAKE THE FILLING. Chop the nut meats until they are smaller than a pea. In a bowl, mix the nuts, marmalade, raisins, chèvre, and mascarpone.

2 CORE THE APPLES. Stand the apple on the table and if the stem is not pointing straight up, with a paring knife cut off a thin layer of the bottom to make it stand level. Now insert a paring knife at an angle into the top, stem end, and cut out a shallow crater about 1½ inches wide. Then use a spoon to dig out

the seeds and core almost all the way to the bottom, but don't go all the way. A serrated grapefruit spoon works especially well. Spoon the filling into the cavity and top the filling with the brown sugar.

3 FIRE UP. Set up the grill for 2-zone cooking.

4 BAKE. Place a pan over the direct heat side of the grill, add water until it is about ½ inch deep. Sit the apples in the water in the pan and close the lid. Keep an eye on the pan and when the water is almost gone chances are the apples are done. Insert a thermometer probe or the blade of a knife, and if it slides in like an ice skater, you are good to go.

5 BRÛLÉ. Now we want to brûlé the sugar. You can do this by placing the apples in front of or under the rotisserie burner if your grill has one, or under your indoor broiler, or with a torch. Be vigilant, they brown very quickly.

PINEAPPLE FOSTER

Bananas Foster is a spectacular flaming dessert prepared tableside in many white tablecloth restaurants. It was created at Brennan's restaurant in the French Quarter of New Orleans in the 1950s and named for a regular customer, Dick Foster. When I was in Hawaii I dined at the excellent Michael Mina Stripsteak in Waikiki Beach. They did their riff on the classic by substituting pineapple for the bananas, and frankly, not only was it much better tasting, it was also more beautiful, and more spectacular.

Since fire and smoke is the theme of this book, I thought for this last recipe I would go out with a flash—flambé. In order to pull this off you need to take some precautions. If you have long hair, it is ponytail time. If you have long sleeves, roll them up. Use a long-handled lighter or match. Close windows and turn off the fan. Have a fire extinguisher handy for this and all other recipes in this book.

MAKES *2 servings*
TAKES *30 minutes*

SPECIAL TOOLS *Long-handled lighter, large skillet with a lid*

- 1 ripe pineapple
- 3 tablespoons (1½ ounces) unsalted butter
- ¼ cup dark brown sugar
- ¼ cup dark rum
- ¼ teaspoon vanilla extract
- 3 tablespoons heavy cream
- 2 scoops vanilla ice cream
- **OPTIONAL.** Toasted macadamia nuts, for garnish

1 PREP. Cut the pineapple in half from top to bottom. Remove the hard core. Scoop out the center of the pineapple, leaving about ¼-inch walls. Cut the stuff you scooped out into ½-inch cubes.

2 MAKE THE SAUCE. In a large frying pan, melt the butter over medium heat over the grill or the side burner. If you have a portable butane or propane burner, you can do this tableside. Add the brown sugar. Stir until it bubbles. Add the pineapple cubes and coat thoroughly. Cook on low until it thickens, 4 to 5 minutes.

3 FIRE UP. In a small bowl, stir together the rum and vanilla and add it to the pan. Stir. Make sure there is nothing within 5 feet overhead. Place the tip of a long-handled lighter about ¼ inch above the liquid and light it. It should catch and flame for a few minutes until the alcohol vapors have mostly burned off.

4 CARAMELIZE. Place a lid over the pan for about 30 seconds to make sure there are no flames hiding. Remove the lid, add the cream, and stir several minutes until the liquid browns and thickens.

5 SERVE. Spoon the pineapple chunks and liquid back into the hollowed-out pineapple halves and top each with a scoop of vanilla ice cream. Garnish with toasted macadamia nuts if you wish. Aloha!

APPENDIX

B elow are useful weights, measures, and equivalents. Go to AmazingRibs.com/mm for a link to nifty interactive calculators for converting one salt to another, calculating the fat percentage of ground meats for sausages and burgers, for determining how much nitrite to add to meat for curing, as well as metric conversions for volume, temperature, length, and weight. A lot of these equivalents are approximate and that's what this symbol ≈ means.

TEMPERATURE

The Warp Scale

Cooking is all about temperature control. When talking about cooking in the indirect convection heat zone, I specify a temperature such as 225°F. That's easy to measure with a thermometer. But because infrared in the direct zone is measured in watts per square meter, I use a number scale I made up: The Warp Scale. Warp 10 is maximum IR, "Give 'er all she's got, Scottie."

USEFUL TEMPERATURES

0°F (−18°C) Best freezer temperature.

25°F (−4°C) Meat freezes.

32°F (0°C) Water freezes.

34° to 39°F (1° to 4°C) Best refrigerator temperature.

41° to 130°F (5° to 54°C) Danger zone in which many pathogenic bacteria grow.

131°F (55°C) Hardiest pathogenic bacteria begin to die. Minimum sous vide temp.

135°F (57°C) Sous vide eggs for 90 minutes to pasteurize them.

135°F (57°C) Connective tissues in meat begin to contract and squeeze out pink juice (myoglobin).

150° to 165°F (66° to 74°C) Large cuts cooking at low temps stall and do not rise for hours.

165°F (71° to 74°C) Instant kill temp. Most pathogens die in seconds.

160° to 205°F (71° to 96°C) Collagens melt, form gelatin, making meat succulent.

173°F (78°C) Alcohol begins to boil at sea level.

180° to 185°F (82° to 85°C) Water begins to simmer at sea level.

183°F (84°C) Pectin begins to break down.

208° to 210°F (100°C) Baked potatoes are fluffy.

212°F (100°C) Water boils at sea level. Subtract 2°F for every 1,000 feet above.

225°F (107°C) Recommended air temp for low and slow roasting tough cuts of meats.

230° to 234°F (110° to 112°C) Sucrose (granulated white sugar) melts and makes syrup. Fructose starts to caramelize.

310°F (154°C) Maillard browning accelerates.

325°F (163°C) Recommended minimum cooking temp for browning and crisping poultry skins.

338°F (170°C) Sucrose (granulated white sugar) caramelizes.

425°F (218°C) Teflon thermometer cables can melt.

450°F (232°C) Teflon pans can emit toxic gases.

500° to 700°F (299° to 399°C) Hardwoods gassify and smoke.

700° to 1,000°F (399° to 538°C) Hardwood gases produce flame.

FATS AND OILS

85°F (29°C) Butter starts to melt.

130°F (54°C) Beef and other fats begin to liquefy.

300°F (149°C) Butter starts to smoke.

325° to 375°F (163° to 191°C) Extra-virgin olive oil begins to smoke.

350° to 375°F (177° to 191°C) Best oil temp for most deep-frying (I like 375°F).

361°F (183°C) Some animal fats begin to smoke.

400° to 450°F (205° to 230°C) Vegetable oil begins to smoke.

440°F (227°C) Inexpensive olive oil and sunflower oil begin to smoke.

450°F (232°C) Peanut oil, corn oil, soybean oil begin to smoke.

482°F (250°C) Clarified butter begins to smoke.

510°F (265°C) Safflower oil begins to smoke.

600° to 700°F (316° to 371°C) Flashpoint: Smoke from burning fat can burst into flame.

DRY MEASUREMENTS

These are all leveled at the top of the measuring spoon.

1 tablespoon = 3 teaspoons

1 cup = 16 tablespoons

1 ounce = 28 grams

1 pound = 16 ounces = 454 grams

1 kilogram = 2.2 pounds

1 pinch ≈ 1/16 teaspoon ≈ the amount you can hold between your thumb and a finger

SALT

These are conversions only if you are measuring by volume such as teaspoons or cups. Within a box of salt there is some slight variation. There is also variation from batch to batch.

1.0 volume Morton Salt and Morton Iodized Salt equals

1.0 volume Morton Canning & Pickling Salt

1.0 volume Morton All-Purpose Natural Sea Salt

1.1 volumes Baleine Fine Sea Salt

1.2 volumes Morton Coarse Kosher Salt

1.6 volumes Diamond Crystal Kosher Salt

1.9 volumes Maldon Sea Salt Flakes

BEANS

1 can of beans is usually 15 ounces undrained ≈ about 10 ounces drained ≈ 1½ cups cooked beans ≈ ¼ pound dried beans

1 pound dried beans ≈ 4 cans of beans

1 cup dried beans ≈ 3 cups cooked beans

1 pound dried beans ≈ about 2 cups dried beans ≈ makes about 6 cups cooked beans

CHARCOAL BRIQUETS

1 quart ≈ 16 Kingsford briquets

1 Weber chimney ≈ 5 quarts ≈ 80 Kingsford briquets

CHEESE

1 pound cheese ≈ 4½ cups shredded

1 cup shredded cheese ≈ a little more than 4 ounces

1 cup cottage cheese ≈ 8 ounces

FATS

1 stick butter = 8 tablespoons = 4 ounces = ¼ pound = 113 grams = ½ cup

To convert unsalted butter to salted, add ⅓ teaspoon Morton Coarse Kosher Salt for every 4 ounces of butter

If you are using salted butter, remove ⅓ teaspoon Morton Coarse Kosher Salt for every 4 ounces of butter in the recipe

1 pound solid fat (lard or shortening) ≈ 2 cups

Butter averages about 82% fat

FLOUR

1 cup unsifted all-purpose flour or cake flour ≈ 4¼ ounces by weight ≈ 120 grams

1 tablespoon all-purpose flour for thickening ≈ ½ tablespoon cornstarch

GARLIC (MINCED OR PRESSED)

Small clove ≈ ½ teaspoon

Medium clove ≈ 1 teaspoon ≈ ⅛ teaspoon garlic powder

Large clove ≈ 1½ teaspoons

Extra-large clove ≈ 2 teaspoons

GROUND BEEF

Ground chuck usually is 15% fat

Ground round usually is about 10% fat

Ground sirloin usually is about 5% fat

MUSHROOMS

1 pound trimmed fresh mushrooms

≈ 5 cups chopped

≈ one 8-ounce can sliced mushrooms, drained

≈ 3 ounces dried mushrooms, rehydrated

≈ 2 cups sautéed

PASTA AND RICE

1 pound dried pasta ≈ 4 servings

4 ounces dried macaroni ≈ 1 cup ≈ 3 cups cooked and drained

4 ounces dried spaghetti or other noodles ≈ 1.75 cups ≈ 4 to 5 cups cooked and drained

1 cup uncooked white rice ≈ ½ pound cooked

1 cup uncooked white rice + 2 cups boiling water ≈ 3 cups cooked

1 cup brown whole-grain rice ≈ ½ pound or 4 cups cooked rice

SUGARS

1 cup granulated white sugar ≈ 7 ounces ≈ 198 grams

1 cup packed dark brown sugar ≈ 7 ounces ≈ 198 grams

1 cup packed dark brown sugar ≈ 1 cup granulated white sugar + 2 tablespoons unsulfured molasses

1 cup packed light brown sugar ≈ 1 cup granulated white sugar + 1 tablespoon unsulfured molasses

1 cup honey ≈ ¾ cup granulated white sugar + ¼ cup water

1 cup corn syrup ≈ 1 cup granulated white sugar +¼ cup water

Simple syrup ≈ 1 part granulated white sugar to 1 part water by volume; thoroughly dissolved

CALORIES

1 gram water contains 0 calories

1 gram carbohydrates contains 4 calories

1 gram protein contains 4 calories

1 gram alcohol contains 7 calories

1 gram fat contains 9 calories

OTHER

APPLE. 1 medium apple ≈ ½ cup slices or chopped

BAKING SODA. To improve browning, use about ¼ teaspoon to 1 square foot of surface or mixed with 1 cup (8 fluid ounces) of water

BREADCRUMBS. 1 slice dried bread ≈ ½ cup bread crumbs

COFFEE. 1 coffee scoop = 2 tablespoons

HERBS. 2 to 3 parts fresh herbs ≈ 1 part dried (most of the time)

MSG. ½ teaspoon Ac'cent or Aji-no-moto per pound of food

MUSTARD. 1 tablespoon prepared mustard ≈ 1 teaspoon mustard powder

ONION. 1 large onion ≈ 4 inches diameter ≈ 1¼ cups chopped

POPCORN. ¼ cup popcorn ≈ 5 cups popped

TOMATOES. 1 pound tomatoes ≈ 1½ cups chopped

YEAST. 1 envelope active dry yeast = 2¼ teaspoons

ZEST. Depending on its size, an orange will yield 1½ to 2 tablespoons of fresh grated zest and less than 1 tablespoon dried. A lemon or lime will yield about 1 tablespoon of fresh zest and ½ teaspoon dried.

DRY BRINE

Sprinkle the meat with about ½ teaspoon of Morton Coarse Kosher Salt per trimmed pound and refrigerate for 2 to 12 hours depending on thickness.

WET MEASUREMENTS

Wet volumetric measurements like tablespoons and teaspoons allow for a little depression in the center, but the edges of the liquid should meet the edges of the spoon. In cups it is the center of the meniscus that you measure to. The meniscus is the upward slope where the liquid contacts the sides.

BASIC WET BRINE

Making brines is easiest with metric measurements:

6% brine: 1 liter water + 60 grams of any salt

4% brine: 1 liter water + 40 grams of any salt

6.4% brine: Add 1 cup of hot water to a 2-cup measuring cup. Then pour in salt, *any salt*, until the water line reaches 1½ cups.

Wet brine meat in a ratio of 2½ parts 6.4% brine to 1 part meat, so for a 3-pound chicken use 7½ pounds of brine.

½-inch-thick meat should be brined for about ½ hour in the refrigerator

1-inch-thick meat should be brined for about 1 hour in the refrigerator

2-inch-thick meat should be brined for about 4 hours in the refrigerator

3-inch-thick meat should be brined for about 12 hours in the refrigerator

1 dash water ≈ 3 drops ≈ 1/16 teaspoon

1 cup water = 8 fluid ounces = 16 tablespoons = 48 teaspoons = 237 milliliters and grams

1 pint water = 2 cups = 1 pound (a pint's a pound the world around)

1 quart water = 2 pints = 4 cups = 32 fluid ounces = 0.95 liters

1 gallon water = 4 quarts = 128 fluid ounces = 3.785 liters = 3,785 cubic centimeters

1 liter water = 100 centiliters = 1,000 milliliters = 34 fluid ounces = 1.06 quarts

1 #10 can = 0.75 gallons

1 fluid ounce water = 1.043 dry ounces = 0.0652 pounds = 29.574 grams

1 gram water = 0.0022 pounds = 0.035 ounces

1 milligram water = 0.000035 ounces

CREAM

Light cream ≈ 18% butterfat

Light whipping cream ≈ 26–30% butterfat

Heavy cream (whipping cream) ≈ 36% or more butterfat

Double cream = clotted or Devonshire cream ≈ 42% butterfat

1 cup heavy cream ≈ ¾ cup milk + ¼ cup unsalted butter (use only in cooking)

1 cup buttermilk ≈ 1 cup plain yogurt ≈ 1 cup milk + 1 tablespoon distilled white vinegar or lemon juice

1 cup half-and-half ≈ ½ cup heavy cream + ½ cup milk

JUICES

Tomato juice ≈ ½ cup tomato sauce + ½ cup water

1 small orange ≈ 3 tablespoons juice ≈ 1½ ounces

1 medium orange ≈ 4 tablespoons juice ≈ 2 ounces

1 large orange ≈ 5 tablespoons juice ≈ 2½ ounces

1 small lemon ≈ 2 tablespoons juice ≈ 1 ounces

1 medium lemon ≈ 3 tablespoons juice ≈ 1½ ounces

1 large lemon ≈ 4 tablespoons juice ≈ 2 ounces

1 lime ≈ 2 tablespoons juice ≈ 1 ounce

EGGS

1 large egg ≈ 2 ounces by volume ≈ 4 tablespoons ≈ 2 ounces by weight

1 large egg white ≈ 2 tablespoons

1 large egg yolk ≈ 1 tablespoon

1 cup ≈ 4 to 6 whole eggs without shells

WINE AND WHISKEY

1 shot whiskey = 2 tablespoons

1 jigger whiskey = 3 tablespoons

1 bottle wine = 750 ml = 25.3 fluid ounces

1 barrel wine ≈ 60 gallons ≈ 300 bottles

1 ton grapes ≈ 2 barrels of wine

1 barrel bourbon ≈ 53 gallons ≈ 200 liters per barrel ≈ 267 bottles

1 fifth whiskey = 25.6 fluid ounces

SPECIAL THANKS

"Cooking for others is an act of love, and who is in the chairs is more important that what is on the plates."

I stand on the shoulders of a lot of talented people without whom this book would have been impossible.

BRIGIT BINNS is my culinary consultant, advisor, confidant, and muse. We spent months in exhilarating conversations concocting about fifty playful, fun recipe concepts some of which are used in this book. She is the author of more than thirty cookbooks (she has lost count) and has served as editor-developer-doctor for countless others. Some of her most popular titles are *The Cook and The Butcher*, *The New Wine Country Cookbook: Recipes from California's Central Coast*, and *Sunset: Eating Up the West Coast*. Her memoir of growing up in Hollywood is titled *Rotten Kid*. Brigit has taught me much and

corrected my stupidities, and the recipes that we have developed together always bowl me over.

She and her husband, the actor Casey Biggs, operate a very comfy "bed and bottle" named Refugio near wine country around Paso Robles, California (refugiopasorobles.com).

PROFESSOR GREG BLONDER, PHD. Greg is an amazing polymath and my science advisor. Greg attended MIT as an undergraduate and did his PhD in physics at Harvard. He has served as chief technical advisor in AT&T's legendary Bell Labs in its heyday, holds more than one hundred patents, and is currently a professor of design and manufacturing at Boston University. His love of food, especially barbecue, brought him to food science research. More of his kitchen-science research can be found at GenuineIdeas.com. He has taught me so much.

CLINT CANTWELL is the president and general manager of Meathead's AmazingRibs.com. He is also a heckuva creative recipe writer and editor and I have shared some of his recipes herein. Clint is the pitmaster of the Smoke in da Eye competition team, he was the winner of the Travel Channel's *American Grilled* nationwide cooking competition series, and he was named one of the "10 Faces of Memphis Barbecue" by *Memphis Magazine*. He has lived in such barbecue hotspots as Texas, Kansas City, New Orleans, New York City, and now Memphis.

MY TEAM AT Meathead's AmazingRibs.com. They busted butt and that allowed me the time to write this tome. In alphabetical order: Alvaro Arteaga, senior developer and server wrangler; John "Spinaker" Bowlsby, moderator; Jerod Broussard, moderator and poultry expert; Chris Coppieters, programmer; Max Good, vice president of product reviews and keeper of the flame; David Joachim, former editorial director and current pizza maven; Ray Johnson, webmaster; Aaron "Huskee" Lyons, pit boss of the Pitmaster Club; Casey Markee, SEO consultant; Bill McGrath, thermometer tester; Jeff McNear, WordPress developer; and Charlotte Wager, marketing consultant and social media director. Special thanks to Jim Maivald, my webmaster for fifteen years until he died in November 2023.

LISA KOLEK. Her illustrations help me explain concepts so well both in this book, my last book, and on the website.

RICK LONGHI, a Cordon Bleu–trained chef and food stylist who worked by my side for many days cooking, styling sets for recipe photos, and critiquing the food and pictures.

HAROLD ROSS. For many of the recipes I used an unusual photographic technique called "light painting." To achieve the painterly look I wanted I first had to black out my studio. For each exposure I used small handheld lights often lighting each element on the set individually and assembling them in Photoshop. The shutter of my Nikon DSLR was open for as long as 30 seconds. Harold is a brilliant and inspirational artist and he taught me how to make art with light painting.

JERRY N. UELSMANN, JOHN PAUL CAPONIGRO, SONIA LANDY SHERIDAN, and **DAVE BLACK** have been my mentors and inspirations in art over the years.

CHEF RYAN UDVETT was my right hand on my last book, and some of the recipes and methods we developed together are in this book. I hired him right out of culinary school and he has gone on to great things as a product developer for US Foods.

CHEF ETIENNE MERLE, a brilliant intuitive French chef of the late great L'Auberge du Cochon Rouge in Ithaca, New York. He let me work in his kitchen and he paid me with knowledge.

FIRST READERS. When I thought I had written a good book I gave the manuscript to what the trade calls "first readers." They promptly let me know that I'm not as smart as I thought I was. I am so grateful to them, in alphabetic order: Brigit Binns; Professor Blonder; Clint Cantwell; Max Good; Tamar Haspel, *Washington Post* columnist; Aaron "Huskee" Lyons; Richard Longhi; and M. L. Tortorello, PhD microbiologist and retired chief of the Food Technology Branch of the Division of Food Processing Science & Technology for the US Food and Drug Administration.

ALL 15,000 MEMBERS OF OUR PITMASTER CLUB, whose financial support made this book possible and whose feedback on early drafts helped polish the content. Without their financial support I would still be stocking shelves in liquor stores.

SALLY EKUS FROM THE EKUS GROUP, agent extraordinaire. Sally has been my advisor, confidant, consultant, shrink, negotiator, drover, and favorite jalapeño.

MY EDITOR, SARAH KWAK, AND THE TEAM AT HARPERCOLLINS: JACQUELINE QUIRK, SHELBY PEAK, MARK MCCAUSLIN, YEON KIM, AND ESPECIALLY DESIGNER TAI BLANCHE. I have been a full-time writer and editor for about fifty years, so here's a lesson for any aspiring writer: Listen to your editors. They will make you look better than you are.

J. KENJI LÓPEZ-ALT of SeriousEats.com, author of *The Food Lab: Better Home Cooking Through Science*, *The Wok*, and the foreword to my last book. He is an inspiration. He thinks about food the same way I do, never accepting conventional wisdom without testing it first. We are brothers by different mothers only he is smarter and better looking.

CHRISTOPHER KIMBALL, creator of Christopher Kimball's Milk Street, a brilliant culinary magazine, TV show, radio show, cooking school, and store. He was the first to teach the science of cooking alongside the art.

HAROLD MCGEE, PHD, food scientist, author, inspiration. His seminal volume *On Food and Cooking: The Science and Lore of the Kitchen* has taught me a lot, and when I had a vexing question, he has been generous in answering my emails.

ALTON BROWN. My hero. I have a shrine to him in my office.

JERRY AND NORMA GOLDWYN. Dad lit the fire in me. He grilled and he let me watch. He was also a serial entrepreneur, with a degree in food technology, so in more ways than one, I am a chip off the old block. He died in 2005 and I know he'd be proud of my work. Growing up, most of the meals were made by Mom and I cannot recall her cooking anything that wasn't delicious.

MY DOCTORS at Loyola University Health System. I am alive because of them.

SO MANY OTHERS. I have benefitted from the help of scores of people more knowledgeable than I am, and I could fill pages naming them all. Here are a few: **STERLING BALL, ARDIE "REMUS POWERS" DAVIS, CHEF RICK GRESH,**

THE LATE ANTONIO MATA, MARIETTA SIMS, CHEF BARRY SORKIN, CHEF PAUL VIRANT, CAROLYN WELLS, and the **MEMBERS OF THE BARBECUE HALL OF FAME,** who allowed me into their exclusive ranks.

YOU. There are so many more to thank and you know who you are. I am especially indebted to the readers of Meathead's AmazingRibs.com and my last book who have questioned, commented, and criticized. You have helped me fill the gaps in my writing with your questions. As my Dad said, "Praise is cheap but criticism is priceless."

And thanks to **THE GODS OF GRAPE, GRAIN, AND FIRE** who have watched over me so far.

Take a Deep Dive

Hungry for more of my methods, recipes, tips, and techniques? Download my six Deep Dive Guide e-books for $3.99 from your favorite e-book store. Or get all six e-books free as a paid member of the AmazingRibs.com Pitmaster Club. And don't forget to read this book's predecessor, *Meathead: The Science of Great Barbecue and Grilling* (see page xvi).

INDEX